Mathematical
Surveys
and
Monographs

Volume 169

Connective Real *K*-Theory of Finite Groups

Robert R. Bruner
J. P. C. Greenlees

American Mathematical Society
Providence, Rhode Island

2010 *Mathematics Subject Classification.* Primary 19L41, 19L47, 19L64, 55N15, 55N91, 13D45; Secondary 13D02, 20C15, 20G05, 20J06, 53C21, 55T15, 55U20.

For additional information and updates on this book, visit
www.ams.org/bookpages/surv-169

Library of Congress Cataloging-in-Publication Data

Bruner, R. R. (Robert Ray), 1950–
Connective real K-theory of finite groups / Robert R. Bruner, J.P.C. Greenlees.
p. cm. — (Mathematical surveys and monographs ; v. 169)
Includes bibliographical references and index.
ISBN 978-0-8218-5189-0 (alk. paper)
1. K-theory. 2. Finite groups. 3. Algebraic topology. 4. Homology theory. I. Greenlees, J. P. C. (John Patrick Campbell), 1959– II. Title.

QA612.33.B78 2010
512′.66—dc22

2010032901

---

10 9 8 7 6 5 4 3 2 1 15 14 13 12 11 10

Contents

CHAPTER 1

Introduction

The aim of this book is to study the connective real K-cohomology $ko^*(BG)$ as a ring and the associated homology $ko_*(BG)$ as a module over the ring, for a range of compact Lie groups G, with both general results and a wide range of specific calculations. We believe that these calculations are only the beginning of what can be done with these methods, and we hope others will find them useful in many more cases.

1.1. Motivation

There are general reasons for interest that would not justify a project of this size. On the one hand, calculations of connective K-theory of a space are illuminating for its homotopy theory; on the other, classifying spaces BG provide both embodiments of group theory and especially accessible infinite spaces.

A more compelling argument comes from three processes for deducing subtle information from cruder information: uncompletion, descent and unlocalization. All three occur rather widely, and most fashionably in the study of elliptic cohomology, but they still have considerable substance even in the more accessible case of K-theory. To explain further, we imagine we live in a perfect world but with imperfect understanding. Thus, although there is an equivariant form of connective real K-theory, we begin with a knowledge of non-equivariant, periodic, complex K-theory alone. Since the Atiyah-Segal theorem states that $KU^*(BG)$ is a completion of KU_G^*, the process of going from a non-equivariant theory to the equivariant theory can be viewed as uncompletion. Since KO is the fixed point spectrum of KU, the process of recovering real from complex K-theory is a form of descent. Finally, the periodic theory KO is obtained from the connective theory ko by inverting the Bott element, so the recovery of ko from KO is an unlocalization. In the forms just described, these processes are well understood theoretically, but if we ask for equivariant forms of descent and unlocalization there is more to be said, and similarly if there is a space involved as well. Finally, we might actually want to use this theoretical understanding to give calculations. The present book could be viewed as a case study in these matters.

Finally, it is worth explaining why we *actually* began this project. When G is a discrete group, $ko_n(BG)$ is the obstruction group for the existence of a positive scalar curvature metric on a spin manifold of dimension $n \geq 5$ and fundamental group G. We will give a brief description of work on the Gromov-Lawson-Rosenberg (GLR) conjecture, referring to [**88**] for a more detailed guide. First, for a spin manifold M of dimension $n \geq 5$, the existence of a positive scalar curvature metric depends only on the bordism class $[M]$, and there are maps

$$\Omega_n^{spin}(BG) \xrightarrow{D} ko_n(BG) \xrightarrow{p} KO_n(BG) \xrightarrow{A} KO_n(C^*_{red}G),$$

with the composite vanishing on $[M]$ when it admits a positive scalar curvature metric, by the Lichnerowicz method. The GLR conjecture for G is that this necessary condition is also sufficient; this is known to be false in general [**89, 34**], but there are no known finite groups for which it fails. In any case, Jung and Stolz [**66, 95**] have shown that the existence of a positive scalar curvature metric depends only on $D([M]) \in ko_n(BG)$, and we write $ko_n^+(BG)$ for the subgroup of $ko_n(BG)$ consisting of indices of manifolds M admitting a metric of positive scalar curvature. The conjecture states that the containment

$$ko_n^+(BG) \subseteq \ker(Ap)$$

is actually an equality. To verify the conjecture for G, it suffices to find a small set of generators of $\ker(Ap)$ and realize them as indices of manifolds with psc metric.

Evidently it is essential to have a precise calculational understanding of the module $ko_*(BG)$, with as much additional structure as possible to minimize the difficult geometric constructions. As a result of our calculations it has been possible to verify the conjecture in a number of new cases, and details will appear elsewhere [**63, 78, 79**]. This gives a compelling reason for interest in $ko_*(BG)$ for discrete groups G; on the other hand the importance of characteristic classes of representations in giving comprehensible answers shows that we should consider compact Lie groups G as well.

1.2. Forms of K-theory

We found ourselves led into considering several different forms of K-theory, and it seems helpful to introduce them by explaining how they came in.

Periodic K-theory. A template for the project comes from Atiyah's periodic theories. Periodic complex K-theory is 2-periodic with coefficient ring

$$KU_* = \mathbb{Z}[v, v^{-1}],$$

where v is the Bott element of degree 2. Periodic real K-theory is 8-periodic, with coefficient ring

$$KO_* = \mathbb{Z}[\alpha, \beta, \beta^{-1}, \eta]/(2\eta, \eta^3, \alpha^2 = 4\beta)$$

where β is the degree 8 Bott element, α is of degree 4, and η is the image of the Hopf map in degree 1. The study of the periodic theories may be viewed as classical.

The connective theories are obtained by taking connective covers. These have coefficient rings

$$ku_* = \mathbb{Z}[v]$$

and

$$ko_* = \mathbb{Z}[\alpha, \beta, \eta]/(2\eta, \eta^3, \alpha^2 = 4\beta)$$

The complex case. In fact our earlier work [**28**] on ku was originally undertaken as a preliminary study, to see the likely shape of the answer, and to identify effective methods in a simpler case. This turned out to have been very wise, since phenomena familiar from the case of ku recur for ko, but often in a significantly more complicated form: it is doubtful that we would have found our way through without the experience gained by looking at ku. However we have been delighted to find that it is not just the methods, but also the calculations themselves that are useful.

ko-cohomology. There is a Bockstein spectral sequence (see Chapter 4) which takes $ku^*(BG)$ as an input, and collapses at E_4 to give $ko^*(BG)$ as an output. Furthermore, the differentials have almost algebraic descriptions, so that it is usually a mechanical and combinatorial process to calculate them. Finally, the filtration on $ko^*(BG)$ only has three terms, and filtrations 1 and 2 lie in the $\mathbb{Z}$-torsion submodule of $ko^*(BG)$. Our experience is that the Bockstein spectral sequence is more effective than a direct Adams spectral sequence calculation of $ko^*(BG)$, especially when coupled with representation theoretic information.

ko-homology. To obtain $ko_*(BG)$, there are two methods, each of which has been useful. On the one hand we may apply the local cohomology spectral sequence (see Chapter 2) to $ko^*(BG)$. On the other hand we may apply the Bockstein spectral sequence to $ku_*(BG)$. The local cohomology spectral sequence has the advantage that it displays the homotopy Gorenstein property of the cohomology ring. This implies a fascinating and precise connection between filtrations of representation rings and the action of the Steenrod algebra on ordinary group cohomology.

Equivariant forms. Anyone repeating our calculations will swiftly fall into the habit of using representation theoretic names for elements and making character theoretic calculations. This proves very effective, despite the fact that representation theory in $ko^*(BG)$ only enters through a completion. For example, the Atiyah-Segal completion theorem [**14**] shows

$$ko^0(BG) = KO^0(BG) = RO(G)^\wedge_J,$$

where $RO(G)$ is the real representation ring and $J = \ker(RO(G) \longrightarrow RO(1) = \mathbb{Z})$ is the augmentation ideal. The fact that uncompleted calculations are effective suggests that there is an uncompleted object to which they apply: in the periodic case, *equivariant* KO-theory gives the uncompleted version of the Borel theory, and in particular

$$KO^*(BG) = (KO^*_G)^\wedge_J.$$

Again KO_G^{-i} has the purely representation theoretic description in terms of the real, complex and symplectic representation rings:

$$RO(G),\ RO(G)/RU(G),\ RU(G)/RSp(G),\ 0,\qquad RSp(G),\ RSp(G)/RU(G),\ RU(G)/RO(G),\ 0$$

for $i \equiv 0, 1, 2, 3, 4, 5, 6, 7 \bmod 8$, so we obtain the desired representation theoretic description.

To make the best use of our calculations we therefore begin by defining an appropriate equivariant form of ko-theory. Those averse to equivariance can ignore this and complete all calculations, but we believe that they will soon find this sufficiently tiresome to overcome the aversion. Although the equivariant form of ko sees representation theory in uncompleted form, the v-torsion is unchanged. Indeed, the only method we have for calculating this torsion in ko^*_G is to look in $ko^*(BG)$. Hence the necessity for even equivariant fanatics to resort to classical homotopy theory. Using the uncompleted form, we have maps

$$\begin{array}{ccc} ko^*_G & \longrightarrow & KO^*_G \\ \downarrow & & \downarrow \\ ko^*(BG) & \longrightarrow & KO^*(BG), \end{array}$$

and we prove that the square is a pullback (the analogous complex case was proved in [**46**]).

Relationship to periodic forms. In the complex case we found it useful to consider the short exact sequence

$$0 \longrightarrow TU \longrightarrow ku_*^G \longrightarrow QU \longrightarrow 0$$

where

$$QU = \mathrm{im}(ku_*^G \longrightarrow KU_*^G)$$

and TU is the v-power torsion. Remarkably often QU is generated by $1, v$ and the Chern classes (so that it is the *modified Rees ring*) and TU is detected in ordinary cohomology. Our calculations in the real case show that something similar happens: it is useful to consider the exact sequence

$$0 \longrightarrow ST \longrightarrow ko_*^G \longrightarrow QO \longrightarrow 0$$

where

$$QO = \mathrm{im}(ko_*^G \longrightarrow KO_*^G)$$

and ST is the β-power torsion. Remarkably often QO is generated by elements in degrees ≥ 0 together with characteristic classes lifting to ko. Remarkably often ST is detected in ordinary cohomology. It is also sometimes convenient to consider a $\mathbb{Z}$-torsion free version of the representation theoretic part. For this we consider the exact sequence

$$0 \longrightarrow T \longrightarrow ko_*^G \longrightarrow \overline{QO} \longrightarrow 0$$

where

$$\overline{QO} = \mathrm{im}(ko_*^G \longrightarrow KU_*^G = RU(G)[v, v^{-1}]).$$

In particular, T also contains the submodule τ of η-multiples, and we have an exact sequence

$$0 \longrightarrow \tau \longrightarrow T \longrightarrow TO \longrightarrow 0,$$

and in all the examples we calculate, $TO = ST$.

Complex conjugation. In fact our calculations lead us one further step into the equivariant world. In studying the Bockstein spectral sequence we need to consider the action of complex conjugation on $ku^*(BG)$. In other words we are forced to consider equivariant K-theory with reality (in the sense of Atiyah [**12**]). The version of $K\mathbb{R}$ considered by Atiyah-Segal [**14**] is equivariant for a semidirect product $G \rtimes Q$ where Q is a group of order 2 corresponding to complex conjugation, but experience shows ([**37**], [**73**]) that the need to consider subgroups of $G \rtimes Q$ makes it desirable to allow non-split extensions $G.Q$ as in Karoubi [**68**]. We include a brief summary in a form convenient for our use.

Now that we have set the context, we describe our results in more detail.

1.3. The complex case

Before tackling the real case considered in this book, the reader should be generally familiar with the complex case as described in [**28**]; it should be sufficient to look at Pages 1-18. We summarize the contents briefly, so the reader knows where to look for further details.

In [**28**, Chapter 1] there are a number of general results giving a description of the coarser features of $ku^*(BG)$ as a ring in terms of the group theory of G. As might be expected, the answer is that the variety is a mixture of that of ordinary cohomology at various integer primes and that of K-theory. Quillen proved that the mod p cohomology ring has Krull dimension equal to the p-rank of G, whilst periodic K-theory is one dimensional. It follows that $ku^*(BG)$ has dimension equal to the rank of G if G is non-trivial, and the variety is formed by sticking the variety of $KU^*(BG)$ (with components corresponding to conjugacy classes of cyclic subgroups of G) to those of $H^*(BG;\mathbb{F}_p)$ (with components corresponding to conjugacy classes of maximal elementary abelian p-subgroups of G).

In [**28**, Chapter 2], there are calculations of the cohomology $ku^*(BG)$ as a ring for a number of groups of small rank. The additive structure largely follows from the Adams spectral sequence

$$\mathrm{Ext}_{\mathcal{A}}^{*,*}(H^*(ku;\mathbb{F}_p), H^*(BG;\mathbb{F}_p)) \Rightarrow [BG_+, ku_p^\wedge]^* = ku^*(BG)_p^\wedge.$$

The mod p cohomology of ku is free over $\mathcal{A}//E(Q_0,Q_1)$, where $\mathcal{A}$ is the mod p Steenrod algebra and $E(Q_0,Q_1)$ is the exterior algebra on the Milnor primitives Q_0 and Q_1, the E_2-term, so the E_2-term can be calculated from $H^*(BG)$ as a module over $E(Q_0,Q_1)$. Amongst the rank 1 groups treated are the cyclic and generalized quaternion groups. Those of rank 2 are the Klein 4-group, the dihedral groups and the alternating group A_4.

For rank 1 groups, $ku^*(BG)$ is a subring of $KU^*(BG)$, so that once we have identified the generators of $ku^*(BG)$ using the Adams spectral sequence, we can use $RU(G)$ (via equivariant K-theory) to determine the relations. Among higher rank groups, the calculation is complicated by the presence of (p,v)-torsion; although the Adams spectral sequence is less effective, it often shows that there is no (v)-torsion in positive Adams filtration. Accordingly, ordinary mod p cohomology together with representation theory determines the multiplicative structure.

In any case, the outcome of this process gives a short exact sequence

$$0 \longrightarrow TU \longrightarrow ku^*(BG) \longrightarrow QU^\wedge \longrightarrow 0,$$

where QU is the image in periodic K-theory. The image QU contains the Modified Rees ring generated by $1, v$ and the Chern classes, and is often equal to it. The kernel TU is often detected in ordinary cohomology. The equivariant cohomology ring itself can then be calculated from the pullback square

$$\begin{array}{ccc} ku_G^* & \longrightarrow & KU_G^* \\ \downarrow & & \downarrow \\ ku^*(BG) & \longrightarrow & KU^*(BG), \end{array}$$

from [**46**]. This has the effect of replacing the exact sequence above by an uncompleted version

$$0 \longrightarrow TU \longrightarrow ku_G^* \longrightarrow QU \longrightarrow 0,$$

where QU is the image in periodic *equivariant* K-theory (the torsion part TU is unaffected).

In [**28**, Chapter 3] the homology $ku_*(BG)$ is deduced from the cohomology ring $ku^*(BG)$ using the local cohomology spectral sequence:

$$E_2^{*,*} = H_J^*(ku^*(BG)) \Rightarrow ku_*(BG),$$

where H_J^* is Grothendieck's local cohomology functor [**54**]. Because the local cohomology vanishes above the rank of the group, this is a finite spectral sequence. In fact, the appropriate commutative algebra shows that the $ku^*(BG)$ and $ku_*(BG)$

contain the same information up to duality. The local cohomology spectral sequence together with the universal coefficient theorem establishes a remarkable duality property of the ring $ku^*(BG)$. This is a somewhat more elaborate counterpart of the fact that for ordinary mod p cohomology the corresponding duality implies that a Cohen-Macaulay cohomology ring is automatically Gorenstein. The implications for Tate cohomology are also considered.

Finally, [**28**, Chapter 4] gives an account of the group theoretically benign case of elementary abelian 2-groups V. It is chastening that both $ku^*(BV)$ and $ku_*(BV)$ are nonetheless so intricate. Again we consider the extension

$$0 \longrightarrow TU \longrightarrow ku^*(BV) \longrightarrow QU^{\wedge} \longrightarrow 0$$

of $ku^*(BV)$-modules, where TU is the v-power torsion (in this case annihilated by (p,v)). We view TU as a subset of $H^*(BV;\mathbb{F}_2) = \mathbb{F}_2[x_1, x_2, \ldots, x_r]$; it is the PC-submodule generated by the elements $q_S = Q_1Q_0(\prod_{s\in S} x_s)$ with S a subset of $\{1, 2, \ldots, r\}$, where PC is the polynomial ring $\mathbb{F}_2[y_1, y_2, \ldots, y_r]$ generated by the Chern classes $y_i = x_i^2$. However, the commutative algebra of $ku^*(BV)$ is rather complicated. It turns out that there is a direct sum decomposition $TU = TU_2 \oplus TU_3 \oplus \cdots \oplus TU_r$, and TU_i is of dimension r and depth i. Remarkably, the local cohomology of TU_i is concentrated in degrees i and r. More startling still is the duality: the subquotients of $H^1_I(Q)$ under the 2-adic filtration are the modules $H^i_I(TU_i)$, and the differentials in the local cohomology spectral sequence give the isomorphism. Furthermore the top local cohomology groups pair up $H^r_I(TU_i)^\vee = \Sigma^{-r+4}TU_{r-i+2}$ for $i = 2, 3, \ldots, r-1$ (the exceptional behaviour of TU_r is exactly what is required to lead to a clean final duality statement): there is a natural exact sequence

$$0 \longrightarrow \Sigma^{-4}TU^\vee \longrightarrow ku_*(BV) \longrightarrow \mathbb{Z}[v] \oplus \Sigma^{-1}(2^{r-1}H^1_I(QU)) \longrightarrow 0.$$

After this summary of results for complex connective K-theory, we may turn to the contents of the present book.

1.4. Highlights of Chapter 2

Since we need to consider both real and complex K-theory, we are led to K-theory with reality in the sense of Atiyah. We are really concerned with the K-theory of classifying spaces, but as exemplified by the Atiyah-Segal completion theorem, these are best understood through the representation theory of G. We therefore begin in Chapter 2 by describing the relevant background in representation theory. We want to understand the ring and module structure of KO, so we need a little more detail in the theory of Clifford algebras than is explicit in the literature, so we spend some time describing the theory in a convenient and explicit form. Nonetheless, the periodic theory is well-known, and we are just giving a summary. When it comes to the connective forms, the second author has given a construction in the complex case [**46**], and we present here the obvious counterpart in the real case. Fausk [**38**] has pointed out that there is a t-structure for which this is indeed the connective cover, thereby making the construction seem more natural. In fact we go a little further by generalizing to the case with reality.

This gives the entire array of theories we need. The construction provides a Hasse square for calculating the coefficients, and by using characteristic classes we can show that these coefficient rings are Noetherian. Having written down the equivariant forms, we need to show that their relationship to the homology and cohomology of classifying spaces is as expected, and we do this by proving a completion theorem and a local cohomology theorem for the theory. This is ideally suited to the study of manifolds with positive scalar curvature, since the local cohomology calculation of homology highlights exactly the subgroup one needs to realize to prove the GLR conjecture.

Finally, we end the chapter by showing that for $i \geq -7$, there is a representation theoretic description of ko_i^G. We give examples to show that this regularity vanishes in lower degrees. These results are proved by using the classical Atiyah-Hirzebruch spectral sequence together with the generalization to ko of Bott's original sequence of Ω-spaces defining the spectrum KO.

1.5. Highlights of Chapter 3

In Chapter 3 we turn to twisting, periodicity and descent. In the periodic case, the behaviour was described by Atiyah in terms of his K-theory with reality. Continuing with our convention that Q is the group of order 2 giving complex conjugation, all theories are 8-fold periodic, and K-theory with reality is periodic under suspension by $\mathbb{C} = 1 \oplus \xi$, where ξ is the non-trivial representation of the group Q of order 2 on the real line; this is equivalent to a 4ξ-periodicity equivalence $\Sigma^{4\xi} K\mathbb{R} \simeq \Sigma^4 K\mathbb{R}$. The twisting by other representations give different theories, but they are well understood. When it comes to describing the relationship between real and complex K-theory, the descent theorem is familiar in the periodic case. It states that the homotopy fixed points of complex K-theory under complex conjugation is real K-theory so that there is a spectral sequence $H^*(C_2; KU^*(X)) \Rightarrow KO^*(X)$. We give a brief summary of this behaviour.

The situation in the connective case, for twisting, periodicity and descent is considerably more complicated, and we spend some time making it explicit. It turns out that $k\mathbb{R}$ is connectively 8-fold periodic, but not quite 4ξ-periodic. On the other hand, the Borel theory is 4ξ-periodic, so it is still quite practical to make calculations. One of the most useful facts for us is that there is an equivalence

$$(S^{-\xi} \wedge k\mathbb{R})^Q \simeq \Sigma ko$$

even though the corresponding statement is false before passage to fixed points (note the negative suspension on the left and the positive suspension on the right). We provide a chart of the entire $RO(Q)$-graded homotopy of $k\mathbb{R}$, and describe all twists of the descent spectral sequence. It is illuminating to see how these periodicities emerge from successive truncations of the descent spectral sequence.

1.6. Highlights of Chapter 4

The results of Chapter 3 show that the descent spectral sequence does not give ko from ku. It turns out that this is really because the wrong truncation was chosen in moving from the periodic case. Indeed, the η-Bockstein spectral sequence does give a method, taking the form

$$E^1_{*,*} = ku_*(X)[\tilde{\eta}] \Rightarrow ko_*(X);$$

we describe it in detail in Chapter 4. Perhaps the most familiar construction begins from a cofibre sequence

$$\Sigma ko \xrightarrow{\eta} ko \xrightarrow{c} ku,$$

but after the construction through K-theory with reality, it is in some ways more natural to give an equivariant approach through the cofibre sequence

$$F(S^{\xi}, k\mathbb{R}) \longrightarrow k\mathbb{R} \longrightarrow F(Q_+, k\mathbb{R})$$

of Q-spectra. We describe both approaches and show that they agree up to the expected completion, both in the connective and periodic cases. All of this has a G-equivariant version.

In any case, since $\eta^3 = 0$ in K-theory, the Bockstein spectral sequences all collapse at the E_4-page. If we apply the spectral sequences for the periodic theories to the spaces BG, the complete behaviour of the spectral sequence is known. This gives a convenient packaging of the relationship between self-conjugate and anti-self-conjugate K-theory on the one hand, and real, complex and quaternionic K-theory on the other. The fact that this spectral sequence agrees with the descent spectral sequence in the periodic case is also interesting; for example, the fact that the descent spectral sequence collapses at E_4 is not obvious a priori from its construction.

Turning to the connective case, since the construction is compatible with the Bott element, the periodic case gives extensive information about the connective case. In particular, we may consider the exact sequence

$$0 \longrightarrow TU \longrightarrow ku_G^* \longrightarrow QU \longrightarrow 0$$

as an exact sequence of chain complexes under d^1. The known behaviour in the periodic case means that the d^1 differential on the QU part is $1 \pm \tau$, with the sign depending on the degree mod 4. When TU is detected by mod 2 cohomology, d^1 is given by Sq^2 on the TU part. The connecting homomorphism appears on the charts as a d^1, and we typically display Bockstein spectral sequence charts at the $E^{1\frac{1}{2}}$-page (i.e., after the d^1 on the TU and QU parts, but before the connecting homomorphism has been taken into account as a late d^1).

1.7. Highlights of Chapter 5

Our primary focus is on finite groups. Nonetheless, when calculating group cohomology, characteristic classes of representations are extremely valuable. Since representation theory is relatively easy to understand, characteristic classes give a large collection of cohomology classes we can name and manipulate. In particular, we can follow them through change of groups homomorphisms and changes of cohomology theory. In the best behaved cases (exemplified by Chern classes in complex oriented theories) we obtain formulae for the values on sums of representations, or even for tensor products. Of course the basis of this is knowing the values of a cohomology theory on the classical groups, and obtaining universal formulae for the structure maps.

Accordingly, we spend Chapter 5 studying ko_G^* for the groups $O(n)$, $U(n)$ and $Sp(n)$, which serve as universal examples for characteristic classes of real, complex and quaternionic representations. The Yoneda Lemma tells us that the value of $ko_{G(n)}^*$ or $ku_{G(n)}^*$ gives all characteristic classes in ko^* or ku^* for $G(n)$ representations.

The symplectic groups give Pontrjagin classes: these are the best behaved, and we have a complete calculation. Indeed,

$$ko_{Sp(1)^n}^* = ko^*[z_1, \ldots, z_n]$$

where z_i is the first Pontrjagin class of the natural representation of the ith factor, and

$$ko^*_{Sp(n)} = ko^*[z_1, \dots, z_n]^{\Sigma_n} = ko^*[p_1, \dots, p_n].$$

This allows us to define Pontrjagin classes of symplectic representations and gives a symplectic splitting principle. Accordingly, we obtain numerous useful formulae and we can easily relate the classes to counterparts in periodic theories and in representation theory.

We have made one departure from convention in our definition of Pontrjagin classes. Classically, $p_i \mapsto \sigma_i(y_1^2, \dots, y_n^2)$ under the restriction $H^*(BSp(n); \mathbb{Z}) \longrightarrow H^*(BT^n; \mathbb{Z})$ to the maximal torus. Noting that this factors through restriction to $Sp(1)^n$, and that the generators z_i of $E^*Sp(1)^n$ (for complex oriented E) map to $y_i[-1](y_i)$ we contend that, in general, we should have

$$p_i \mapsto \sigma_i(y_1[-1](y_1), \dots, y_n[-1](y_n)).$$

For cohomology, this specializes to $p_i \mapsto \sigma_i(-y_1^2, \dots, -y_n^2)$ so that our p_i is the traditional $(-1)^i p_i$. This simplifies a number of formulae and has the merit of being the $v = 0$ case of a formula which has clear meaning in terms of the formal group law for K-theory.

The unitary groups are well behaved for ku, but have quite complicated relations and torsion for ko. The orthogonal groups are more complicated, and we only give complete calculations in small dimensions.

1.8. Highlights of Chapter 6

In Chapter 6 we make the first calculations of $ko^*(BG)$ for finite groups G. Naturally we begin with the easiest groups, which are those of rank 1, namely cyclic and generalized quaternion, together with the Klein 4-group and A_4. Even in these cases, there is some complexity.

Our method has already been layed out. We take ku^*_G from **[28]** or elsewhere, and use the Bockstein spectral sequence. Working from the exact sequence

$$0 \longrightarrow TU \longrightarrow ku^*_G \longrightarrow QU \longrightarrow 0,$$

we must find the E^2-term by calculating d^1. The work falls naturally into two parts. First, a calculation of the self-conjugate (QU_{sc}) and anti-self-conjugate (QU_{asc}) parts of QU, together with $H = QU_{sc}/v^{-1}(1+\tau)QU$ gives the representation theoretic contibution. Second, a calculation of the Sq^2-cycles and homology of TU gives the torsion contribution. The connecting homomorphism may be treated as a late d^1 in the spectral sequence.

Knowing the answer in positive degrees agrees with KO^*_G with a known representation theoretic description, we may usually deduce the remaining differentials (since d^2 and d^3 must cancel everything above the 3-line). Finally, we attempt to give a representation theoretic description of the d^2 and d^3-cycles in QU_{sc}: in many cases (all except A_4 in this chapter) we observe it is generated by Chern and and Pontrjagin classes.

In any case, we describe our conclusions in terms of the short exact sequence

$$0 \longrightarrow ST \longrightarrow ko^*_G \longrightarrow QO \longrightarrow 0.$$

Of course QO agrees with KO^G_* in degrees ≥ 0. In negative degrees, the torsion free quotient is often generated by characteristic classes of self-conjugate representations in degrees which are multiples of 4, together with anti-self-conjugate classes in degrees which are 2 mod 4. The η-multiples are eventually 8-periodic, but may or may not be bounded below. The torsion TO is zero above degree -8 and it is elementary abelian in our examples, with growth equal to the rank of the group.

1.9. Highlights of Chapter 7

In Chapter 7 we apply the local cohomology spectral sequence to determine $ko_*(BG)$ for the same collection of groups as in Chapter 6. This takes the form

$$H_I^*(ko^*(BG)) \Rightarrow ko_*(BG)$$

where H_I^* denotes Grothendieck's local cohomology with respect to the augmentation ideal $I = \ker(ko^*(BG) \longrightarrow ko^*)$. Because the local cohomology vanishes above the rank of the group, this is a finite spectral sequence.

The first step here is to calculate the local cohomology modules. Using the exact sequences

$$0 \longrightarrow T \longrightarrow ko_G^* \longrightarrow \overline{QO} \longrightarrow 0$$

and

$$0 \longrightarrow \tau \longrightarrow T \longrightarrow TO \longrightarrow 0,$$

this is broken into three parts: calculations for TO, for τ and for $\overline{QO}$. The calculation of $H_I^*(TO)$ tends to be straightforward, since TO is finite over a polynomial subring of $H^*(BG; \mathbb{F}_2)$, often one generated by Pontrjagin classes: by observation TO is always of depth ≥ 2, so the local cohomology is in degrees 2 and above.

If τ is bounded below, then $H_I^*(\tau) = H_I^0(\tau) = \tau$. If τ is unbounded, its local cohomology is in degrees 0 and 1. In large degrees, it is very regular, but there are often edge effects near degree 0. Calculations can be made with characters, using the connection between QO and representation theory.

Finally, there is $H_I^*(\overline{QO})$, which tends to be the most laborious. First, it is easy to see that $H_I^0(\overline{QO})$ consists of characters supported at the identity, which is therefore $\mathbb{Z}$ in degrees divisible by 4 and ≥ 0. Next, we have found in every example we have calculated that the augmentation ideal is the radical of a principal ideal (p_*), so that local cohomology may be calculated using the ideal (p_*). Very often p_* can be taken to be the Pontrjagin class of some representation of G, and often its degree d is 4 or 8. Accordingly local cohomology vanishes above cohomological degree 1 and we have an exact sequence

$$0 \longrightarrow H_I^0(\overline{QO}) \longrightarrow \overline{QO} \longrightarrow \overline{QO}[\frac{1}{p_*}] \longrightarrow H_I^1(\overline{QO}) \longrightarrow 0.$$

One finds that multiplication by p_* becomes an isomorphism $\overline{QO}_n \longrightarrow \overline{QO}_{n-d}$ for sufficiently negative n, giving a calculation of $\overline{QO}[\frac{1}{p_*}]$. Thus, to calculate local cohomology in degree n, we choose k large enough that $p_*^k : \overline{QO}_n \longrightarrow \overline{QO}_{n-kd}$ has codomain at the stable value of $\overline{QO}[\frac{1}{p_*}]_n$ in this congruence class mod d, and then uses

$$H_I^1(\overline{QO})_n = \mathrm{cok}(p_*^k : \overline{QO}_n \longrightarrow \overline{QO}_{n-kd}).$$

This gives the E_2-term of the local cohomology spectral sequence. In rank 1 there are no differentials (except for the d^1 corresponding to the connecting homomorphism), and in rank 2 there is only one. This is very often forced by the fact that $ko_*(BG)$ is connective, together with the module structure.

In very general terms, all the answers have the following form. The torsion free part consists of a copy of $\mathbb{Z}$ in each degree $4k \geq 0$. Corresponding to the η-multiples in cohomology there is a copy of RO/RU in large degrees 1 and 2 mod 8 (and a subgroup of this in low degrees), and similarly a copy of RSp/RU in large degrees 5 and 6 mod 8 (and a subgroup of it in low degrees). Next, in odd degrees there is a sum of cyclic groups in odd degrees with orders increasing as the degree increases by 8. In degrees $4k-1$ the number of factors is one less than the rank of RU_{sc}, and in degrees $4k-3$ the number of factors is the rank of RU_{asc}. Finally, there is a an elementary abelian summand, with growth rate equal to the rank of the group.

1.10. Highlights of Chapter 8

In Chapter 8 we take a step into a more complicated realm by moving to an infinite class of groups of rank more than 1: we apply our methods to calculate $ko^*(BD)$ and $ko_*(BD)$ where D is a dihedral 2-group. It is convenient to relate these to the answers for the group $O(2)$, because it neatly encodes the maps induced by the inclusions of groups, and because it highlights the role of representation theory and characteristic classes. We used a more ad hoc basis in [**28**], so the present approach to $ku^*(BD)$ and $ku_*(BD)$ gives new information.

From $ku^*(BD)$ we deduce $ko^*(BD)$ using the Bockstein spectral sequence, but there is a single additive extension in degree -6 for which we needed to appeal directly to the Adams spectral sequence. We then apply the local cohomology theorem to give $ku_*(BD)$ and $ko_*(BD)$. In the complex case, we have short exact sequences

$$0 \longrightarrow TU \longrightarrow ku^*_D \longrightarrow QU \longrightarrow 0$$

where QU is the Modified Rees ring and TU is a free module over the double of $H^*(BD;\mathbb{F}_2)$ on a class of degree -6. Similarly in the real case we have a short exact sequence

$$0 \longrightarrow TO \longrightarrow ko^*_D \longrightarrow QO \longrightarrow 0$$

so that $ST = TO$. It turns out that in this case, TO is a proper submodule of the Sq^2-cyles ZTU; indeed there is a short exact sequence

$$0 \longrightarrow TO \longrightarrow ZTU \longrightarrow \Sigma^{-12}\mathbb{F}_2[p_2] \longrightarrow 0.$$

We have $QO = \overline{QO} \oplus \tau$ and it is notable that the group τ of η-multiples is unbounded. Since the size of TU is independent of the order of D, it is the representation theory that is hardest to understand explicitly. In cohomology this means we do not give the Chern filtration precisely, but in homology it means that, even for the dihedral group of order 8, we only give the order of $ko_i(BD)$ if i is 3 or 7 mod 8, and do not determine the exact structure.

1.11. Highlights for elementary abelian groups

Finally, we devote four entire chapters (Chapters 9 to 12) to elementary abelian 2-groups $V(r)$, where r is the rank. This may seem excessive, but they give the simplest family of increasing rank, and therefore provide an essential test of our methods as well as a basis for future calculations. In fact, we give a complete calculation of the ring $ko^*(BV(r))$ and the module $ko_*(BV(r))$. Additively the homology calculation is due to Cherng Yih Yu [**99**], but our method gives more information about naturality and the module structure. Those familiar with the complex case from Chapter 4 of [**28**] will be prepared for some complexity, but the intricate additional complications in the real case may still be some surprise.

On the other hand, the entire calculation is done from the complex case purely algebraically. Indeed, we begin with the short exact sequence

$$0 \longrightarrow TU \longrightarrow ku^*(BV(r)) \longrightarrow QU^\wedge \longrightarrow 0,$$

where QU is the Rees ring, and $TU \subseteq H^*(BV(r))$ is a submodule over the polynomial ring PC on Chern classes, understood in detail from [**28**]. No other input is required, though this relies on some numerical good fortune.

1.11.A. Highlights of Chapter 9. In Chapter 9 we use the Bockstein spectral sequence to deduce $ko^*(BV(r))$ from $ku^*(BV(r))$: perhaps surprisingly, it is the combinatorial complexity of the action of Sq^2 on TU that needs most work, and the details necessary to give precise numerical values for dimensions are not worked through until Chapter 11. The chapter ends with Bockstein spectral sequence charts for ranks ≤ 8, but the summary of general conclusions in Section 9.12 may be of particular interest. In effect we identify the exact sequence

$$0 \longrightarrow ST \longrightarrow ko_*^{V(r)} \longrightarrow QO \longrightarrow 0.$$

We find in this case that $QO \cong \overline{QO} \oplus \tau$, and $ST = TO = ZTU = \ker(Sq^2 : TU \longrightarrow TU)$, where τ consists of the η-multiples (and is bounded below by roughly $-2r$) and $\overline{QO}$ is generated by $1, v$, the Pontrjagin classes and JU^4. It is notable that $ko^G_{-8} = JU^4$ and $ko^G_{-9} = 0$, which contrasts with the fact that in the complex case ku^G_{-6} has torsion for $r \geq 2$ and $ku^G_{-7} \neq 0$ for $r \geq 3$.

1.11.B. Highlights of Chapter 10. In Chapter 10 we apply the Bockstein spectral sequence to deduce $ko_*(BV(r))$ from $ku_*(BV(r))$ (the spectral sequences are displayed in a range for $r \leq 6$). It is rather surprising that it is possible to understand the entire behaviour, and indeed, we only came to this method at a late stage because we incorrectly expected it to be impractical. It is only possible because of the detailed structural information we have about the complex case. In fact the additive extensions in degrees congruent to 3 mod 8 involve special consideration, and we only resolve them by using the local cohomology spectral sequence and some arithmetic good fortune. Once again the answers are made numerically explicit by appeal to Chapter 11. In the complex case we have the exact sequence

$$0 \longrightarrow \mathrm{Start}(2)TU^\vee \longrightarrow \widetilde{ku}_*(BV(r)) \longrightarrow \mathrm{Start}(1)2^{r-1}H^1_I(QU) \longrightarrow 0$$

In the real case it is a little more complicated, but we do have an isomorphism

$$ko_*(BV(r))/\eta \cong ku_{\equiv 0(4)} \oplus D_{\equiv 1(8)} \oplus D_{\equiv 2(8)} \oplus \mathrm{Start}(2)(TU/Sq^2)^\vee \oplus \mathrm{Start}(1)2^{r-1}H^1_I(QU)_{\equiv 0(4)},$$

where D is concentrated in degrees 1 and 2 mod 8, where it is elementary abelian and of rank $2^r - 1$ in large degrees. The η-multiples are also identified explicitly.

1.11.C. Highlights of Chapter 11. Chapter 11 is a combination of combinatorics and commutative algebra. It makes explicit how Sq^2 acts on TU, and shows how it is controlled by the combinatorics of pairs of subsets of $\{1, 2, \ldots, r\}$. In order to understand the Hilbert series explicitly, and to make calculations of local cohomology possible, we also give explicit resolutions of $TO = ZTU = \ker(Sq^2 : TU \longrightarrow TU)$ as modules over the polynomial algebra PP on the Pontrjagin classes. The process by which these are formed, starting from the PC-resolutions of TU, and putting a compatible action of Sq^2, passing to cycles and then adding generators to remove homology may be of some interest. Due to the combinatorial intricacy we will not attempt a detailed summary here.

1.11.D. Highlights of Chapter 12. Finally, in Chapter 12 we calculate $ko_*(BV(r))$ using the local cohomology spectral sequence for $r = 1, 2, 3$ and 4. This alternative calculation gives additional information, and makes certain duality properties explicit. Unfortunately it seems unlikely this method can be extended to arbitrary rank: although Chapter 11 reduces the local cohomology calculation to combinatorics and linear algebra, it still involves calculating with rather large matrices with entries in the polynomial ring PP. In rank 3 the calculation could be done by hand, but computer assistance seems essential even in rank 4, and the complexity it reveals suggests this is unavoidable. On the other hand, the calculations do reveal striking behaviour: firstly the E^∞ term is concentrated on the H^0_I, H^1_I and H^r_I columns, and secondly almost all differentials come from the H^1_I column, with the finitely many exceptions coming from the H^0_I column. Furthermore the differentials from the H^1_I all seem to be of maximal rank. This conclusion shows $ko_*(BV(r))$ has a filtration of length 2: the top quotient comes from the coefficients, the middle section is purely representation theoretic and the bottom piece comes from mod 2 homology. The last piece is by far the most complicated, and is the only piece really new in rank r, but the relationship to mod 2 homology makes it rather accessible. The way in which differentials are forced by the bare requirement that $ko_*(BV(r))$ is connective is astonishing, and the required cancellation establishes remarkable relationships between local cohomology groups.

1.12. Conclusions

The specific calculations feeding into other applications, and the specific structural results are discussed at the appropriate point. Here, we record the general conclusion, that $ku^*(BG), ku_*(BG), ko^*(BG)$ and $ko_*(BG)$ are remarkably computable. Indeed, providing both the group cohomology and representation theory are accessible, there is a good chance of a complete calculation of all four.

Perhaps more interesting is that in most cases the only homotopical input required is $ku^*(BG)$, with the rest being encoded by Sq^2 alone, together with representation theory and commutative algebra. As for methods of calculation for $ko_*(BG)$, it seems that for groups of small rank it is best to go via $ko^*(BG)$, but that for larger rank it is best to go via $ku_*(BG)$.

More striking perhaps are the remarkable phenomena that the calculations illustrate. Firstly, we were constantly struck by the intimacy of the relationship between ordinary cohomology and representation theory that is necessary for the calculations to be consistent. Secondly, the Gorenstein-type duality given by combining the local cohomology theorem with the Universal Coefficient Theorem continues to be very striking for ko.

1.13. History and comparisons with other methods

The following historical remarks are intended to give a sense of the way in which the subject has developed, but are in no sense a comprehensive survey of all relevant work. Notably omitted is the interesting work on operations and cooperations in connective K-theories. Our goal here is simply to put the present volume into context.

Historically, the first substantial calculation involving connective K-theory was Adams's 1961 determination of $H^*(ku)$ in his work on the integrality of Chern classes [**1**], although this was formulated as the calculation of the cohomology of Postnikov covers of BU in the stable range. In 1963, Stong [**97**] completed the calculation and also calculated the cohomology of the Postnikov covers of BO. At this point, it would have been possible to make easy Adams spectral sequence calculations of connective real and complex K-theory in a number of simple cases. Nonetheless, the first explicit appearance of this method came in Don Davis's Stanford thesis [**31, 32**]. Davis credits Milgram with having explained to him the relevance of Stong's results to the calculation of connective K-theory [**33**]. Although no mention of connective K-theory can be found there, the 1968 paper by Gitler, Mahowald and Milgram [**40**] contains Adams spectral sequence charts for the real connective K-theory of stunted projective spaces. Mahowald [**76**] said that "By the late 60s this sort of thing was well known." However, the absence of any explicit reference in the literature to Adams spectral sequence methods for calculating connective K-theory until [**31, 32, 77**] suggests that it was only well known to a small circle of experts.

The rest of the work involving connective K-theory until the middle 1970s was concerned with the relation to complex cobordism and the beginnings of what we now recognize as chromatic phenomena, as in the work of Conner and Floyd [**29**], Novikov [**86**], Conner and Smith [**30**], Smith [**94**] and Johnson and Smith [**64**].

In the 1970s and into the 1980s quite a lot of work appeared using connective K-theory to study the immersion problem for projective spaces. This involved computing, among other things, $ko_*(BC_2)$ and $ko_*(BC_2 \times BC_2)$, as well as the ko homology of subquotients of these classifying spaces. Results in the odd primary case at around this time were proved by Hashimoto [**55**]. In this period, homology theories had come to be accepted as the preferred algebraic context, in order to avoid the complications arising from the inevitable completions and topological algebra occuring in the generalized cohomology of infinite complexes. (Adams, [**2**, pp. 50-55], gives a very nice account of the reasons for this switch in emphasis.) An exception to this preference for ko_* over ko^* is in the 1981 work of Kane [**67**]. Another, in 1988, was Ossa's [**87**] calculation of $ku^*(BV)$ and $ko^*(BV)$, where V is elementary abelian, using the Adams spectral sequence, the earlier calculations for cyclic groups and the relation to representation theory.

In the early 1990s, the role of $ko_*(BG)$ as the recipient of the $\widehat{A}$-genus obstruction to the existence of metrics of positive scalar curvature (Sections 1.1 and 2.7) brought the problem of calculating $ko_*(BG)$ for many more groups to the attention of the first author. The work by his student Bayen [**17**] started by calculating $ko_*(BG)$ using the Adams spectral sequence. However, they soon realized [**18**] that the ring structure of $ko^*(BG)$ and $ku^*(BG)$ was crucial. In particular, for periodic groups, the Tate theory and the ko and ku homology are straightforward to deduce from the cohomology by periodicity, while the cohomologies $ko^*(BG)$ and $ku^*(BG)$ can be calculated by using the ring structure to propagate the relations deduced from periodic K-theory. For non-periodic groups, once the ring structure is known, we may use the local cohomology spectral sequence to give a more structured understanding of the homology groups $ko_*(BG)$ and $ku_*(BG)$. The Adams spectral sequence alone gives very little information about the ring structure, since the relations involve terms of mixed Adams filtration in all but the simplest cases. However, the Adams spectral sequence often implies that the crucial relations occur in dimensions in which the Bott map acts monomorphically and this shows that the relations are accessible through the representation rings [**28**]. This gives us the ability to determine the exact ring structure, and often provides the simplest derivation of Adams spectral sequence differentials as well. Already for C_{p^2}, D_8 or Q_8 there are multiplicative extensions which are obscure in the Adams spectral sequence [**28**]. Even for a group as simple as Q_{16}, there are additive extensions which are unclear in the Adams spectral sequence.

In the end, representation theoretic means have supplanted the use of the Adams spectral sequence in most parts of our work. However, the Adams spectral sequence still remains a useful way to do some accounting: it can tell us when we have enough generators and what sort of relations to look for. It also provides a different filtration than either the Bockstein spectral sequence or the local cohomology spectral sequence, so can help resolve extensions which are obscure in one or another of these. As is always the case, having multiple methods of getting information about the objects of interest gives a check on each and together they can give more information than any one of them alone; we have corrected many errors by these comparisons.

In the present century, more work has used equivariant methods, for example in the work of the second author [**45, 46**] and in the present volume. It is worth also noting Dugger's equivariant Atiyah-Hirzebruch spectral sequence [**35**].

Finally, we want to point out that, conceptually, the Atiyah-Hirzebruch and Bockstein spectral sequences can be seen as associated graded versions of the Adams spectral sequence, drawing in the structure in stages, rather than all at once, as the Adams spectral sequence does. To be precise, we can think of the passage from mod 2 cohomology to ko as taking place by three Bockstein spectral sequences:

(i) the Q_0, or p-Bockstein spectral sequence, computes integral cohomology,
(ii) the Q_1, or v-Bockstein spectral sequence, then gives complex connective K-theory, and finally
(iii) the Sq^2, cR or η-Bockstein spectral sequence then gives real connective K-theory.

If we combine the first two steps, we have the Adams spectral sequence proceeding from the mod 2 cohomology as a module over $E(Q_0, Q_1)$ directly to ku theory, while if we combine all three, we use the $\mathcal{A}(1)$-module structure to go directly to ko theory. It is no surprise that all these approaches taken together yield better information than any one alone can.

1.14. Comparisons with other theories

One might also compare the calculations of the connective K-theory of BG with what we know for other theories. The simplest answers are for periodic K-theory in the real and complex form. There one has a purely algebraic answer in terms of representation theory. The connective theories are a hybrid of this and the less predictable ordinary cohomology, an infinitesimal extension of the modified Rees ring.

The next comparison is with bordism. On the one hand connective K-theory is Noetherian and of finite Krull dimension, but on the other it usually has $\mathbb{Z}$-torsion and nilpotents. This should be contrasted with bordism, which is not Noetherian, but is often $\mathbb{Z}$-torsion free.

The usual approach to get better approximations to bordism whilst keeping Noetherian rings is to use higher chromatic approximations. The periodic forms are the Johnson-Wilson theories $E(n)$ or their completed Lubin-Tate counterparts. The results of Hopkins-Kuhn-Ravenel [**57, 58**] give a version of character theory and suggest one may hope for a higher version of representation theory. As for K-theory, one would expect this to come from an equivariant version of the theory. In higher Krull dimensions, there are many connective theories with the same periodic theory and one needs to understand the geometry to see which connective form is appropriate. In any case, at the superficial level one will again expect a hybrid of the representation theory and cohomology, perhaps again an infinitesimal extension of a modified Rees ring. Work of Strickland [**98**] on $BP\langle n\rangle$ for abelian groups already shows the complexity one may expect.

In forming K-theory with reality we used the group Q of order 2 since the only global automorphism of the multiplicative group is negation. For higher chromatic levels, this still gives an automorphism, so one may again seek versions with reality. This was done by Landweber [**74**] and Araki [**8, 9**] for bordism. Hu and Kriz [**60, 61**] used this to introduce real oriented versions of the Johnson-Wilson theories and Kitchloo and Wilson have investigated them further [**70, 71, 72**]. For completed or localized theories we expect more automorphisms of the associated geometry, and hence a variety of analogues of the real form.

As usual in higher dimensions, one cannot proceed by simple analogy: one needs to be guided by a detailed geometric understanding. Accordingly, the correct picture is only likely to emerge when we are able to follow the pattern established by Hopkins, Lurie and others in the construction of tmf. The complex periodic form comes from some family of geometric objects, which gives a moduli problem, and one hopes to construct a moduli stack in the world of E_∞-ringed spaces. In the case of K-theory there is only one multiplicative group, corresponding to KU. By considering abelian compact Lie groups one has a model for the abelian equivariant theory. From these one assembles a general equivariant theory from the category of subgroups. In the case of K-theory this gives the familiar equivariant KU. Finally, one has the global sections of this E_∞-ringed moduli stack. In the case of K-theory this gives KO. The connective versions are obtained by careful examination of the geometry. Thus the meaning of 'reality' will depend on the automorphisms of various objects, and the meaning of 'connective' on the geometry of the moduli stack.

In any case, one may hope that one may be able to adapt the methods of this book to extend the work of Hill [**56**] on $tmf^*(B\Sigma_3)$ a little further, though the complexity of the calculation of the coefficient ring of tmf shows that this will be a major undertaking.

1.15. Prerequisites

Although we have made some effort to summarize things about K-theory that we need, we have assumed that the reader is familiar with classifying spaces of compact Lie groups and the usual tools of ordinary cohomology, including cohomology operations and the Steenrod algebra. Similarly, we have assumed the reader is familiar with basic representation theory of finite groups in characteristic zero, has some knowledge of the classical Lie groups. We use a little of the language of commutative algebra, and we assume the reader is fluent with the language of spectral sequences. This standard material is available in many places. May's "Concise Course" [**82**] is a convenient source for the algebraic topology, Serre's book [**93**] on representation theory, Bröcker and tom Dieck [**22**] for the structure and representation theory of compact Lie groups. For commutative algebra we recommend the books of Matsumura [**81**] and Bruns and Herzog [**26**], and on spectral sequences McCleary's book [**84**].

Going further, there are several places where we have used stable homotopy theory in the sense that cohomology theories are represented. We have also used the Adams spectral sequence as a tool for calculating connective K-theory. All this is well covered in Adams's Chicago Lectures [**3**]. There are many useful techniques for using and displaying the Adams spectral sequence; the first author is preparing a systematic account [**25**], but for the present it is necessary to explain as we go along. In particular, we often need to discuss modules over the $\mathcal{A}(1) = \langle Sq^1, Sq^2\rangle$ and the exterior algebra $E(1)$, and we summarize the essentials in Appendix A.

There are also a few places where we have used the technology of highly structured ring spectra. The main point is that this gives a triangulated homotopy category of modules over a commutative ring spectrum with a symmetric monoidal tensor product. The second author has attempted to give an introduction to this in [**48**], but the original sources of [**36, 59**] should be consulted for details.

1.16. Reading this book

Assuming the prerequisites described above and some familiarity with the case of *complex* connective K-theory, the reader should read Chapter 2 to set the context. It is also almost essential to be familiar with the Bockstein spectral sequence as described in Chapter 4. Apart from this, the reader should be able to select the calculations of interest. In particular, it is not necessary to read Chapter 5 on characteristic classes before reading about the calculations for finite groups, though its contents permit a more coherent overview.

For calculations of ko homology or cohomology, the reader should first be familiar with the corresponding calculation with ku. For calculations of ko-homology the reader should look at the corresponding calculation of ko-cohomology first. It would be wise do look first at Chapter 6 to see how things work out in fairly simple examples.

Many readers interested in elementary abelian groups will not need to read Chapter 12, and readers only interested in small ranks or low degrees may be happy to check results from first principles rather than read Chapter 11 in detail.

In some of the diagrams, there is more information than fits easily into the space available. Rather than omit large numbers, we have sometimes left them in, hoping that the fact that they overflow the bounds of boxes will be a vivid indication of their size. Unfortunately this means that beauty and readability are sometimes compromised.

1.17. Thanks

This book and its predecessor [**28**] have been written over the period 1998-2010, and has benefitted from opportunities to meet (especially in Edale, Sheffield and Detroit). We are grateful for grants from EPSRC and the Midwest Topology Network that have made this possible.

The authors would like to thank various people for interest, advice and instruction including Neil Dummigan, Michael Joachim, Moty Katzman, Stephan Stolz, Don Davis and Neil Strickland. They would also like to thank Dilip Bayen, Arjun Malhotra, Kijti Rodtes and Nguyen Phuc Son for critical reading of earlier drafts, and for persistence in asking questions. Finally, we thank Mike Catanzaro for his TEX and *xy* expertise and the staff at the AMS for their adroit technical assistance.

CHAPTER 2

K-Theory with Reality

This chapter provides a number of the basic tools we need in the rest of the book. Much of this summarizes and extends classical material in a convenient form. Thus Sections 2.1 and 2.2 recall basic representation theory and periodic K-theory with an involution in the sense of Atiyah and Karoubi. In Section 2.3 we give an extended summary of the facts we need from Clifford algebras.

This equips us to describe in Section 2.4 the adaption of the construction of a fully equivariant version of the connective theory [**46**] to the case with reality. This provides the basis for the fully equivariant (uncompleted) version of our calculations. Sections 2.5 and 2.6 prove and exploit the local cohomology and completion theorems which provide the basic connections to non-equivariant theories. In Section 2.7 we show how well suited this is to the study of the Gromov-Lawson-Rosenberg conjecture on the existence of positive scalar curvature metrics on spin manifolds.

Finally, we consider the first few negative coefficient groups of the theory. By using the classical results of Bott (recalled in Section 2.8) and Atiyah-Hirzebruch we are able to show in Section 2.9 that ko^G_n in terms of representation theory for $n \geq -7$.

2.1. Representation theory

We summarize here the standard representation theory that we need. See [**4, 22**] for further details.

The basic structures are linear representations.

DEFINITION 2.1.1. When $K = \mathbb{R}, \mathbb{C}$ or $\mathbb{H}$ a representation of G over K is a K-vector space V with a K-linear action of G on V.

For the purpose of comparison, and for the study of multiplicative structure, it is worth giving an alternative approach.

DEFINITION 2.1.2. (i) A *real* representation of G is a representation V of G over $\mathbb{C}$ with a conjugate linear map $J : V \longrightarrow V$ with $J^2 = 1$.
(ii) A *quaternionic* representation of G is a representation V of G over $\mathbb{C}$ with a conjugate linear map $J : V \longrightarrow V$ with $J^2 = -1$.
(iii) A *complex* representation of G is the same as a representation over $\mathbb{C}$.

WARNING 2.1.3. To avoid confusion later, we emphasize that a real representation is a different thing to a representation over $\mathbb{R}$. Similarly a quaternionic representation is a different thing to a representation over $\mathbb{H}$.

LEMMA 2.1.4. *(i) There is an equivalence of categories between representations over $\mathbb{R}$ and real representations. A representation U over $\mathbb{R}$ gives rise to the real*

representation $\mathbb{C} \otimes_{\mathbb{R}} U$*, where* J *is given by complex conjugation on the* $\mathbb{C}$ *factor. Conversely, given a real representation* $(\hat{U}, J)$ *the* $(+1)$*-eigenspace of* J *gives a representation* $\hat{U}_+$ *over* $\mathbb{R}$.
(ii) There is an equivalence of categories between representations over $\mathbb{H}$ *and quaternionic representations. A representation* W *over* $\mathbb{H}$ *gives* W *the structure of a quaternionic representation, by taking* J *to act as* j*. Conversely a quaternionic representation* (W, J) *can be viewed as a representation over* $\mathbb{H}$ *where* j *acts as* J*.* □

We work largely with real, complex and quaternionic representations and we write $RO(G)$, $RU(G)$ and $RSp(G)$ for the Grothendieck groups of finite dimensional real, complex and quaternionic representations.

Forgetting the structure map J gives natural transformations

$$RO(G) \xrightarrow{c} RU(G) \xleftarrow{\tilde{c}} RSp(G),$$

both called *complexification.* There are also maps

$$RO(G) \xleftarrow{r} RU(G) \xrightarrow{q} RSp(G),$$

called *realification* and *quaternionification*; these are easily deduced from the evident constructions giving representations over $\mathbb{R}$ or $\mathbb{H}$. Finally there is *conjugation*

$$\tau : RU(G) \longrightarrow RU(G).$$

The composites of these maps have certain obvious properties.

LEMMA 2.1.5. *(i)* $rc = 2$, $cr = 1 + \tau$
(ii) $q\tilde{c} = 2$, $\tilde{c}q = 1 + \tau$.
(iii) Every real or quaternionic representation is self-conjugate: $\tau c = c$ *and* $\tau\tilde{c} = \tilde{c}$*. Also,* $r\tau = r$ *and* $q\tau = q$*.* □

Note that $RO(G)$ and $RU(G)$ are rings and $c : RO(G) \longrightarrow RU(G)$ is a ring homomorphism. However the product of quaternionic representations is real, and the product of a real and a quaternionic representation is quaternionic. For calculational purposes, use c to embed $RO(G)$ in $RU(G)$ and $\tilde{c}$ to embed $RSp(G)$ in $RU(G)$; all products are then compatible with the usual product on $RU(G)$ which can be calculated using characters. This is simply expressed by the commutativity of the diagrams

$$\begin{array}{ccc} RO \otimes RO & \longrightarrow & RO \\ \downarrow{\scriptstyle c\otimes c} & & \downarrow{\scriptstyle c} \\ RU \otimes RU & \longrightarrow & RU \end{array} \qquad \begin{array}{ccc} RO \otimes RSp & \longrightarrow & RSp \\ \downarrow{\scriptstyle c\otimes\tilde{c}} & & \downarrow{\scriptstyle \tilde{c}} \\ RU \otimes RU & \longrightarrow & RU \end{array} \qquad \begin{array}{ccc} RSp \otimes RSp & \longrightarrow & RO \\ \downarrow{\scriptstyle \tilde{c}\otimes\tilde{c}} & & \downarrow{\scriptstyle c} \\ RU \otimes RU & \longrightarrow & RU \end{array}$$

in which the horizontal maps are tensor product of representations.

Each simple representation is either real, complex or quaternionic. Simple complex representations come in non-isomorphic conjugate pairs. We suppose $\{U_i\}_{i\in I}$ is the list of simple real representations, $\{V_j, \tau V_j\}_{j\in J}$ is the list of simple complex representations, and $\{W_i\}_{i\in I}$ is the list of simple quaternionic representations. In this case, the additive bases of the representation groups are as follows:

- $RO(G) = \mathbb{Z}\{U_i, rV_j, r\tilde{c}W_k \mid i \in I, j \in J, k \in K\}$

- $RU(G) = \mathbb{Z}\{cU_i, V_j, \tau V_j, \tilde{c}W_k \mid i \in I, j \in J, k \in K\}$
- $RSp(G) = \mathbb{Z}\{qcU_i, qV_j, W_k \mid i \in I, j \in J, k \in K\}$

2.2. The periodic case

Suppose given a group $\widehat{G}$ in a short exact sequence

$$1 \longrightarrow G \longrightarrow \widehat{G} \longrightarrow Q \longrightarrow 1,$$

where Q is of order 2. We write G^* for the non-identity coset of G. This notation will be fixed throughout Chapters 2, 3 and 4.

For example if G has an automorphism τ of order 2 we may take $\widehat{G}$ to be the split extension; Atiyah and Segal [**14**] restrict themselves to this case, but Karoubi [**68**] permits the general case. In fact the only case of relevance in the present work will be $\widehat{G} = G \times Q$, but the structure is clarified by appropriate generality.

When an extension $\widehat{G}$ is specified, we may make the following definition.

DEFINITION 2.2.1. A *real* G-representation is a $\mathbb{C}$-vector space V with an action of $\widehat{G}$ so that G acts through $\mathbb{C}$-linear maps and G^* acts through $\mathbb{C}$-antilinear maps. A morphism of real G-representations is a $\mathbb{C}$-linear map commuting with $\widehat{G}$-actions. The category of real G-representations is written Rep$\mathbb{R}(G)$; it is symmetric monoidal under direct sum.

REMARK 2.2.2. The terminology is due to Atiyah, and leads to little difficulty provided one always maintains the distinction between a real representation and a representation over $\mathbb{R}$. Indeed, the classical case corresponds to $\widehat{G} = G \times Q$. In this case, a real representation of G in the sense of Definition 2.2.1 is a complex representation of G and $1 \times Q$ supplies a conjugate linear map J with $J^2 = 1$ so as to obtain a real representation in the sense of 2.1.2. We have already seen that real representations in the sense of 2.1.2 correspond to represention over $\mathbb{R}$.

In the classical case, a representation U over $\mathbb{R}$ gives rise to a real representation $\mathbb{C} \otimes_{\mathbb{R}} U$. This construction admits a generalization.

CONSTRUCTION 2.2.3. A representation U of G over $\mathbb{R}$ gives an induced representation $\hat{U} = \mathbb{R}\widehat{G} \otimes_{\mathbb{R}G} U$ of $\widehat{G}$. This admits the structure of a real representation where J is $+1$ on the identity coset $\mathbb{R}G \otimes_{\mathbb{R}G} U$, and -1 on the G^* coset $\mathbb{R}G^* \otimes_{\mathbb{R}G} U$.

DEFINITION 2.2.4. (i) A *real* G-space is the same thing as a $\widehat{G}$-space.

(ii) A *real* G-vector bundle over a real G-space X is a real $\widehat{G}$-vector bundle $\xi \longrightarrow X$ so that the fibres of ξ are complex vector spaces, and so that $g : \xi_x \longrightarrow \xi_{gx}$ is $\mathbb{C}$-linear for $g \in G$ and $\mathbb{C}$-antilinear if $g \in G^*$. The category of real G-vector bundles over X is written Vect$\mathbb{R}_G(X)$.

(iii) Real K-theory is defined by

$$K\mathbb{R}^0_G(X) = Gr\mathrm{Vect}\mathbb{R}_G(X).$$

THEOREM 2.2.5. *(Karoubi* [**68**, 3.9]*)* $K\mathbb{R}^0_G(\cdot)$ *extends to a cohomology theory on* $\widehat{G}$*-spaces with suspension isomorphisms for representations of* $\widehat{G}$*. The theory is Bott periodic for representations* $\hat{U}$ *obtained from representations* U *of* G *over* $\mathbb{R}$*, in the sense that there is a Thom class giving an isomorphism*

$$K\mathbb{R}^{\alpha}_G(X) \cong K\mathbb{R}^{\alpha+\hat{U}}_G(X)$$

for all $\alpha \in RO(\widehat{G})$*.*

Proof: The suspension isomorphism allows one to define

$$\widetilde{KR}_G^{-V}(X) = \widetilde{KR}_G^0(S^V \wedge X)$$

for all representations V.

Karoubi establishes Bott periodicity in the form

$$KR_G^{\hat{U}}(X) \cong KR_G^0(X).$$

Since representations of this form are cofinal amongst all representations of $\widehat{G}$, this can be used to extend to a theory graded over $RO(\widehat{G})$ which is periodic for representations $\hat{U}$. □

REMARK 2.2.6. The standard notation for a $\widehat{G}$-equivariant cohomology theory would have a subscript $\widehat{G}$. We have nonetheless retained the classical notation in the present section $K\mathbb{R}_G^*$ since its use is universal, and justify it on the grounds that, as reflected in the terminology, we are thinking of the cohomology theory as defined on a category of G-spaces with additional structure. For ease of comparison we have also used this notation in Section 2.4.

When arguing with representing $\widehat{G}$-spectra it is impossible to retain this classical notation since $K\mathbb{R}$ would need to be treated differently from other spectra. Accordingly, we usually use the standard notation (i.e., with the group referring to the equivariance), for example in Sections 2.5 and 2.6 and throughout Chapter 3. We have attempted to ensure that it is always clear which of the two notational conventions is in force at any time.

COROLLARY 2.2.7. *The $\widehat{G}$-equivariant cohomology theory $K\mathbb{R}_G^*(\cdot)$ is represented by a $\widehat{G}$-spectrum $K\mathbb{R}$.* □

LEMMA 2.2.8. *(i) As a G-spectrum $K\mathbb{R}$ coincides with Atiyah-Segal G-equivariant complex K-theory KU.*
(ii) If $\widehat{G} = G \times Q$ then the Lewis-May Q-fixed point spectrum represents Atiyah-Segal G-equivariant orthogonal K-theory:

$$K\mathbb{R}^Q = KO.$$

Proof: Part (i) amounts to the statement that a real vector bundle over an extended $\widehat{G}$-space $X = \widehat{G} \times_G Y$ is uniquely determined by a complex vector bundle over Y.

Part (ii) amounts to the statement that a real vector bundle over a Q-fixed $\widehat{G}$-space is equivalent to an $\mathbb{R}$-vector bundle. The proof is the same as in 2.1.4: complexification and taking $(+1)$-eigenspaces give the equivalence. □

REMARK 2.2.9. Note that the restriction to $\widehat{G} = G \times Q$ is necessary to obtain a G-equivariant conclusion. We need to know that Q is a normal subgroup of $\widehat{G}$ with quotient G.

We can describe the coefficients KO_G^* and KU_G^* conveniently in terms of the notation from the end of the preceding section. First, we recall the coefficients of the nonequivariant theories:

$$KU^* = \mathbb{Z}[v, v^{-1}]$$

with $v \in KU^{-2}$, and

$$KO^* = \mathbb{Z}[\eta, \alpha, \beta]/(2\eta, \eta^3, \eta\alpha, \alpha^2 - 4\beta)$$

with $\eta \in KO^{-1}$, $\alpha \in KO^{-4}$, and $\beta \in KO^{-8}$.

We will find it useful to note the following.

LEMMA 2.2.10. $\tau v = -v$

Proof: This is simply the fact that the conjugate (dual) of the Hopf bundle on $\mathbb{C}P^1$ is its negation [**11**, p. 115]. □

It is also useful to exhibit the maps induced by complexification and realification which we will deduce in Lemmas 2.3.10 and 2.3.11.

LEMMA 2.2.11. *Complexification $KO^* \xrightarrow{c} KU^*$ is the ring homomorphism given by $c(\eta) = 0$, $c(\alpha) = 2v^2$ and $c(\beta) = v^4$. Realification $KU^* \xrightarrow{r} KO^*$ is the KO^*-module homomorphism $r(1) = 2$, $r(v) = \eta^2$, $r(v^2) = \alpha$ and $r(v^3) = 0$.*

Proof: This is an immediate consequence of Lemmas 2.3.10 and 2.3.11 using the identities satisfied by r,c,q and $\tilde{c}$ (Lemma 2.1.5). For example, KO^{-4} is generated by α, represented by $qc(1)$, where 1 is the trivial 1-dimensional real representation. By Lemma 2.3.10, $c(\alpha)$ is then represented by $\tilde{c}qc(1) = 2c(1)$, which is $2v^2$. □

THEOREM 2.2.12. (i) $KU_G^* = RU(G)[v, v^{-1}]$.

(ii) $KO_G^* = RO^*(G)[\beta, \beta^{-1}]$ *where*

$$\begin{array}{lcllcl}
RO^0(G) & = & RO(G) & \cong & \mathbb{Z}\{U_i, rV_j, r\tilde{c}W_k \mid i \in I,\, j \in J,\, k \in K\} \\
RO^{-1}(G) & = & RO(G)/RU(G) & \cong & \mathbb{F}_2\{U_i \mid i \in I\} \\
RO^{-2}(G) & = & RU(G)/RSp(G) & \cong & \mathbb{F}_2\{cU_i \mid i \in I\} \oplus \mathbb{Z}\{\overline{V}_j \mid j \in J\} \\
RO^{-3}(G) & = & 0 & & \\
RO^{-4}(G) & = & RSp(G) & \cong & \mathbb{Z}\{qcU_i, qV_j, W_k \mid i \in I,\, j \in J,\, k \in K\} \\
RO^{-5}(G) & = & RSp(G)/RU(G) & \cong & \mathbb{F}_2\{W_k \mid k \in K\} \\
RO^{-6}(G) & = & RU(G)/RO(G) & \cong & \mathbb{Z}\{\overline{V}_j \mid j \in J\} \oplus \mathbb{F}_2\{\tilde{c}W_k \mid k \in K\} \\
RO^{-7}(G) & = & 0 & &
\end{array}$$

In $RO^{-2}(G)$ and $RO^{-6}(G)$, $\overline{V} = [V] = -[\tau V]$ is the equivalence class of V in

$$\mathbb{Z}\{\overline{V}_j \mid j \in J\} = \mathbb{Z}\{V_j, \tau V_j \mid j \in J\}/(\{V_j + \tau V_j \mid j \in J\}).$$

Multiplication by η gives the evident epimorphisms $RO^{4n} \twoheadrightarrow RO^{4n-1}$ and the evident monomorphisms $RO^{4n-1} \rightarrowtail RO^{4n-2}$. Multiplication by α is $qc : RO^0 \longrightarrow RO^{-4}$, $r\tilde{c} : RO^{-4} \longrightarrow RO^{-8} \cong RO^0$, zero on RO^{4n-1}, zero on the torsion in RO^{4n-2} and multiplication by 2 on the torsion-free summands of RO^{4n-2}.

Proof: The values of KU_G^* and KO_G^* follow immediately from the Atiyah-Segal theorems, as summarized in Theorem 2.3.9. The action of η can be deduced by tensoring with the Clifford module representing it (see Subsection 2.3.D). In principle, the action of α could be calculated in the same way, but it is simpler to compute this by comparison with the complex case. The proofs that multiplication by η and

α are as stated are given in Propositions 2.3.13 and 2.3.14. □

2.3. Clifford modules

Atiyah, Bott and Shapiro [**13**] used the language of Clifford modules to give a systematic account of periodicity properties of KO-theory. We need to prove some properties that they did not make explicit, so it is necessary to give a brief summary of results from [**13**]. An account in the non-equivariant setting can be found in Bröcker and tom Dieck [**22**] or Husemoller [**62**].

2.3.A. Clifford algebras. Given a real vector space V with a quadratic form Q, the Clifford algebra $C(Q)$ is the quotient of the tensor algebra $T(V)$ by the ideal generated by by the elements $x \otimes x - Q(x) \cdot 1$ for $x \in V$. Since the ideal is generated by even degree elements, $C(Q)$ is $\mathbb{Z}/2$ graded, with $C(Q)^0$ generated by the images of even tensors and $C(Q)^1$ generated by odd tensors. In particular, we may take V to be $\mathbb{R}^k$ with the basis $e_1, e_2, \ldots, e_k$ and consider the diagonal negative definite quadratic form. We write C_k for the associated Clifford algebra, so that it is generated by symbols $e_1, \ldots, e_k$ which anticommute and $(e_i)^2 = -1$. Similarly, we write C'_k for the Clifford algebra associated to the positive definite diagonal quadratic form, which is generated by symbols $e'_1, \ldots, e'_k$ which anticommute and $(e'_i)^2 = +1$.

There are some places where we find it convenient to be very explicit. For brevity we write $K(n)$ for the algebra of $n \times n$ matrices over K.

EXAMPLE 2.3.1. (a) We have $C_0 \cong \mathbb{R}$.
(b) We have $C_1 \cong \mathbb{C}$ where i corresponds to e_1.
(c) $C_2 \cong \mathbb{H}$ where i corresponds to e_1, j to e_2, and k to $e_1 e_2$.

EXAMPLE 2.3.2. (a) We have $C'_0 \cong \mathbb{R}$
(b) We have $C'_1 \cong \mathbb{R} \times \mathbb{R}$ where the idempotents are $f_{11} = (1 + e'_1)/2$ and $f_{22} = (1 - e'_1)/2$.
(c) $C'_2 \cong \mathbb{R}(2)$ where we take f_{11} to be $\begin{pmatrix} 1 & 0 \\ 0 & 0 \end{pmatrix}$, f_{22} to be $\begin{pmatrix} 0 & 0 \\ 0 & 1 \end{pmatrix}$, $f_{12} = (e'_2 + e'_1 e'_2)/2$ to be $\begin{pmatrix} 0 & 1 \\ 0 & 0 \end{pmatrix}$ and $f_{21} = (e'_2 - e'_1 e'_2)/2$ to be $\begin{pmatrix} 0 & 0 \\ 1 & 0 \end{pmatrix}$.

This is enough for us to identify all the C_k and C'_k using the following lemma, which is the root of all periodicity phenomena.

LEMMA 2.3.3. [**13**, 4.2] *There are isomorphisms* $C_{k+2} \cong C'_k \otimes_{\mathbb{R}} C_2$ *and* $C'_{k+2} \cong C_k \otimes_{\mathbb{R}} C'_2$.

Proof: The first isomorphism is given by $\psi(e_i) = 1 \otimes e_i$ for $i = 1, 2$ and $\psi(e_i) = e'_{i-2} \otimes e_1 e_2$ for $i \geq 3$. The second is obtained by exchanging the roles of the e_is and the e'_is. □

Using these, we may complete the following table. Since $C_2 \otimes C'_2 \cong \mathbb{H}(2)$, we have $C_{k+4} \cong C_k \otimes \mathbb{H}(2)$ and $C'_{k+4} \cong C'_k \otimes \mathbb{H}(2)$. Finally, the isomorphism $\mathbb{H} \otimes_{\mathbb{R}} \mathbb{H} \cong \mathbb{R}(4)$ (see below for the definition) gives the Morita equivalence which gives Bott periodicity,

$$C_{k+8} \cong C_k \otimes \mathbb{R}(16), \text{ and}$$

$$C'_{k+8} \cong C'_k \otimes \mathbb{R}(16).$$

k	C_k	C'_k
0	$\mathbb{R}$	$\mathbb{R}$
1	$\mathbb{C}$	$\mathbb{R} \times \mathbb{R}$
2	$\mathbb{H}$	$\mathbb{R}(2)$
3	$\mathbb{H} \times \mathbb{H}$	$\mathbb{C}(2)$
4	$\mathbb{H}(2)$	$\mathbb{H}(2)$
5	$\mathbb{C}(4)$	$\mathbb{H}(2) \times \mathbb{H}(2)$
6	$\mathbb{R}(8)$	$\mathbb{H}(4)$
7	$\mathbb{R}(8) \times \mathbb{R}(8)$	$\mathbb{C}(8)$
8	$\mathbb{R}(16)$	$\mathbb{R}(16)$

Here we have used the well-known facts that $\mathbb{C} \otimes_{\mathbb{R}} \mathbb{C} \cong \mathbb{C} \times \mathbb{C}$, $\mathbb{H} \otimes_{\mathbb{R}} \mathbb{C} \cong \mathbb{C}(2)$ and $\mathbb{H} \otimes_{\mathbb{R}} \mathbb{H} \cong \mathbb{R}(4)$. These are best expressed by giving algebra isomorphisms ω_F, $F = \mathbb{C}$ or $\mathbb{H}$:

$$\begin{array}{ccc} \mathbb{C} \otimes_{\mathbb{R}} \mathbb{H} & \xrightarrow{\omega_{\mathbb{C}}} & \mathrm{Hom}_{\mathbb{C}}(\mathbb{H}, \mathbb{H}) \cong \mathbb{C}(2) \\ \downarrow & & \downarrow \\ \mathbb{H} \otimes_{\mathbb{R}} \mathbb{H} & \xrightarrow{\omega_{\mathbb{H}}} & \mathrm{Hom}_{\mathbb{R}}(\mathbb{H}, \mathbb{H}) \cong \mathbb{R}(4) \end{array}$$

where $\omega_F(x_1 \otimes x_2)(x) = x_1 x \overline{x}_2$. Restricting to $\mathbb{R} \otimes_{\mathbb{R}} \mathbb{H}$ and $\mathbb{R} \otimes_{\mathbb{R}} \mathbb{C}$ we obtain familiar embeddings of $\mathbb{H}$ in $\mathbb{C}(2)$ and $\mathbb{C}$ in $\mathbb{R}(2)$:

$$\alpha + \beta j \mapsto \begin{pmatrix} \overline{\alpha} & \overline{\beta} \\ -\beta & \alpha \end{pmatrix}$$

and

$$a + bi \mapsto \begin{pmatrix} a & -b \\ b & a \end{pmatrix}.$$

Further, restricting $\omega_{\mathbb{C}}$ to $\mathbb{C} \otimes_{\mathbb{R}} \mathbb{C}$ gives the isomorphism $\omega'(\alpha \otimes \beta) = (\alpha\overline{\beta}, \alpha\beta)$:

$$\begin{array}{ccc} \mathbb{C} \otimes_{\mathbb{R}} \mathbb{C} & \xrightarrow{\omega'} & \mathbb{C} \times \mathbb{C} \\ \downarrow & & \downarrow \\ \mathbb{C} \otimes_{\mathbb{R}} \mathbb{H} & \xrightarrow{\omega_{\mathbb{C}}} & \mathbb{C}(2) \end{array}$$

Using these isomorphisms, we find that the standard inclusions of $\mathbb{R}^k$ in $\mathbb{R}^{k+1}$, induce the inclusions of C_k in C_{k+1} one would expect. More specifically, the expected maps are built from

- the standard inclusions $\mathbb{R} \longrightarrow \mathbb{C} \longrightarrow \mathbb{H}$,
- the inclusions $F \longrightarrow F \times F$ of the diagonal,
- the inclusion $F \times F \longrightarrow F(2)$ of diagonal matrices, and
- the embeddings $H \longrightarrow \mathbb{C}(2)$ and $\mathbb{C} \longrightarrow \mathbb{R}(2)$ above.

We may also deduce the even parts.

LEMMA 2.3.4. [**13**, 5.4] *For $k \geq 0$ there is an isomorphism $C_k \cong C^0_{k+1}$ induced by $\phi(e_i) = e_i e_{k+1}$.* □

REMARK 2.3.5. The analogous result for $(C'_{k+1})^0$ breaks the symmetry: we have $(C'_{k+1})^0 \cong C_k$, not C'_k.

REMARK 2.3.6. We will also need to refer to Clifford algebras for complex vector spaces, and we write $C_k(\mathbb{C})$ for the complex analogue of $C_k = C_k(\mathbb{R})$.

Evidently, $C_k(\mathbb{C}) = C_k(\mathbb{R}) \otimes_{\mathbb{R}} \mathbb{C}$, and we see $C_1(\mathbb{C}) = \mathbb{C} \times \mathbb{C}$ and $C_2(\mathbb{C}) \cong \mathbb{C}(2)$. Since the positive and negative diagonal quadratic forms are equivalent over $\mathbb{C}$ we have $C(\mathbb{C})_k \cong C(\mathbb{C})'_k$, and thus we have 2-periodicity

$$C_{k+2}(\mathbb{C}) \cong C_k(\mathbb{C}) \otimes_{\mathbb{C}} \mathbb{C}(2).$$

2.3.B. Clifford modules. Now consider G-equivariant graded modules over C_k, and let $M_G(C_k)$ denote the Grothendieck group of these. Similarly, let $N_G(C_k^0)$ denote the Grothendieck group of (ungraded) equivariant modules over C_k^0.

LEMMA 2.3.7. *For $k \geq 1$, extension of scalars gives an isomorphism*

$$N_G(C_k^0) \cong M_G(C_k).$$

REMARK 2.3.8. A graded C_0 module M consists of two unrelated $\mathbb{R}$-modules M^0 and M^1. If we adopt the convention that $C_0^0 = \mathbb{R} \times \mathbb{R}$, then the above equivalence holds also for $k = 0$, although the equivalence is no longer simply extension of scalars.

Using Morita equivalences, we find the following table for $M_G(C_k)$ and the maps induced by inclusions $C_k \longrightarrow C_{k+1}$, together with their cokernels.

k	$M_G(C_k)$	$\operatorname{cok}(M_G(C_{k+1}) \longrightarrow M_G(C_k))$
0	$RO(G) \times RO(G)$	$RO(G)$
	$\uparrow \Delta$	
1	$RO(G)$	$RO(G)/RU(G)$
	$\uparrow r$	
2	$RU(G)$	$RU(G)/RSp(G)$
	$\uparrow \tilde{c}$	
3	$RSp(G)$	0
	$\uparrow \nabla$	
4	$RSp(G) \oplus RSp(G)$	$RSp(G)$
	$\uparrow \Delta$	
5	$RSp(G)$	$RSp(G)/RU(G)$
	$\uparrow q$	
6	$RU(G)$	$RU(G)/RO(G)$
	$\uparrow c$	
7	$RO(G)$	0
	$\uparrow \nabla$	
8	$RO(G) \oplus RO(G)$	$RO(G)$
	$\uparrow \Delta$	

Finally, the connection with K-theory uses the construction which takes a graded module $M = (M^0, M^1)$ over C_k and forms the KO-theory class $\chi(M)$ consisting of the pair of bundles $(D^k \times M^0, D^k \times M^1)$ over $D^k = D(\mathbb{R}^k)$ with the isomorphism of their restriction to S^{k-1} given by $(x, v) \mapsto (x, xv)$.

THEOREM 2.3.9. [**14**, Section 8] *For $k \geq 1$ there is an exact sequence*

$$M_G(C_{k+1}) \longrightarrow M_G(C_k) \xrightarrow{\chi} KO_G(D^k, S^{k-1}) \longrightarrow 0$$

giving

$$KO_k^G \cong KO_G(D^k, S^{k-1})$$

in terms of the representation theory of Clifford algebras. □

2.3.C. Complexification. We will need to compare the real case to the much simpler complex case.

Using Morita equivalences, we find the following table for $M_G(C_k(\mathbb{C}))$ and the inclusions induced by $C_k(\mathbb{C}) \longrightarrow C_{k+1}(\mathbb{C})$, together with their cokernels.

k	$M_G(C_k(\mathbb{C}))$	$\mathrm{cok}(M_G(C_{k+1}(\mathbb{C})) \longrightarrow M_G(C_k(\mathbb{C})))$
0	$RU(G) \times RU(G)$	$RU(G)$
	$\uparrow \Delta$	
1	$RU(G)$	0
	$\uparrow \nabla$	
2	$RU(G) \times RU(G)$	$RU(G)$
	$\uparrow \Delta$	

The χ-construction of bundles works as in the real case, and we have the following compatibility statement.

LEMMA 2.3.10. *Complexification induces a diagram*

$$\begin{array}{ccccccc} M_G(C_{k+1}) & \longrightarrow & M_G(C_k) & \xrightarrow{\chi} & KO_G(D^k, S^{k-1}) & \longrightarrow & 0 \\ \otimes_{\mathbb{R}}\mathbb{C}\downarrow & & \otimes_{\mathbb{R}}\mathbb{C}\downarrow & & \otimes_{\mathbb{R}}\mathbb{C}\downarrow & & \\ M_G(C_{k+1}(\mathbb{C})) & \longrightarrow & M_G(C_k(\mathbb{C})) & \xrightarrow{\chi} & KU_G(D^k, S^{k-1}) & \longrightarrow & 0 \end{array}$$

and hence complexification induces $c : RO(G) \longrightarrow RU(G)$ in degree 0, and $\tilde{c} : RSp(G) \longrightarrow RU(G)$ in degree 4. In degrees 2 and 6, the maps $RU(G)/RSp(G) \longrightarrow RU(G)$ and $RU(G)/RO(G) \longrightarrow RU(G)$ are both induced by $1 - \tau : RU(G) \longrightarrow RU(G)$.

Proof: Most of the lemma is clear, but it is worth commenting on behaviour in degrees 2 and 6. For degree 2, the real case gives

$$\mathbb{C} = C_2^0 \longrightarrow C_3^0 = \mathbb{H},$$

and the corresponding complex case is

$$\mathbb{C} \times \mathbb{C} = \mathbb{C} \otimes_{\mathbb{R}} \mathbb{C} = C_2^0 \longrightarrow C_3^0 = \mathbb{H} \otimes_{\mathbb{R}} \mathbb{C} = \mathbb{C}(2).$$

A module M over $C = C_2^0$ is complexified to $M \oplus \tau M = M \otimes_{\mathbb{R}} \mathbb{C}$; working modulo modules $N \oplus N$ (which are restrictions of modules over $\mathbb{C}(2)$), we obtain the stated formula. The argument in degree 6 is precisely analogous. □

Similarly, we can consider the restriction.

LEMMA 2.3.11. *Restriction along $C_k \longrightarrow C_k(\mathbb{C})$ induces a diagram*

$$\begin{array}{ccccccc} M_G(C_{k+1}(\mathbb{C})) & \longrightarrow & M_G(C_k(\mathbb{C})) & \xrightarrow{\chi} & KU_G(D^k, S^{k-1}) & \longrightarrow & 0 \\ \downarrow & & \downarrow & & \downarrow & & \\ M_G(C_{k+1}) & \longrightarrow & M_G(C_k) & \xrightarrow{\chi} & KO_G(D^k, S^{k-1}) & \longrightarrow & 0 \end{array}$$

and hence restriction induces realification $r : RU(G) \longrightarrow RO(G)$ in degree 0, and quaternionification $q : RU(G) \longrightarrow RSpG)$ in degree 4. In degrees 2 and 6, the *maps $RU(G) \longrightarrow RU(G)/RSp(G)$ and $RU(G) \longrightarrow RU(G)/RO(G)$ are the natural quotient maps.*

Proof: This is straightforward except in degree 4, where it is the assertion that the Morita equivalence $N_G(\mathbb{C}) \longrightarrow N_G(\mathbb{C}(2))$ composed with restriction along $\mathbb{H} \longrightarrow \mathbb{C}(2)$ is the same as induction along $\mathbb{C} \longrightarrow \mathbb{H}$. This follows because the left $\mathbb{H}$ module $\mathbb{C}^2 \otimes_{\mathbb{C}} M$ is isomorphic to $\mathbb{H} \otimes_{\mathbb{C}} M$. □

2.3.D. Explicit representatives. We may now describe explicit representatives for $\eta \in KO_1$, $\alpha \in KO_4$ and $\beta \in KO_8$. Indeed, we see that $C_1 \cong \mathbb{C}$ as a module over itself generates the cokernel of $M(C_2) \longrightarrow M(C_1)$, and hence that $\chi(C_1)$ represents η. Similarly, the left factor $\mathbb{H}_L = C_4^0 f_{11}$ of $C_4^0 \cong \mathbb{H} \times \mathbb{H}$ generates the cokernel of $N(C_5^0) \longrightarrow N(C_4^0)$, and it corresponds to the left hand column module $LC_{\mathbb{H}} = C_4 f_{11}$ over C_4. A representative for α is thus $\chi(LC_{\mathbb{H}})$. Again, the left factor $\mathbb{R}(8)_L$ of $C_8^0 \cong \mathbb{R}(8) \times \mathbb{R}(8)$ generates the cokernel of $N(C_9^0) \longrightarrow N(C_8^0)$, and it corresponds to the left hand column module $LC_{\mathbb{R}(8)}$ over C_8. A representative for β is thus $\chi(LC_{\mathbb{R}(8)})$.

We can now finish the derivation of the action of the coefficients KO^* on KO_G^*. The action of η is induced by tensoring graded $C_k[G]$ modules with C_1 and treating the result as a $C_{k+1}[G]$ module using the the isomorphism $C_1 \otimes C_k \cong C_{k+1}$. Equivalently, using ungraded C_k^0 modules, it is given by the composite

$$N_G(C_k^0) \cong M_G(C_k) \xrightarrow{C_1 \otimes -} M_G(C_{k+1}) \cong N_G(C_{k+1}^0).$$

LEMMA 2.3.12. *Multiplication by $\eta : N_G(C_k^0) \longrightarrow N_G(C_{k+1}^0)$ is induction along the inclusion $C_k^0 \longrightarrow C_{k+1}^0$:*

$$\eta M = C_{k+1}^0 \otimes_{C_k^0} M$$

Proof: It is sufficient to show that the composite displayed before the lemma sends C_k^0 to C_{k+1}^0. Indeed, the graded module C_k^0 corresponds to the graded module $(C_k^0, e_1 C_k^0)$. Tensoring with $C_1 = (\mathbb{R}, e_{k+1}\mathbb{R})$ and taking the even part gives

$$\mathbb{R} \otimes C_k^0 \oplus e_{k+1}\mathbb{R} \otimes e_1 C_k^0 = C_k^0 \oplus e_{k+1} e_1 C_k^0 = C_{k+1}^0.$$

□

PROPOSITION 2.3.13. *Multiplication by η induces the natural quotient maps $RO(G) \longrightarrow RO(G)/RU(G)$ and $RSp(G) \longrightarrow RSp(G)/RU(G)$ in degrees* 0 *and* 4 *modulo* 8*, and induces complexification $RO(G)/RU(G) \longrightarrow RU(G)/RSp(G)$ and $RSp(G)/RU(G) \longrightarrow RU(G)/RO(G)$ in degrees* -1 *and* -5 *modulo* 8.

Proof: Clearly it suffices to calculate the homomorphism $N_G(C_k^0) \longrightarrow N_G(C_{k+1}^0)$ induced by η. We treat the four cases separately.

Degree 0 is exceptional (see Remark 2.3.8). The map $M_G(C_0) \longrightarrow M_G(C_1)$ sends a graded C_0-module (M^0, M^1) to the graded C_1 module $(M^0 \oplus M^1, e_1(M^0 \oplus M^1)$. Modulo restrictions of C_1 and C_2 modules, respectively, this is exactly the quotient map $RO(G) \longrightarrow RO(G)/RU(G)$.

In degree -1, the preceding lemma says that we have the map induced by passage to quotients from complexification $RO(G) \longrightarrow RU(G)$.

In degree -4, as in degree 0, we get the map induced on quotients by direct sum $RSp(G) \times RSp(G) \longrightarrow RSp(G)$. Precisely, we induce up along $\mathbb{H} \times \mathbb{H} \longrightarrow \mathbb{H}(2)$ and then use the Morita equivalence between $\mathbb{H}(2)$ modules and $\mathbb{H}$ modules. This sends the $\mathbb{H} \times \mathbb{H}$ module $M_0 \oplus M_1$ (in which the left and right factors $\mathbb{H}$ act as usual on M_0 and M_1, respectively, and act trivially upon the other summand) to

$$\mathbb{H}^2 \otimes_{\mathbb{H}(2)} \mathbb{H}(2) \otimes_{\mathbb{H}\times\mathbb{H}} M_0 \oplus M_1 \cong \mathbb{H}^2 \otimes_{\mathbb{H}\times\mathbb{H}} M_0 \oplus M_1 \cong M_0 \oplus M_1$$

with the diagonal $\mathbb{H}$ action.

Finally, in degree -5, we have induction along $\mathbb{H}(2) \longrightarrow \mathbb{C}(4)$ composed with Morita equivalences on each end. This is equivalent to restriction along $\mathbb{C} \longrightarrow \mathbb{H}$ by the isomorphisms

$$\begin{aligned} \mathbb{C}^4 \otimes_{\mathbb{C}(4)} \mathbb{C}(4) \otimes_{\mathbb{H}(2)} \mathbb{H}^2 \otimes_{\mathbb{H}} M &\cong \mathbb{C}^4 \otimes_{\mathbb{H}(2)} \mathbb{H}^2 \otimes_{\mathbb{H}} M \\ &\cong \mathbb{H}^2 \otimes_{\mathbb{H}(2)} \mathbb{H}^2 \otimes_{\mathbb{H}} M \\ &\cong \mathbb{H} \otimes_{\mathbb{H}} M \\ &\cong M, \end{aligned}$$

thought of as a $\mathbb{C}$ module now. Here we are making the identification of $\mathbb{C}^4$ with $\mathbb{H}^2$ as a left $\mathbb{C}$, right $\mathbb{H}(2)$ module. Thus, η multiplication $N_G(C_5^0) \longrightarrow N_G(C_6^0)$ is identified with complexification $RSp(G) \longrightarrow RU(G)$ which is induced by restriction of scalars from $\mathbb{H}$ to $\mathbb{C}$. $\square$

To compute the effect of multiplication by α, we use the comparison with complex K-theory.

PROPOSITION 2.3.14. *Multiplication by α induces*

- $qc : RO(G) \longrightarrow RSp(G)$ *in degrees* 0 *modulo* 8
- $0 : RO(G)/RU(G) \longrightarrow RSp(G)/RU(G)$ *in degrees* -1 *modulo* 8
- $2 : RU(G)/RSp(G) \longrightarrow RU(G)/RO(G)$ *in degrees* -2 *modulo* 8
- $r\tilde{c} : RSp(G) \longrightarrow RO(G)$ *in degrees* -4 *modulo* 8
- $0 : RSp(G)/RU(G) \longrightarrow RO(G)/RU(G)$ *in degrees* -5 *modulo* 8
- $2 : RU(G)/RO(G) \longrightarrow RU(G)/RSp(G)$ *in degrees* -6 *modulo* 8

REMARK 2.3.15. Thus, $\alpha : RO(G) \longrightarrow RSp(G)$ is an isomorphism on the summands corresponding to real representations, is multiplication by 2 on the summands corresponding to complex representations, and is multiplication by 4 on the summands corresponding to quaternionic representations (see Theorem 2.2.12). The roles are reversed for $\alpha : RSp(G) \longrightarrow RO(G)$. This is the factorization of $\alpha^2 = 4\beta$ in degrees 0 modulo 4.

Proof: Since $\alpha\eta = 0$, and every element in $RO(G)/RU(G)$ or $RSp(G)/RU(G)$ is in the image of η, degrees -1 and -5 are clear.

In the remaining cases we use the fact that realification is a KO^*-module homomorphism, and the fact that α acts as $2v^2$ on KU^*. In degrees -2 modulo 4, this is simply multiplication by 2, since realification induces the natural quotient maps $RU \twoheadrightarrow RU/RSp$ and $RU \twoheadrightarrow RU/RO$.

In degree 0, we have

$$\begin{array}{ccc} RU & \xrightarrow{r} & RO \\ {\scriptstyle 2v^2}\downarrow & & \downarrow{\scriptstyle \alpha} \\ RU & \xrightarrow{q} & RSp \end{array}$$

Thus, multiplication by α sends $rcU_i = 2U_i$ to $2qcU_i$, and hence U_i to qcU_i, sends rV_j to $2qV_j = qcrV_j$, and sends $r\tilde{c}W_k$ to $2q\tilde{c}W_k = qcr\tilde{c}W_k$. That is, $\alpha = qc : RO \longrightarrow RSp$.

The analogous calculation in degree -4, shows that $\alpha = r\tilde{c} : RSp \longrightarrow RO$.

□

2.4. The connective case

First note that the structure maps c, q, r, $\tilde{c}$ and τ of Section 2.1 are $RO(G)$-module homomorphisms. For example, r is an $RO(G)$-module homomorphism viewing $RU(G)$ as an $RO(G)$-module through c. Similarly, the map R lifts to the connective case, to give a ko-module map $R : ku \longrightarrow \Sigma^2 ko$ (see Lemma 4.1.2).

Following [**46**] we want to define $k\mathbb{R}$ by a pullback square. However it seems that we must assume $\widehat{G} = G \times Q$. In this case G is a quotient of $\widehat{G}$ (as well as being a subgroup), and so we may view $EG = E(\widehat{G}/Q)$ as a $\widehat{G}$-space. In this case $K\mathbb{R}$ is relatively split in the sense that there is a map

$$\inf_Q^{\widehat{G}} K\mathbb{R} \longrightarrow K\mathbb{R}$$

of $\widehat{G}$-spectra, which is a Q-equivalence. Indeed, when working over G-fixed spaces, we may view a real vector bundle as a G-fixed G-equivariant real vector bundle, and this gives a Q-map $K\mathbb{R} \longrightarrow K\mathbb{R}^G$, which is clearly a Q-equivalence. Accordingly we have an equivalence

$$F(EG_+, K\mathbb{R}) \simeq F(EG_+, \inf_Q^{\widehat{G}} K\mathbb{R})$$

of $\widehat{G}$-spectra.

In the non-equivariant case, the crudest possible construction works. More precisely, when $G = 1$ we take the Q-spectrum $k\mathbb{R}$ to be the connective cover of $K\mathbb{R}$, and we then use this to define the general case.

DEFINITION 2.4.1. The $\widehat{G}$-spectrum $k\mathbb{R}$ is defined by the homotopy pullback square

$$\begin{array}{ccc} k\mathbb{R} & \longrightarrow & K\mathbb{R} \\ \downarrow & & \downarrow \\ F(EG_+, \inf_Q^{\widehat{G}} k\mathbb{R}) & \longrightarrow & F(EG_+, \inf_Q^{\widehat{G}} K\mathbb{R}) \end{array}$$

In the bottom left position, the Q-spectrum $k\mathbb{R}$ is the connective cover of the Q-spectrum $K\mathbb{R}$, the right hand vertical is the map

$$K\mathbb{R} \longrightarrow F(EG_+, K\mathbb{R}) \simeq F(EG_+, \inf_Q^{\widehat{G}} K\mathbb{R}),$$

and the bottom horizontal is induced by the Q-map $k\mathbb{R} \longrightarrow K\mathbb{R}$.

Before we can proceed we need to make explicit the definition of equivariant ko. The definition is implicit in [**46**].

DEFINITION 2.4.2. The equivariant version of ko is defined by the homotopy pullback square

$$\begin{array}{ccc} ko & \longrightarrow & KO \\ \downarrow & & \downarrow \\ F(EG_+, \mathrm{inf}_1^G ko) & \longrightarrow & F(EG_+, \mathrm{inf}_1^G KO) \end{array}$$

We are now in a position to give equivariant versions of familiar relationships between $k\mathbb{R}$, ku and ko.

LEMMA 2.4.3. *(i) $k\mathbb{R}$ is G-equivariantly ku.*
(ii) $k\mathbb{R}^Q$ is G-equivariantly ko.

Proof: Part (i) is clear from Definition 2.4.1; indeed, regarding the defining pullback square G-equivariantly we obtain the defining pullback square for ku.

Part (ii) is similarly obtained by taking Q-fixed points of the defining pullback square. At the top right, we obtain $KO = (K\mathbb{R})^Q$ in the usual way. For the bottom row we argue as follows. First note that for any $\widehat{G}$-spectrum Y, $F(EG_+, Y)^Q \simeq F(EG_+, Y^Q)$. We are interested in the cases $Y = \mathrm{inf}_Q^{\widehat{G}} k\mathbb{R}$ and the corresponding periodic case. For definiteness, we discuss the case with $k\mathbb{R}$. Here Y is Q-equivalent to $k\mathbb{R}$, and hence has the same Q-fixed points non-equivariantly, namely ko. More precisely, since EG_+ is G-free we may calculate over a Q-fixed universe, where the conclusion is straightforward. □

LEMMA 2.4.4. *(i) The $\widehat{G}$-spectrum $k\mathbb{R}$ is relatively split, so that $k\mathbb{R}_*^G$ is a ko_*-module.*
(ii) Inverting the Bott element we obtain Atiyah-Segal equivariant K-theory with reality: $k\mathbb{R}[1/\beta] = K\mathbb{R}$

Proof: Part (ii) is immediate from Part (i) since inverting β makes the lower horizontal an equivalence (using the analogue of Lemma 1.1.1 of [**28**]).

For Part (i), we note that the maps

$$\mathrm{inf}_Q^{\widehat{G}} k\mathbb{R} \longrightarrow F(EG_+, \mathrm{inf}_Q^{\widehat{G}} k\mathbb{R})$$

and

$$\mathrm{inf}_Q^{\widehat{G}} k\mathbb{R} \longrightarrow \mathrm{inf}_Q^{\widehat{G}} K\mathbb{R} \longrightarrow K\mathbb{R}$$

are compatible in $F(EG_+, \mathrm{inf}_Q^{\widehat{G}} K\mathbb{R})$. This shows there is a map $\mathrm{inf}_Q^{\widehat{G}} k\mathbb{R} \longrightarrow k\mathbb{R}$. To see that any choice is a Q-equivalence, note that Q-equivariantly, the defining pullback square for $k\mathbb{R}$ has both horizontals $k\mathbb{R} \longrightarrow K\mathbb{R}$. To ensure we have a ring map, take the homotopy pullback square in $\mathbb{S}$-algebras. □

LEMMA 2.4.5. *The ring ko_*^G is Noetherian.*

Proof: First, by [**46**], ku_*^G is a Noetherian ring. In Chapter 4 we will give a Bockstein spectral sequence for calculating ko_*^G from ku_*^G. This starts from an E_1 term in which each row is ku_*^G, and converges to $E_4 = E_\infty$ which is concentrated entirely on rows 0, 1 and 2. We argue that each of these three rows is a finitely generated module over a finitely generated subalgebra of ko_*^G. It follows that ko_G^* is itself a finitely generated module over a finitely generated subalgebra, and it is therefore a Noetherian ring as required.

The Bockstein spectral sequence is a spectral sequence of modules over the 0-line at E_∞, so it suffices to find a large enough subalgebra of this 0-line. In fact we choose a faithful representation of G, in $Sp(n)$ for some n. Then $Sp(n)/G$ is a finite $Sp(n)$-complex and therefore $ku^*_G = ku^*_{Sp(n)}(Sp(n)/G)$ is a finitely generated module over $ku^*_{Sp(n)} = \mathbb{Z}[v, p_1, p_2, \ldots, p_n]$ [**46**]. This in turn is finitely generated as a module over $\mathbb{Z}[v^4, p_1, p_2, \ldots, p_n]$, and we will shown in Chapter 5 that this is in the image of $ko^*_{Sp(n)}$ under complexification. Thus ku^*_G is a finitely generated module over $ko^*_{Sp(n)}$ through $V^* : ko^*_{Sp(n)} \longrightarrow ko^*_G$, and hence over a subalgebra of the 0-line at E_∞ of the Bockstein spectral sequence as required. □

2.5. The local cohomology and completion theorems

We will need a special case of the completion theorem for $K\mathbb{R}$ and $k\mathbb{R}$. However the inductive nature of the proof means we need to prove it for an arbitrary family and without the assumption that the extension $1 \longrightarrow G \longrightarrow \widehat{G} \longrightarrow Q \longrightarrow 1$ is a product.

Accordingly, we begin with a family $\mathcal{F}$ of subgroups of $\widehat{G}$, and define

$$J\mathcal{F} = \ker(k\mathbb{R}^{\widehat{G}}_* \longrightarrow \prod_{K\in\mathcal{F}} k\mathbb{R}^K_*).$$

The particular case of concern to us is the family $[\cap G = 1]$ of subgroups intersecting G trivially. If $\widehat{G} = G \times Q$ is a product, we may view EG as a $\widehat{G}$ space and in that case $E[\cap G = 1] = EG$.

Throughout this section, because we are working with spectra and based spaces, all homology and cohomology is reduced.

PROPOSITION 2.5.1. *The local cohomology and completion theorems hold for $k\mathbb{R}$. More precisely, for any family $\mathcal{F}$ of subgroups of $\widehat{G}$ there are spectral sequences*

$$H^*_{J\mathcal{F}}(k\mathbb{R}^{\widehat{G}}_*(X)) \Rightarrow k\mathbb{R}^{\widehat{G}}_*(X \wedge E\mathcal{F}_+)$$

and

$$H^{J\mathcal{F}}_*(k\mathbb{R}^*_{\widehat{G}}(X)) \Rightarrow k\mathbb{R}^*_{\widehat{G}}(X \wedge E\mathcal{F}_+),$$

where $H^*_{J\mathcal{F}}$ is Grothendieck local cohomology [**54**] *and $H^{J\mathcal{F}}_*$ is local homology* [**50**].

If X is a finite complex, the second spectral sequence collapses to an isomorphism

$$k\mathbb{R}^*_{\widehat{G}}(X)^\wedge_{J\mathcal{F}} \cong k\mathbb{R}^*_{\widehat{G}}(X \wedge E\mathcal{F}_+).$$

Proof: The proof follows the pattern of [**41**], but to justify the steps, we should run through the argument. The final clause comes from the fact that local homology calculates the left derived functors of completion [**50**]. Since $k\mathbb{R}^{\widehat{G}}_*$ is Noetherian (2.4.5), the Artin-Rees lemma shows completion is exact on finitely generated modules.

Since $k\mathbb{R}$ is a strictly commutative ring spectrum, for any $k\mathbb{R}$-module M we may form local cohomology modules $\Gamma_{J\mathcal{F}}M$. Indeed, if $x \in k\mathbb{R}^{\widehat{G}}_*$ we take

$$\Gamma_x M = \text{fibre}(M \longrightarrow M[1/x]),$$

and if $J = (x_1, \ldots, x_n)$ we take

$$\Gamma_J M = \Gamma_{x_1} M \wedge_{k\mathbb{R}} \ldots \wedge_{k\mathbb{R}} \Gamma_{x_n} M.$$

Up to equivalence, this is independent of the generators $x_1, \ldots, x_n$, and only depends on the radical of J.

Now consider the two maps

$$k\mathbb{R} \wedge E\mathcal{F}_+ \longleftarrow \Gamma_{J\mathcal{F}} k\mathbb{R} \wedge E\mathcal{F}_+ \longrightarrow \Gamma_{J\mathcal{F}} k\mathbb{R},$$

and argue separately that the two maps are equivalences. The local cohomology theorem then follows from the evident spectral sequence for calculating the homotopy of $\Gamma_{J\mathcal{F}} k\mathbb{R} \wedge X$. The completion theorem from the evident spectral sequence for calculating the homotopy of $F_{k\mathbb{R}}(\Gamma_{J\mathcal{F}} k\mathbb{R}, F(X, k\mathbb{R}))$.

The equivalence is straightforward for the left-hand map. Indeed, $E\mathcal{F}_+$ is made up of cells $\widehat{G}/H_+$ for $H \in \mathcal{F}$ so we may consider the map

$$k\mathbb{R} \wedge \widehat{G}/H_+ \longleftarrow \Gamma_{J\mathcal{F}} k\mathbb{R} \wedge \widehat{G}/H_+.$$

This is induced from the H-equivariant map

$$k\mathbb{R} \longleftarrow \mathrm{res}^{\widehat{G}}_H \Gamma_{J\mathcal{F}} k\mathbb{R},$$

which is an equivalence because $\mathrm{res}^{\widehat{G}}_H J\mathcal{F} = 0$.

The right hand map needs a little more work. It is sufficient to show the mapping cone, $\Gamma_{J\mathcal{F}} k\mathbb{R} \wedge \tilde{E}G$ is contractible. For this we argue by induction on the size of G, supposing the result is already proved for all proper subgroups. This is legitimate for compact Lie groups since there are no infinite descending chains of subgroups. Now consider the cofibre sequence

$$E\mathcal{P}_+ \longrightarrow S^0 \longrightarrow \tilde{E}\mathcal{P}:$$

note that $E\mathcal{P}_+$ is built from cells $\widehat{G}/K$ for proper subgroups K, and

$$\tilde{E}\mathcal{P} \simeq \bigcup_{V^{\widehat{G}}=0} S^V.$$

Smashing with $\Gamma_{J\mathcal{F}} k\mathbb{R} \wedge \tilde{E}G$, we obtain the cofibre sequence

$$\Gamma_{J\mathcal{F}} k\mathbb{R} \wedge \tilde{E}G \wedge E\mathcal{P}_+ \longrightarrow \Gamma_{J\mathcal{F}} k\mathbb{R} \wedge \tilde{E}G \longrightarrow \Gamma_{J\mathcal{F}} k\mathbb{R} \wedge \tilde{E}G \wedge \tilde{E}\mathcal{P}.$$

We aim to prove the middle term is contractible, so it suffices to show the other two terms are contractible.

Consider first the term $\Gamma_{J\mathcal{F}} k\mathbb{R} \wedge \tilde{E}G \wedge \tilde{E}\mathcal{P}$; this is K-contractible for all proper subgroups, so it suffices to show that its $\widehat{G}$-equivariant homotopy groups are zero. Here we note that it suffices to allow the representations V with $V^{\widehat{G}} = 0$ in the model for $\tilde{E}\mathcal{P}$ to run through those of form $\hat{U}$ for $U^G = 0$. Indeed, the trivial summand does not occur, and all non-trivial real representations of $\widehat{G}$ occur as a summand in some representation of the form $\hat{U}$. By 2.2.5, these representations are orientable, and hence have associated Euler classes. Thus, for any $k\mathbb{R}$-module M we have

$$\pi^{\widehat{G}}_*(M \wedge \tilde{E}\mathcal{P}) = \mathcal{E}^{-1}\pi^{\widehat{G}}_*(M),$$

where $\mathcal{E}$ is the multiplicative set of Euler classes of representations $\hat{U}$ where $U^G = 0$. If we take $M = \Gamma_{J\mathcal{F}} k\mathbb{R}$ there is a spectral sequence

$$H^*_{J\mathcal{F}}(k\mathbb{R}^{\widehat{G}}_*) \Rightarrow \pi^{\widehat{G}}_*(\Gamma_{J\mathcal{F}} k\mathbb{R}).$$

Next, we claim that inverting $\mathcal{E}$ annihilates the E_2-term, showing

$$\pi^{\widehat{G}}_*(\Gamma_{J\mathcal{F}} k\mathbb{R} \wedge \tilde{E}\mathcal{P}) = \mathcal{E}^{-1}\pi^{\widehat{G}}_*(\Gamma_{J\mathcal{F}} k\mathbb{R}) = 0$$

as required. If G is finite, we may simply use the representation $\hat{U}$ where U is the reduced regular representation of G since the restriction of $\hat{U}$ to any proper subgroup of $\widehat{G}$ has a trivial real summand and therefore a trivial Euler class: thus $\chi(\hat{V}) \in J\mathcal{F}$. For a compact Lie group one may give a similar argument elementwise: any homotopy element which is $J\mathcal{F}$-power torsion is annihilated by some $\chi(\hat{U})$.

Finally, we turn to $\Gamma_{J\mathcal{F}} k\mathbb{R} \wedge \tilde{E}G \wedge E\mathcal{P}_+$. Since $E\mathcal{P}_+$ is built from cells $\widehat{G}/K$ for proper subgroups K, it suffices to show that $\Gamma_{J\mathcal{F}} k\mathbb{R} \wedge \tilde{E}G \wedge \widehat{G}/K \simeq *$ for all proper subgroups K. However this is induced from

$$\mathrm{res}^{\widehat{G}}_{K}(\Gamma_{J\mathcal{F}} k\mathbb{R} \wedge \tilde{E}G) \simeq \Gamma_{\mathrm{res}^{\widehat{G}}_{K} J\mathcal{F}} k\mathbb{R} \wedge \widetilde{E}K.$$

Now, since $k\mathbb{R}^*_K$ is Noetherian [**44**] or by explicit calculation, we have

$$\sqrt{\mathrm{res}^{\widehat{G}}_{K} J\mathcal{F}} = J\mathrm{res}^{\widehat{G}}_{K} F.$$

Hence,

$$\Gamma_{\mathrm{res}^{\widehat{G}}_{K} J\mathcal{F}} k\mathbb{R} \wedge \widetilde{E}K \simeq \Gamma_{J\mathrm{res}^{\widehat{G}}_{K} F} k\mathbb{R} \wedge \widetilde{E}K \simeq *$$

as required. □

REMARK 2.5.2. We wil often use the special case when $\widehat{G} = G \times Q$ and $E\mathcal{F}_+ = EG_+$ with trivial Q action, and $X = S^0$. The spectral sequence then reads

$$H^*_J(ko^*_G) \Rightarrow ko^G_*(EG_+ \wedge S^{ad(G)}) \cong ko_*(BG^{ad(G)}).$$

If G is a finite p-group the proof may be somewhat simplified, since G acts freely on a product $S(V_1) \times \cdots \times S(V_N)$ of unit spheres of ko-orientable representations $V_1, \cdots, V_N$. We may then take the model

$$EG = S(\infty V_1) \times \cdots \times S(\infty V_N),$$

so that EG_+ has an evident filtration as a product of the evident 1-step filtrations of the terms

$$S(\infty V_i)_+ = \mathrm{fibre}(S^0 \longrightarrow S^{\infty V_i}).$$

Furthermore, it is well known [**15**, III.1.1] that the JO-adic filtration agrees with the p-adic filtration on JO, and that the JO-adic and JU-adic filtrations agree on $RU(G)$ [**14**, 6.1].

2.6. Calculations

Finally we have the ingredients we need to show how to calculate equivariant connective K-theory with reality from rings more directly accessible. We revert to our usual assumption that $\widehat{G} = G \times Q$, and that homology and cohomology is unreduced.

THEOREM 2.6.1. *The defining homotopy pullback square induces a pullback when applied to any finite complex X*

$$\begin{array}{ccc} k\mathbb{R}^*_{\widehat{G}}(X) & \longrightarrow & K\mathbb{R}^*_{\widehat{G}}(X) \\ \downarrow & & \downarrow \\ k\mathbb{R}^*_Q(EG \times_G X) & \longrightarrow & K\mathbb{R}^*_Q(EG \times_G X) \end{array}$$

In particular, with $X = pt$ we obtain the pullback square

$$\begin{array}{ccc} ko_*^G & \longrightarrow & KO_*^G \\ \downarrow & & \downarrow \\ ko^*(BG) & \longrightarrow & KO^*(BG) \end{array}$$

REMARK 2.6.2. Other special cases are also of interest. If $X = Q = S(\xi)$, we obtain the pullback square for calculating ku_*^G as in [**46**, Proposition 6.1]. If we take $X = S(2\xi)$, the two left-hand rings are self-conjugate K-theory, and $ko_{\widehat{G}}^*(S(2\xi) \times Y)$ could be viewed as an equivariant form of connective self-conjugate K-theory.

Proof: There is a long exact Mayer-Vietoris sequence associated to the homotopy pullback square, and we must argue that it splits into short exact sequences.

Let $M(V) = k\mathbb{R}_*^{\widehat{G}}(S^V)$, and note that $M(V)[1/\beta] = K\mathbb{R}_*^{\widehat{G}}(S^V)$. Similarly, by the completion theorem, since $M(V)$ is Noetherian, $k\mathbb{R}_Q^*(BG^V)$ is the completion of $M(V)$ with respect to the augmentation ideal. Thus the square is of the form

$$\begin{array}{ccc} M(V) & \longrightarrow & M(V)[1/\beta] \\ \downarrow & & \downarrow \\ M(V)_J^\wedge & \longrightarrow & M(V)^\wedge[1/\beta]. \end{array}$$

Furthermore, the completions can be viewed as over the ideal generated by β. Indeed, from the Bockstein spectral sequence, this follows from the result for $ku_G^*(X)$ proved in [**46**].

As a matter of algebra [**43**], this is a pullback square provided we are working with finitely generated modules over a Noetherian ring. The result therefore follows from Lemma 2.4.5.

□

2.7. The local cohomology theorem and positive scalar curvature

As described in the introduction, our calculations may be applied to study the Gromov-Lawson-Rosenberg (GLR) conjecture [**88**] for finite groups. When G is a discrete group, $ko_n(BG)$ is the obstruction group for the existence of a positive scalar curvature metric on a Spin manifold of dimension $n \geq 5$ and fundamental group G.

There are maps

$$\Omega_n^{spin}(BG) \xrightarrow{D} ko_n(BG) \xrightarrow{p} KO_n(BG) \xrightarrow{A} KO_n(C_{red}^* G),$$

The existence of a positive scalar curvature metric depends only on the ko-index $D([M]) \in ko_n(BG)$ [**66, 95**], and we write $ko_n^+(BG)$ for the subgroup of $ko_n(BG)$ consisting of indices of spin manifolds M admitting a metric of positive scalar curvature. The Lichnerowicz method gives an inclusion

$$ko_n^+(BG) \subseteq \ker(Ap),$$

and the conjecture states that this is actually an equality. To verify the conjecture for G, it suffices to identify $\ker(Ap)$ and to realize a set of generators as indices of manifolds with psc metric.

For a finite group G, we find $KO_*(C_{red}^*(G)) = KO_*^G$, so that the kernel we must identify is that of the composite

$$ko_*(BG) \xrightarrow{p} KO_*(BG) \xrightarrow{A} KO_*^G.$$

Of course the two maps are the effect in homotopy groups of the maps

$$ko \wedge EG_+ \xrightarrow{p} KO \wedge EG_+ \xrightarrow{\pi} KO$$

of G-spectra. From the construction of the local cohomology spectral sequence we see that π induces the edge homomorphism of the local cohomology spectral sequence.

LEMMA 2.7.1. *The image of Ap is isomorphic to the 0-column at E_∞ of the local cohomology spectral sequence for $ko_*(BG)$.*

The kernel of Ap has a filtration with subquotients given by the higher columns at E_∞.

Proof: The direction of the filtration is such that the 0-column is a quotient of $ko_*(BG)$. The higher columns map to the submodule of higher columns for $KO_*(BG)$, and these evidently map to zero under the edge homomorphism. The 0-column in the periodic case is entirely in degrees ≥ 0, and ko^G_* agrees with KO^G_* in positive degrees so p induces a monomorphism on the 0-column. □

It is sometimes helpful to check when the 1-column is detected in representation theory. Indeed, if ST has depth ≥ 2, the short exact sequence $0 \longrightarrow ST \longrightarrow ko^*_G \longrightarrow QO \longrightarrow 0$ shows the map $ko^*_G \longrightarrow QO$ induces a monomorphism on first local cohomology; since KO^G_*/QO is zero in positive degrees, the map $QO \longrightarrow KO^G_*$ is also a monomorphism of first local cohomology in positive degrees. Since the GLR conjecture is true for cyclic groups [**21**], all classes induced from cyclic subgroups of G are realized by psc spin manifolds. The quotient of the 1-column at E_∞ by classes induced from cyclic subgroups need not be zero (e.g. for the semidihedral group of order 16 [**79**]), but it is often very small so that few classes need to be realized by psc spin manifolds.

Finally, it sometimes happens (for example with all the groups we consider) that the only other column remaining at E_∞ is the rth where r is the 2-rank of G. From the behaviour of the local cohomology theorem for mod 2 cohomology, one checks these columns are detected in $H_*(BG; \mathbb{F}_2)$. Accordingly, to verify the GLR conjecture for n-manifolds, it suffices to find psc spin n-manifolds and calculate $[M] \in H_n(BG; \mathbb{F}_2)$. If these classes span a subspace of the same dimension as the higher columns in degree n, the GLR conjecture is proved in degree n. See [**63, 78**] for numerous examples of this technique.

2.8. Periodic K-theory and maps into classical fibrations

In this section we take an alternative approach to calculating the values of $KO^*(BG)$ using the sequence of loop spaces in the real K-theory spectrum together with the Atiyah-Segal isomorphisms for maps into the classifying spaces BU, BO and BSp. We need this for our comparison between the periodic and connective theories, which uses the values of $[BG, X]$ for various infinite Lie groups and homogeneous spaces. First, recall the Bott periodicity theorem.

THEOREM 2.8.1. (Bott [**19**]) *There are homotopy equivalences which define the real K-theory spectrum:*

$$\begin{array}{lcl} \Omega(BO \times Z) & \simeq & O \\ \Omega(O) & \simeq & O/U \\ \Omega(O/U) & \simeq & U/Sp \\ \Omega(U/Sp) & \simeq & BSp \times \mathbb{Z} \\ \Omega(BSp \times \mathbb{Z}) & \simeq & Sp \\ \Omega(Sp) & \simeq & Sp/U \\ \Omega(Sp/U) & \simeq & U/O \\ \Omega(U/O) & \simeq & BO \times \mathbb{Z} \end{array}$$

Similarly, there are homotopy equivalences which define the complex K-theory spectrum:

$$\begin{array}{lcl} \Omega(BU \times Z) & \simeq & U \\ \Omega(U) & \simeq & BU \times \mathbb{Z} \end{array}$$

Next, we need the Atiyah-Segal isomorphisms ([**14**])

$$\begin{array}{rcl} [BG, BO \times \mathbb{Z}] & \cong & \widehat{RO}(G) \\ [BG, BSp \times \mathbb{Z}] & \cong & \widehat{RSp}(G) \\ [BG, BU \times \mathbb{Z}] & \cong & \widehat{RU}(G) \\ [BG, U] & = & 0 \end{array}$$

Here, $\widehat{(\cdot)}$ denotes completion at the augmentation ideal I of $RO(G)$. For $\widehat{RU}$ we could as well complete at the augmentation ideal of $RU(G)$, since both ideals define the same topology ([**14**]).

Now we can determine the values of $[BG, X]$ for the rest of the infinite loop spaces which figure in the periodic K-theory spectra.

THEOREM 2.8.2. *The values of $[BG, -]$ on the infinite loop spaces above are as follows:*

(i) $[BG, O] = \widehat{RO}(G)/\widehat{RU}(G)$
(ii) $[BG, Sp] = \widehat{RSp}(G)/\widehat{RU}(G)$
(iii) $[BG, U/Sp] = 0$
(iv) $[BG, U/O] = 0$
(v) $[BG, O/U] = \widehat{RU}(G)/\widehat{RSp}(G)$
(vi) $[BG, Sp/U] = \widehat{RU}(G)/\widehat{RO}(G)$

Proof: For Part (i) we see $[BG, O] = \widehat{RO}(G)/\widehat{RU}(G)$ by mapping BG into the fibration sequence

$$\begin{array}{ccccccc} \Omega(U) & \longrightarrow & \Omega(U/O) & \longrightarrow & O & \longrightarrow & U. \\ \| & & \| & & & & \\ BU \times \mathbb{Z} & \longrightarrow & BO \times \mathbb{Z} & & & & \end{array}$$

since $[BG, U] = 0$.

For Part (ii) we see $[BG, Sp] = \widehat{RSp}(G)/\widehat{RU}(G)$ by mapping BG into the fibration sequence

$$\begin{array}{ccccccc} \Omega(U) & \longrightarrow & \Omega(U/Sp) & \longrightarrow & Sp & \longrightarrow & U. \\ \| & & \| & & & & \\ BU \times \mathbb{Z} & \longrightarrow & BSp \times \mathbb{Z} & & & & \end{array}$$

since $[BG, U] = 0$.

For Part (iii) we see $[BG, U/Sp] = 0$ by mapping BG into the fibration sequence

$$U \longrightarrow U/Sp \longrightarrow BSp \times \mathbb{Z} \longrightarrow BU \times \mathbb{Z},$$

since $\tilde{c} : \widehat{RSp}(G) \longrightarrow \widehat{RU}(G)$ is a monomorphism.

For Part (iv) we note $[BG, U/O] = 0$ by mapping BG into the fibration sequence

$$U \longrightarrow U/O \longrightarrow BO \times \mathbb{Z} \longrightarrow BU \times \mathbb{Z},$$

since $c : \widehat{RO}(G) \longrightarrow \widehat{RU}(G)$ is a monomorphism.

For Part (v) we see $[BG, O/U] = \widehat{RU}(G)/\widehat{RSp}(G)$ by mapping BG into the fibration sequence

$$\begin{array}{ccccccccccc} \Omega^2(O/U) & \longrightarrow & \Omega(U) & \longrightarrow & \Omega(O) & \longrightarrow & \Omega(O/U) & \longrightarrow & U & \longrightarrow & O. \\ \| & & \| & & \| & & \| & & & & \\ BSp \times \mathbb{Z} & \longrightarrow & BU \times \mathbb{Z} & \longrightarrow & O/U & \longrightarrow & U/Sp & & & & \end{array}$$

and using Part (iii).

Finally, for Part (vi) we see $[BG, Sp/U] = \widehat{RU}(G)/\widehat{RO}(G)$ by mapping BG into the fibration sequence

$$\begin{array}{ccccccccccc} \Omega^2(Sp/U) & \longrightarrow & \Omega(U) & \longrightarrow & \Omega(Sp) & \longrightarrow & \Omega(Sp/U) & \longrightarrow & U & \longrightarrow & Sp. \\ \| & & \| & & \| & & \| & & & & \\ BO \times \mathbb{Z} & \longrightarrow & BU \times \mathbb{Z} & \longrightarrow & Sp/U & \longrightarrow & U/O & & & & \end{array}$$

and using Part (iv). □

REMARK 2.8.3. The Atiyah-Bott η-c-R sequence $\Sigma KO \xrightarrow{\eta} KO \xrightarrow{c} KU \xrightarrow{R} \Sigma^2 KO$ (see Chapter 4 for extended discussion) combined with the isomorphisms in the preceding remark yields two interesting exact sequences:

$$0 \longrightarrow \widehat{RSp}/\widehat{RU} \xrightarrow{\tilde{c}} \widehat{RU}/\widehat{RO} \xrightarrow{1-\tau} \widehat{RU} \xrightarrow{q} \widehat{RSp} \twoheadrightarrow \widehat{RSp}/\widehat{RU} \longrightarrow 0$$

and

$$0 \longrightarrow \widehat{RO}/\widehat{RU} \xrightarrow{c} \widehat{RU}/\widehat{RSp} \xrightarrow{1-\tau} \widehat{RU} \xrightarrow{r} \widehat{RO} \twoheadrightarrow \widehat{RO}/\widehat{RU} \longrightarrow 0$$

REMARK 2.8.4. For comparisons between the connective and periodic cases, it is useful to note that

$$JU/JO \xrightarrow{\cong} RU/RO \qquad \text{and} \qquad JSp/JU \xrightarrow{\cong} RSp/RU$$

and that there are short exact sequences

$$0 \longrightarrow JO/JU \longrightarrow RO/RU \longrightarrow \mathbb{Z}/2\mathbb{Z} \longrightarrow 0$$

and

$$0 \longrightarrow JU/JSp \longrightarrow RU/RSp \longrightarrow \mathbb{Z}/2\mathbb{Z} \longrightarrow 0$$

In the next section, we shall need to express $[BG, X]$ in terms of representations, for various groups and homogeneous spaces X, including those above and some of their $(n-1)$-connected covers. The general form of these arguments is the following.

(i) Locate X in a fibre sequence $BG_1 \longrightarrow BG_2 \longrightarrow X \longrightarrow B^2G_1 \longrightarrow B^2G_2$, where
(ii) the map $BG_1 \longrightarrow BG_2$ is induced by a known homomorphism $G_1 \longrightarrow G_2$,
(iii) either $[BG, B^2G_1] = 0$ or $[BG, B^2G_1] \longrightarrow [BG, B^2G_2]$ is a monomorphism, for compact Lie G, and
(iv) note that $[BG, BG_i] = \widehat{JG_i}(G)$, the virtual representations of G of 'type G_i' which have virtual dimension 0.

We can then conclude that $[BG, X] = \widehat{JG_1}(G)/\widehat{JG_2}(G)$ and that $\Omega X = G_2/G_1$.

In connection with this last item, Lemma 2.9.1 in the next section shows that JSU, the virtual representations of dimension 0 and determinant 1, can equally be described as JU_2, the ideal generated by second and higher Chern classes, which plays a role in the modified Rees ring. Similarly, we have

$$\widehat{JSO}(G) = [X, BSO] = \ker(\widehat{JO}(G) \xrightarrow{w_1} H^1(BG, \mathbb{Z}/2)), \qquad \text{and}$$
$$\widehat{JSpin}(G) = [X, BSpin] = \ker(\widehat{JSO}(G) \xrightarrow{w_2} H^2(BG, \mathbb{Z}/2)).$$

However, $\widehat{JString}(G) = [X, BString]$ is not necessarily a subset of $\widehat{JSpin}(G)$, just as ku_G^6 is not always a subset of $ku_G^4 = JU_2$ (See Theorem 2.9.5 and Corollary 2.9.4.)

Let us denote the $(n-1)$-connected cover of X by $X\langle n\rangle \longrightarrow X$ and recall that $U\langle 3\rangle = SU$. We first need to establish some facts to use in Step (iii).

COROLLARY 2.8.5. $[BG, SU] = [BG, U\langle 5\rangle] = [BG, SU/Sp] = 0$

Proof: The long exact sequence obtained by applying $[BG, -]$ to the fibration $SU \longrightarrow U \longrightarrow U(1)$ shows that $[BG, SU] = 0$ since $[BG, K(\mathbb{Z}, 0)] = 0 = [BG, U]$. Then, applying $[BG, -]$ to the fibration $U\langle 5\rangle \longrightarrow SU \longrightarrow K(\mathbb{Z}, 3)$ and using the fact that $\Omega SU \longrightarrow \Omega K(\mathbb{Z}, 3)$ is $BU \longrightarrow BU(1)$, which has a section, we get the second equality. Applying $[BG, -]$ to $SU \longrightarrow SU/Sp \longrightarrow BSp \longrightarrow BSU$ shows that $[BG, SU/Sp] = \ker(\widetilde{c} : \widehat{JSp}(G) \longrightarrow \widehat{JSU}(G)) = 0$. □

We can now use the general form of argument outlined above for the covers $SO = O\langle 1\rangle$, $Spin = O\langle 3\rangle$, and $String = O\langle 7\rangle$.

COROLLARY 2.8.6. *For the* 0, 2, *and* 6*-connected covers of* O *and associated homogeneous spaces, we have:*

(i) $\Omega(SO/U) = U/Sp$ *and* $[BG, SO/U] = \widehat{JU}(G)/\widehat{JSp}(G)$
(ii) $\Omega(Spin/SU) = SU/Sp$ *and* $[BG, Spin/SU] = \widehat{JU}_2(G)/\widehat{JSp}(G)$
(iii) $\Omega SO = O/U$ *and* $[BG, SO] = \widehat{JO}(G)/\widehat{JU}(G)$
(iv) $\Omega Spin = SO/U$ *and* $[BG, Spin] = \widehat{JSO}(G)/\widehat{JU}(G)$
(v) $\Omega String = Spin/SU$ *and* $[BG, String] = \widehat{JSpin}(G)/\widehat{JU}_2(G)$

Proof: For Part (i), clearly $\Omega(SO/U) = \Omega(O/U)$ and, by Theorem 2.8.1, this is U/Sp. We therefore have the following diagram of (deloopable) fibre sequences

$$\begin{array}{ccccc}
BSp \times \mathbb{Z} & \longrightarrow & BU \times \mathbb{Z} & \longrightarrow & O/U \\
\| & & \uparrow & & \uparrow \\
BSp \times \mathbb{Z} & \longrightarrow & BU \times 2\mathbb{Z} & \longrightarrow & SO/U \\
\uparrow & & \uparrow & & \| \\
BSp & \longrightarrow & BU & \longrightarrow & SO/U
\end{array}$$

The bottom row proves Part (i), since $[BG, B^2Sp] = [BG, SU/Sp] = 0$ by Corollary 2.8.6.

Part (ii) then follows by killing the bottom homotopy group of SO/U, which is $\pi_2(SO/U) = \mathbb{Z}$. The following diagram shows that the result is $Spin/SU$.

$$\begin{array}{ccccc}
\Sigma H\mathbb{Z} & \longrightarrow & \Sigma H\mathbb{Z}/2 & \xrightarrow{\overline{Q}_0} & \Sigma^2 H\mathbb{Z} \\
\uparrow & & \uparrow & & \uparrow \\
U & \longrightarrow & SO & \longrightarrow & SO/U \\
\uparrow & & \uparrow & & \uparrow \\
SU & \longrightarrow & Spin & \longrightarrow & Spin/SU
\end{array}$$

Now we kill this homotopy group and consider the fibration induced from the bottom row of the display for Part (i).

$$\begin{array}{ccccc}
 & & \Sigma^2 H\mathbb{Z} & = & \Sigma^2 H\mathbb{Z} \\
 & & {\scriptstyle c_1}\uparrow & & \uparrow \\
BSp & \longrightarrow & BU & \longrightarrow & SO/U \\
\| & & \uparrow & & \uparrow \\
BSp & \longrightarrow & BSU & \longrightarrow & Spin/SU
\end{array}$$

The lower fibre sequence shows that we have $\Omega(Spin/SU) = SU/Sp$ and that $[BG, Spin/SU]$ is $\widehat{JU}_2(G)/\widehat{JSp}(G)$, since $[BG, B^2Sp] = [BG, SU/Sp] = 0$ by Theorem 2.8.1 and 2.8.6.

For Part (iii), clearly $\Omega(SO) = \Omega(O)$, while $\Omega(O) = O/U$ by Theorem 2.8.1. We therefore have the following diagram of (deloopable) fibre sequences

$$\begin{array}{ccccc}
BU \times \mathbb{Z} & \longrightarrow & BO \times \mathbb{Z} & \longrightarrow & O \\
\| & & \uparrow & & \uparrow \\
BU \times \mathbb{Z} & \longrightarrow & BO \times 2\mathbb{Z} & \longrightarrow & SO \\
\uparrow & & \uparrow & & \| \\
BU & \longrightarrow & BO & \longrightarrow & SO
\end{array}$$

This proves Part (iii), since $B^2U = SU$ and $[BG, SU] = 0$ by the preceding Corollary.

For Part (iv), we kill the bottom homotopy group of SO to get $Spin$ and use the following diagram to show that $\Omega(Spin) = SO/U$.

$$\begin{array}{ccccc}
 & & \Sigma H\mathbb{Z}/2 & = & \Sigma H\mathbb{Z}/2 \\
 & & \uparrow{\scriptstyle w_1} & & \uparrow{\scriptstyle \Omega w_2} \\
BU & \longrightarrow & BO & \longrightarrow & SO \\
\| & & \uparrow & & \uparrow \\
BU & \longrightarrow & BSO & \longrightarrow & Spin
\end{array}$$

The latter fibre sequence shows that $[BG, Spin] = \widehat{JSO}(G)/\widehat{JU}(G)$ since, again, $[BG, B^2U] = [BG, SU] = 0$.

Finally, for Part (v) we kill the bottom homotopy group of $Spin$, $\pi_3(Spin) = \mathbb{Z}$. The name $String$ has been assigned to this group because of its role in string topology. Composing with the map $BSO \longrightarrow Spin$ from the previous diagram, we get the generator $w_3 = \beta w_2$ of $H^3(BSO, \mathbb{Z})$.

$$\begin{array}{ccccc}
\Sigma^2 H\mathbb{Z} & \longrightarrow & \Sigma^2 H\mathbb{Z}/2 & \xrightarrow{\beta} & \Sigma^3 H\mathbb{Z} \\
\uparrow & & {\scriptstyle w_2}\uparrow & & \uparrow \\
BU & \longrightarrow & BSO & \longrightarrow & Spin \\
\uparrow & & \uparrow & & \uparrow \\
BSU & \longrightarrow & BSpin & \longrightarrow & String
\end{array}$$

The latter fibre sequence shows that $\Omega(String) = Spin/SU$ and that $[BG, String]$ is $\widehat{JSpin}(G)/\widehat{JU}_2(G)$, since $[BG, B^2SU] = [BG, U\langle 5\rangle] = 0$ by Corollary 2.8.5. □

COROLLARY 2.8.7. *The first nine spaces in the spectrum ko and the Moore-Postnikov factorization of $ko \longrightarrow KO$ are*

0: $BO \times \mathbb{Z} \longrightarrow BO \times \mathbb{Z}$
1: $U/O \longrightarrow U/O$
2: $Sp/U \longrightarrow Sp/U$
3: $Sp \longrightarrow Sp$
4: $BSp \longrightarrow BSp \times \mathbb{Z}$
5: $SU/Sp \longrightarrow U/Sp$
6: $Spin/SU \longrightarrow SO/U \longrightarrow O/U$
7: $String \longrightarrow Spin \longrightarrow SO \longrightarrow O$
8: $BString \longrightarrow BSpin \longrightarrow BSO \longrightarrow BO \longrightarrow BO \times \mathbb{Z}$

2.9. Near the edge of periodicity

We have seen that equivariant connective K-theory agrees with the periodic theory in non-negative degrees, and thus is determined entirely by represention theory there. It is possible to give explicit formulas for the first few negative groups in terms of representation theory as well. The results here extend those of [**45**].

First we need a lemma about the filtration we have defined. Recall that for a representation V of dimension n, $\det(V) = \Lambda^n(V) = 1 - c_1^R(V) + \cdots + (-1)^n c_n^R(V)$, where c_i^R denotes the ith representation theory Chern class, $c_i^R = v^i c_i$ ([**28**, Definition 1.3.5]), and that

$$\begin{aligned} ku^*BSU(n) &= ku^*[[c_1, c_2, \ldots, c_n]]/((1-\Lambda^n)/v) \\ &= ku^*[[c_1, c_2, \ldots, c_n]]/(c_1 - vc_2 + \cdots + (-v)^{n-1}c_n) \end{aligned}$$

where the Chern classes c_i and exterior power Λ^n are those of the defining representation (see Sections 5.4 and 5.5).

LEMMA 2.9.1. *$JU_2(G) \subset R(G)$ is the subset consisting of representations of virtual dimension* 0 *and determinant* 1.

REMARK 2.9.2. This is the analogue of the fact that the augmentation ideal $JU(G)$ consists of representations of virtual dimension 0.

Proof: By definition JU_2 is the ideal generated by products of Chern classes together with the Chern classes $c_i(V)$, $i \geq 2$. Let us write JU_2' for the ideal of virtual dimension 0, determinant 1, representations. The virtual dimension 0 requirement is that the (virtual) representation $\alpha : BG \longrightarrow BU \times \mathbb{Z}$ lift to $BU \longrightarrow BU \times \mathbb{Z}$, and the determinant 1 condition is that it lift further over $BSU \longrightarrow BU$.

To see that $JU_2' \subset JU_2$ suppose that $V - W \in JU_2'$ and let $\delta = \det(V) = \det(W)$ and $n = \dim(V) = \dim(W)$. Since δ is a unit in $R(G)$, it is sufficient to show that $V\delta^{-1} - W\delta^{-1} \in JU_2$. Equivalently, we may assume that $\delta = 1$. Since $V - W = (n - W) - (n - V)$, it suffices to show that $c_1(W) = n - W \in JU_2$. But since $\det(W) = 1$, the representation W is pulled back from $SU(n)$. In $ku^*BSU(n)$, $c_1 = vc_2 - v^2c_3 + \cdots \pm v^{n-1}c_n$ and the result follows.

The reverse inclusion requires that we show that any Chern class $c_k(V)$, $k > 1$, and any product $c_1(V)c_1(W)$, has (virtual) dimension 0 and (virtual) determinant 1. These are simple calculations from the definition

$$c_k(V) = \sum_{i=0}^{k} (-1)^i \binom{n-i}{n-k} \Lambda^i(V)$$

for a representation V of dimension n. First,

$$\begin{aligned} \dim(c_k(V)) &= \sum_{i=0}^{k} (-1)^i \binom{n-i}{n-k} \binom{n}{i} \\ &= \binom{n}{k} \sum_{i=0}^{k} (-1)^i \binom{k}{i} \\ &= 0, \end{aligned}$$

since $k > 0$. Next, recall that $\det(\Lambda^i(V)) = (\det(V))^{\binom{n-1}{i-1}}$ if $i > 0$, and $\det(kV) = (\det(V))^k$ for any integer k. Omitting the term $i = 0$ since $\Lambda^0(V) = 1$ has determinant 1, we get

$$\det(c_k(V)) = \prod_{i=1}^{k} (\det V)^{\left((-1)^i \binom{n-i}{n-k}\binom{n-1}{i-1}\right)}$$

and this is $\det(V)$ raised to the power

$$\sum_{i=1}^{k}(-1)^i\binom{n-i}{n-k}\binom{n-1}{i-1} = \binom{n-1}{k-1}\sum_{i=1}^{k}(-1)^i\binom{k-1}{i-1} = 0,$$

since $k-1>0$.

The products of first Chern classes are straightforward. Let $m = \dim(V)$ and $n = \dim(W)$. Then $\dim(c_1(V)c_1(W)) = \dim((m-V)(n-W)) = 0$ and $\det(m-V)(n-W)) = \det(mn - nV - mW + VW) = 1$ since $\det(VW) = (\det(V))^n(\det(W))^m$. □

THEOREM 2.9.3. *In degrees ≥ -5 the coefficients of equivariant connective complex K-theory are entirely representation theoretic, in the sense that $ku^G_* \longrightarrow KU^G_*$ is a monomorphism in this range. The exact values are as follows.*

$$ku^G_i = \begin{cases} 0 & i \geq 0 \quad \text{odd} \\ RU(G) & i \geq 0 \quad \text{even} \\ 0 & i = -1 \\ JU(G) & i = -2 \\ 0 & i = -3 \\ JU_2(G) & i = -4 \\ 0 & i = -5 \end{cases}$$

Proof: We note that the result amounts to proving firstly that the map $ku^G_i \longrightarrow KU^G_i$ is a monomorphism in degrees $i \geq -5$ and secondly identifying the image. The defining pullback square for ku^G_* shows that the fact that for the monomorphic property it suffices to prove that the completed counterpart $ku^{-i}(BG) \longrightarrow KU^{-i}(BG)$ is a monomorphism for $i \geq -5$. The identification of the image in the uncompleted form then follows easily from that of the completed version, again from the pullback square. Accordingly, we may argue entirely in terms of the completed form, which is to say in terms of non-equivariant cohomology of classifying spaces.

We compare the Atiyah Hirzebruch spectral sequences for connective and periodic K-theory:

$$\begin{array}{ccc} H^p(BG, ku_q) & \Longrightarrow & ku^{p-q}(BG) \\ \downarrow & & \downarrow \\ H^p(BG, KU_q) & \Longrightarrow & KU^{p-q}(BG) \end{array}$$

In the range of interest, the only differences between the connective and periodic cases at E_2 are in rows $q = -2$ and -4. This gives the results for $i \geq -1$. The splitting induced by $* = Be \longrightarrow BG \longrightarrow *$ shows that there are no nonzero differentials on $H^0(BG, K_q)$. Hence, in degree -2 we have a short exact sequence

$$0 \longrightarrow ku^2(BG) \longrightarrow KU^2BG \longrightarrow H^0(BG, KU_{-2}) \longrightarrow 0$$

By periodicity, the latter map agrees with the top map in

$$\begin{array}{ccc} R(G)\hat{} = KU^0(BG) & \longrightarrow & H^0(BG, KU_0) \\ \downarrow{\scriptstyle \dim} & & \| \\ R(e)\hat{} = KU^0(*) & = & H^0(*, KU_0) \end{array}$$

The diagram shows that this is the function which computes the dimension of a virtual representation. This proves the result for ku^G_{-2}.

Since G is compact, $H^1(BG, \mathbb{Z}) = 0$. This, and the fact that there are no differentials hitting $H^3(BG, KU_0)$ immediately imply the result in degree -3.

The only differential which can affect the difference between ku^G_* and KU^G_* in degrees $i = -4$ and $i = -5$ is $d_3 : H^2(BG, KU_{-2}) \longrightarrow H^5(BG, KU_0)$. But $H^2(BG, KU_{-2}) = H^2(BG, \mathbb{Z}) = [BG, K(\mathbb{Z}, 2)] = [G, U(1)]$ and every element of $H^2(BG, KU_{-2})$ is the first Chern class of the 1-dimensional representation it corresponds to. The first Chern class in the universal case

$$H^p(BU(1), KU_q) \Longrightarrow KU^{p-q}(BU(1))$$

survives to E_∞, so its images in $H^2(BG, KU_{-2})$ do also. This gives $ku^G_{-5} = 0$ and shows that ku^G_{-4} is the kernel of the map $KU^G_{-4} \longrightarrow H^2(BG; \mathbb{Z}) \otimes H^0(BG, \mathbb{Z})$ induced by the Postnikov section

$$BU \times \mathbb{Z} \stackrel{B\det \times 1}{\longrightarrow} BU(1) \times \mathbb{Z} \simeq K(\mathbb{Z}, 2) \times K(\mathbb{Z}, 0).$$

By Lemma 2.9.1, this is exactly $\widehat{JU}_2(G)$. $\square$

This is best possible, in the sense that ku^G_{-6} contains elements not described by representation theory, since they lie in the kernel of $ku^G_{-6} \longrightarrow KU^G_{-6}$. This occurs for groups as small as $G = V(2) = C_2 \times C_2$ and $G = A_4$ (Sections 6.5 and 6.6). However, the fibre sequence

$$\Sigma^2 ku \stackrel{v}{\longrightarrow} ku \longrightarrow H\mathbb{Z} \stackrel{\overline{Q}_1}{\longrightarrow} \Sigma^3 ku$$

gives the following exact sequences. Recall that $\overline{Q}_1$ is a lift of the Milnor primitive $Q_1 : H\mathbb{F}_2 \longrightarrow \Sigma^3 H\mathbb{F}_2$.

COROLLARY 2.9.4. *There are exact sequences*

$$0 \longrightarrow H^3BG \stackrel{\overline{Q}_1}{\longrightarrow} ku^6BG \stackrel{v}{\longrightarrow} ku^4BG = \widehat{JU}_2(G) \longrightarrow H^4BG \stackrel{\overline{Q}_1}{\longrightarrow} ku^7BG \longrightarrow 0$$

and

$$0 \longrightarrow H^5BG \stackrel{\overline{Q}_1}{\longrightarrow} ku^8BG \stackrel{v}{\longrightarrow} ku^6BG \longrightarrow H^6BG \stackrel{\overline{Q}_1}{\longrightarrow} ku^9BG \longrightarrow \cdots \quad \square$$

THEOREM 2.9.5. *In degrees ≥ -7 the coefficients of equivariant connective complex K-theory are entirely representation theoretic, in the sense that* $ko^G_* \longrightarrow$

KO^G_* is a monomorphism in this range. The exact values are as follows.

$$ko^G_i = \begin{cases} 0 & i = 8k+7 \geq 0 \\ RU(G)/RO(G) & i = 8k+6 \geq 0 \\ RSp(G)/RU(G) & i = 8k+5 \geq 0 \\ RSp(G) & i = 8k+4 \geq 0 \\ 0 & i = 8k+3 \geq 0 \\ RU(G)/RSp(G) & i = 8k+2 \geq 0 \\ RO(G)/RU(G) & i = 8k+1 \geq 0 \\ RO(G) & i = 8k \geq 0 \\ 0 & i = -1 \\ JU(G)/JO(G) & i = -2 \\ JSp(G)/JU(G) & i = -3 \\ JSp(G) & i = -4 \\ 0 & i = -5 \\ JU_2(G)/JSp(G) & i = -6 \\ JSpin(G)/JU_2(G) & i = -7 \end{cases}$$

In degrees -8 and -9 the map $ko^G_i \longrightarrow KO^G_i$ is again often a monomorphism. For ko^G_{-8}, there is an exact sequence

$$JSO \xrightarrow{\beta w_2} H^3(BG;\mathbb{Z}) \longrightarrow ko^G_{-8} \longrightarrow KO^G_{-8} = RO(G)$$

where β is the Bockstein. Similarly, $ko^G_{-9} = 0$ if and only if the two maps

$$JSO \xrightarrow{w_2} H^2(BG;\mathbb{Z}/2) \text{ and } RSpin(G) \xrightarrow{p_1/2} H^4(BG;\mathbb{Z})$$

are both epimorphisms.

REMARK 2.9.6. We are grateful to Alex Gonzalez for giving a specific group where $ko^G_{-9} \neq 0$. We let G be the central extension

$$1 \longrightarrow C_2 \longrightarrow G \longrightarrow C_4 \times C_4 \longrightarrow 1$$

classified by the product of the two classes in $H^1(BG;\mathbb{Z}/2)$ given by projection from the two copies of C_4. The group has sixteen 1-dimensional simple representations and four 2-dimensional simple representations.

One sees that there are no decomposables in the 5-dimensional vector space $H^2(BG;\mathbb{Z}/2)$. Since the subspace generated by w_2s is of dimension ≤ 4 we see that $ko^9(BG) \neq 0$.

Proof: As in the complex case, we note that the result amounts to proving firstly that the map $ko^G_i \longrightarrow KO^G_i$ is a monomorphism in degrees $i \geq -7$ and secondly identifying the image. The defining pullback square for ko^G_* (Theorem 2.6.1) shows that the fact that for the monomorphic property it suffices to prove that the completed counterpart $ko^{-i}(BG) \longrightarrow KO^{-i}(BG)$ is a monomorphism for $i \geq -7$. The identification of the image in the uncompleted form then follows easily from that of the completed version, again from the pullback square. Accordingly, we may argue entirely in terms of the completed form, which is to say in terms of non-equivariant cohomology of classifying spaces.

The Atiyah-Hirzebruch spectral sequence for $KO^*(BG)$ is 0 in rows -1, -2, -3, and -5. It is $H^*(BG,\mathbb{Z})$ in row -4 and is $H^*(BG,\mathbb{Z}/2)$ in rows -6 and -7 (see Figure 2.9.1). When an entry at $E^2_{p,q}$ for KO is replaced by zero for ko there are two effects. On the one hand, its contribution to $KO^{-(p+q)}(BG)$ will be

FIGURE 2.9.1. The Atiyah-Hirzebruch spectral sequence for KO^*BG. The groups giving the difference between ko^7 and KO^7 are enclosed in boxes. (Here $H = H\mathbb{Z}/2$.)

	$\vdots$	$\vdots$	$\vdots$	$\vdots$	$\vdots$	$\vdots$	$\vdots$	$\vdots$	$\vdots$	$\vdots$
$\cdots$	$H\mathbb{Z}^8$	$H\mathbb{Z}^7$	$H\mathbb{Z}^6$	$H\mathbb{Z}^5$	$H\mathbb{Z}^4$	$H\mathbb{Z}^3$	$H\mathbb{Z}^2$	0	$\mathbb{Z}$	8
$\cdots$										7
$\cdots$										6
$\cdots$										5
$\cdots$	$H\mathbb{Z}^8$	$H\mathbb{Z}^7$	$H\mathbb{Z}^6$	$H\mathbb{Z}^5$	$H\mathbb{Z}^4$	$H\mathbb{Z}^3$	$H\mathbb{Z}^2$	0	$\mathbb{Z}$	4
$\cdots$										3
$\cdots$	H^8	H^7	H^6	H^5	H^4	H^3	H^2	H^1	$\mathbb{Z}/2$	2
$\cdots$	H^8	H^7	H^6	H^5	H^4	H^3	H^2	H^1	$\mathbb{Z}/2$	1
$\cdots$	$H\mathbb{Z}^8$	$H\mathbb{Z}^7$	$H\mathbb{Z}^6$	$H\mathbb{Z}^5$	$H\mathbb{Z}^4$	$H\mathbb{Z}^3$	$H\mathbb{Z}^2$	0	$\mathbb{Z}$	0
$\cdots$										-1
$\cdots$										-2
$\cdots$										-3
$\cdots$	$H\mathbb{Z}^8$	$H\mathbb{Z}^7$	$H\mathbb{Z}^6$	$H\mathbb{Z}^5$	$H\mathbb{Z}^4$	$\boxed{H\mathbb{Z}^3}$	$H\mathbb{Z}^2$	0	$\mathbb{Z}$	-4
$\cdots$										-5
$\cdots$	H^8	H^7	H^6	H^5	H^4	H^3	H^2	$\boxed{H^1}$	$\mathbb{Z}/2$	-6
$\cdots$	H^8	H^7	H^6	H^5	H^4	H^3	H^2	H^1	$\boxed{\mathbb{Z}/2}$	-7
$\cdots$	$H\mathbb{Z}^8$	$H\mathbb{Z}^7$	$H\mathbb{Z}^6$	$H\mathbb{Z}^5$	$H\mathbb{Z}^4$	$H\mathbb{Z}^3$	$H\mathbb{Z}^2$	0	$\mathbb{Z}$	-8
	$\vdots$	$\vdots$	$\vdots$	$\vdots$	$\vdots$	$\vdots$	$\vdots$	$\vdots$	$\vdots$	

removed, so $ko^{-(p+q)}(BG)$ will be correspondingly smaller. On the other hand, any differentials from $E^r_{p,q}$, which make $KO^{-(p+q+1)}(BG)$ smaller than the E_2 page will be removed, so that so $ko^{-(p+q)-1}(BG)$ will be correspondingly bigger.

Accordingly, to see that the map

$$ko^{-i}(BG) \longrightarrow KO^{-i}(BG)$$

is a monomorphism, we need only show that the groups $E^2_{p,q}$ with $p+q=i+1$ and $q \leq -4$ consist entirely of infinite cycles. We do not have control of all the differentials, so we show instead that all of $E^2_{p,q}$ lifts to the relevant subgroup of KO^G_{i+1}. For example, the map from KO^G_* to the $p=0$ column is the dimension or the dimension mod 2, and all dimensions are realized by the appropriate virtual representations, so all of the $p=0$ column consists of infinite cycles. The arguments for subsequent columns are more delicate, and we take them in turn, at each stage identifying the filtration on KO^G_i and showing that the maps from them to the groups at E^2 are surjective.

To start with, there is no difference between the periodic and connective cases in degrees -1, -2 and -3. Correspondingly, $JU/JO = RU/RO$ and $JSp/JU = RSp/RU$, since the trivial representation is in both denominators.

Degree -4 is exactly analogous to degree -2 of the complex case. The inclusion $ko^4(BG) \longrightarrow KO^4(BG)$ is induced by $BSp \longrightarrow BSp \times \mathbb{Z}$.

Degree -5 is 0 because the only difference between the connective and periodic cases lies in $H^1(BG, \mathbb{Z}) = 0$.

In degree -6, we have the filtration

$$\begin{array}{ccccc} KO^6_G & \xleftarrow{\supset} & F_1 & \xleftarrow{\supset} & ko^6_G \\ {\scriptstyle \dim}\downarrow & & \downarrow{\scriptstyle c_1} & & \\ \mathbb{Z}/2 & & H^2(BG, \mathbb{Z}) & & \end{array}$$

We observe that $KO^6(BG) = \widehat{RU}(G)/\widehat{RSp}(G) = [BG, O/U]$ and that this comes from the fibre sequence

$$BSp \times \mathbb{Z} \longrightarrow BU \times \mathbb{Z} \longrightarrow O/U$$

(Theorem 2.8.1). Restricting the edge homomorphism $O/U \longrightarrow H\mathbb{Z}/2$ to $BU \times \mathbb{Z}$ gives the dimension modulo 2, by naturality with respect to restriction to the trivial subgroup. This is surjective; taking fibres, we get the fibre sequence

$$BSp \times \mathbb{Z} \longrightarrow BU \times 2\mathbb{Z} \longrightarrow SO/U$$

and we may now replace this by

$$BSp \longrightarrow BU \longrightarrow SO/U,$$

from which it follows that $F_1 = [BG, SO/U] = \widehat{JU}(G)/\widehat{JSp}(G)$ by Corollary 2.8.6. (See also Remark 2.8.4.) The map to the next filtration quotient of the Atiyah-Hirzebruch spectral sequence is $SO/U \longrightarrow \Sigma^2 H\mathbb{Z}$, and the restriction of this to BU is the Chern class c_1^H, since this is the unique nontrivial generator (up to sign), and restriction to the complex case shows that it is a permanent cycle in the universal case $G = U$. This gives $ko^6(BG) = [BG, B(SU/Sp)] = \widehat{JU}_2(G)/\widehat{JSp}(G)$ since the fibre of $c_1 : BU \longrightarrow \Sigma^2 H\mathbb{Z}$ is BSU. Furthermore $c_1 : F_1 - \widehat{JU}(G)/\widehat{JSp}(G) \longrightarrow H^2(BG; \mathbb{Z})$ is surjective.

The argument in degree -7 is similar, though the filtration is a bit longer.

$$\begin{array}{ccccccc} KO^7_G & \xleftarrow{\supset} & F_1 & \xleftarrow{\supset} & F_2 & \xleftarrow{\supset} & ko^7_G \\ {\scriptstyle \dim}\downarrow & & \downarrow{\scriptstyle w_1} & & \downarrow & & \\ \mathbb{Z}/2 & & H^1(BG, \mathbb{Z}/2) & & H^3(BG, \mathbb{Z}) & & \end{array}$$

Here $KO^7(BG) = \widehat{RO}(G)/\widehat{RU}(G) = [BG, O]$ from the fibre sequence

$$BU \times \mathbb{Z} \longrightarrow BO \times \mathbb{Z} \longrightarrow O$$

(Theorem 2.8.1). The edge homomorphism is again the dimension (now over $\mathbb{R}$) modulo 2. Taking fibres of the map to $H\mathbb{Z}/2$, we see that $F_1 = [BG, SO] = \widehat{JO}(G)/\widehat{JU}(G)$, induced by the fibre sequence

$$BU \times \mathbb{Z} \longrightarrow BO \times 2\mathbb{Z} \longrightarrow SO$$

or

$$BU \longrightarrow BO \longrightarrow SO.$$

The map to the next filtration quotient is given by $SO \longrightarrow \Sigma H\mathbb{Z}/2$, and the restriction of this to BO is w_1. This follows by pullback from $G = C_2$, since every element in $H^1(BG, \mathbb{Z}/2)$ is w_1 of the corresponding representation $BG \longrightarrow BO(1)$. For $G = C_2$, we have $KO^7(BC_2) = \widehat{RO}(C_2)/\widehat{RU}(C_2) = (\mathbb{Z}/2)^2$, while $ko^7(BC_2) = 0$ by Theorem 6.3.1. It follows that w_1 is a permanent cycle. Note in passing that $w_1 : \widehat{JO}(G)/\widehat{JU}(G) \longrightarrow H^1(BG; \mathbb{Z}/2)$ is surjective since $H^1(BG; \mathbb{Z}/2) = \mathrm{Hom}(G, \mathbb{Z}/2)$. Next, we have $F_2 = \widehat{JSO}/\widehat{JU}$, realized by $[BG, Spin]$:

$$\begin{array}{ccccc}
 & & \Sigma H\mathbb{Z}/2 & = & \Sigma H\mathbb{Z}/2 \\
 & & {\scriptstyle w_1}\uparrow & & \uparrow{\scriptstyle \Omega w_2} \\
BU & \longrightarrow & BO & \longrightarrow & SO \\
\| & & \uparrow & & \uparrow \\
BU & \longrightarrow & BSO & \longrightarrow & Spin
\end{array}$$

The final step in the filtration is the natural map $Spin \longrightarrow \Sigma^3 H\mathbb{Z}$ killing the bottom homotopy group. This restricts to $w_3 : BSO \longrightarrow \Sigma^3 H\mathbb{Z}$, and this equals βw_2, inducing the map of fibre sequences

$$\begin{array}{ccccc}
\Sigma^2 H\mathbb{Z} & \longrightarrow & \Sigma^2 H\mathbb{Z}/2 & \stackrel{\beta}{\longrightarrow} & \Sigma^3 H\mathbb{Z} \\
{\scriptstyle c_1}\uparrow & & {\scriptstyle w_2}\uparrow & & \uparrow \\
BU & \longrightarrow & BSO & \longrightarrow & Spin \\
\uparrow & & \uparrow & & \uparrow \\
BSU & \longrightarrow & BSpin & \longrightarrow & String
\end{array}$$

This gives the final layer, $ko^7_G = [BG, String] = \widehat{JSpin}/\widehat{JU}_2$ by Corollary 2.8.6.

The argument in degree -8 the filtration is longer again

$$\begin{array}{ccccccccc}
KO^8_G & \stackrel{\supset}{\longleftarrow} & F_1 & \stackrel{\supset}{\longleftarrow} & F_2 & \stackrel{\supset}{\longleftarrow} & F_3 & \stackrel{\supset}{\longleftarrow} & ko^8_G \\
{\scriptstyle \dim}\downarrow & & {\scriptstyle w_1}\downarrow & & {\scriptstyle w_2}\downarrow & & {\scriptstyle p_1/2}\downarrow & & \\
\mathbb{Z} & & H^1(BG, \mathbb{Z}/2) & & H^2(BG, \mathbb{Z}/2) & & H^4(BG, \mathbb{Z}) & &
\end{array}$$

Proceeding as before, we see that $F_1 = \ker(\dim : RO(G) \longrightarrow \mathbb{Z}) = JO$, that $w_1 : JO \longrightarrow H^1(BG; \mathbb{Z}/2)$ is surjective with kernel $F_2 = JSO$. Since $KO^9_G = 0$, this shows that ko^9_G is built from the cokernel of $w_2 : JSO \longrightarrow H^2(BG; \mathbb{Z}/2)$ and the cokernel of $p_1/2 : JSpin \longrightarrow H^4(BG; \mathbb{Z})$. □

CHAPTER 3

Descent, Twisting and Periodicity

The most remarkable property of K-theory is Bott periodicity: for periodic real K-theory this was established by Atiyah and Segal. This can be viewed as a Thom isomorphism for orientable suspensions, and it is also often valuable to know the effect of non-orientable suspensions. When it comes to real and complex K-theory, it is very striking that there is a descent spectral sequence allowing one to reconstruct KO from KU with a suitable action. All of these phenomena have counterparts in the connective case, and we describe them here in some detail.

We continue with the notation of Chapter 2, but restrict ourselves to the case when conjugation (from the group Q of order 2) is independent of the action of G, so that $\widehat{G} = G \times Q$. Indeed, the results are of interest even in the non-equivariant case $G = 1$, and the real content of this chapter is for that case.

Before turning to the work at hand, it is useful to establish notation for some standard constructions.

3.1. Fixed points, homotopy fixed points and geometric fixed points

We recall some standard constructions with fixed points and quotients. These are quite general, but we will be applying them to the group Q of order 2. The two special features of this group are firstly that various filtrations can be made very explicit, and secondly that they are periodic. It is also useful that the trivial subgroup is the only proper subgroup. More specifically, using the nontrivial representation ξ of Q on $\mathbb{R}$, the infinite sphere with antipodal action $S(\infty\xi)$ gives a good model for EQ, and hence $\tilde{E}Q \simeq S^{\infty\xi}$. The specific model plays no real role in the present section, but for later reference we write $S^{\infty\xi}$ instead of $\tilde{E}Q$ throughout.

There are three types of fixed points for a Q-spectrum that will concern us. The Lewis-May (or categorical) fixed points X^Q can be used to define the other two. The homotopy fixed points are defined by $X^{hQ} = F(EQ_+, X)^Q$, and the geometric fixed points by $\Phi^Q X \simeq (\tilde{E}Q \wedge X)^Q = (S^{\infty\xi} \wedge X)^Q$. We will need to use the relationship between them.

The cofibre sequence $EQ_+ \longrightarrow S^0 \longrightarrow S^{\infty\xi}$ gives rise to the cofibre sequence

$$F(EQ_+, X) \longleftarrow X \longleftarrow F(S^{\infty\xi}, X)$$

with fixed point cofibre sequence

$$X^{hQ} \longleftarrow X^Q \longleftarrow F(S^{\infty\xi}, X)^Q.$$

Similarly, we obtain the Tate diagram [**53**]

$$\begin{array}{ccccc} X \wedge EQ_+ & \longrightarrow & X & \longrightarrow & X \wedge S^{\infty\xi} \\ \simeq\downarrow & & \downarrow & & \downarrow \\ F(EQ_+, X) \wedge EQ_+ & \longrightarrow & F(EQ_+, X) & \longrightarrow & F(EQ_+, X) \wedge S^{\infty\xi} \end{array}$$

with fixed points

$$\begin{array}{ccccc} X_{hQ} & \longrightarrow & X^Q & \longrightarrow & \Phi^Q X \\ =\downarrow & & \downarrow & & \downarrow \\ X_{hQ} & \longrightarrow & X^{hQ} & \longrightarrow & t(X)^Q. \end{array}$$

We follow Scandinavian terminology by referring to the top row as the *norm-restriction sequence.*

The point about the homotopy fixed points is that they have homotopy groups which can be calculated by the descent spectral sequence (see Section 3.2). We will therefore approach our calculation of $\pi_*^Q(X) = \pi_*(X^Q)$ through the cofibre sequence $X_{hQ} \longrightarrow X^Q \longrightarrow \Phi^Q X$, calculating $\pi_*(X_{hQ})$ by descent and removing the error term $\pi_*(\Phi^Q X)$. This explains why the calculations of this chapter are all of homotopy orbits or homotopy fixed points.

Taking $\widehat{G} = G \times Q$, we note that $\widehat{G}$ acts on $\widehat{G}/G = Q$ and hence on EQ and $S^{\infty\xi}$, so the above discussion applies to $\widehat{G}$-spaces. The reason for taking $\widehat{G}$ to be a split extension is that we need Q to be a subgroup, and the reason for taking $\widehat{G}$ to be a product is to ensure that Q is a normal subgroup so that $G = \widehat{G}/G$ acts on Q-fixed points and Q-quotients.

3.2. Descent

For any $\widehat{G}$-spectrum X we may consider the Lewis-May Q-fixed point G-spectrum X^Q and the homotopy Q-fixed point G-spectrum

$$X^{hQ} := F(EQ_+, X)^Q,$$

and the comparison map

$$X^Q \longrightarrow F(EQ_+, X)^Q.$$

For the homotopy groups of the homotopy fixed point spectrum we have the descent spectral sequence

$$E_2^{s,t} = H^s(Q; X_G^t) \Rightarrow (X^{hQ})_G^{s+t}.$$

In case the fixed points and homotopy fixed points agree, this lets us use G-equivariant information to calculate $\widehat{G}$-equivariant information.

We want to apply this with $X = K\mathbb{R}$ and $X = k\mathbb{R}$.

LEMMA 3.2.1. *(Atiyah) The Q-fixed points and homotopy fixed points agree for $K\mathbb{R}$, and hence there is a descent spectral sequence*

$$E_2^{s,t} = H^s(Q; KU_G^t) \Rightarrow KO_G^{s+t}.$$

Proof: We must show that the map

$$K\mathbb{R} = F(S^0, K\mathbb{R}) \longrightarrow F(E(\widehat{G}/G)_+, K\mathbb{R})$$

is an equivalence, which is to say that $F(S^{\infty\xi}, K\mathbb{R}) \simeq *$.

Atiyah shows that the Euler class of ξ is η. Thus, using $S(\infty\xi)$ as our model for $E(\widehat{G}/G)$, we find that since $\eta^3 = 0$,

$$[T, F(S^{\infty\xi}, K\mathbb{R})]_*^{\widehat{G}} = 0$$

for any T and the result follows. □

On the other hand, we may calculate that, even for $G = 1$, the corresponding statement is false in the connective case. Before proceeding, it is very useful to

FIGURE 3.2.1. The descent spectral sequence $E^2_{s,t} = H^{-s}(Q; KU_t) \Rightarrow (K\mathbb{R}^{hQ})_{s+t}$. Open and filled dots represent $\mathbb{Z}/2$s, with only the filled dots surviving nontrivially to E^∞. The named elements v^s each generate a copy of $\mathbb{Z}$, with the v^{4i} and $2v^{4i+2}$ surviving to E^∞. The only differential is d^3.

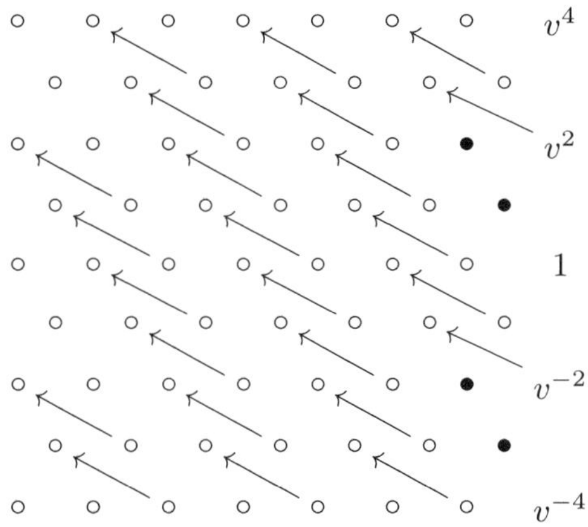

draw a picture (Figure 3.2.1) of the descent spectral sequence in the periodic case, and we will use homological grading, displaying

$$E^2_{s,t} = E_2^{-s,-t} = H^{-s}(Q; KU_t)$$

at the point (s,t). For the present we display the E_2-term, but later we will have reason to look at the E_1-term. Naturality will show that the descent spectral sequence for the connective case is a suitable truncation of that for the periodic case.

EXAMPLE 3.2.2. When $G = 1$ the homotopy fixed points are different to the fixed points.

$$ko = k\mathbb{R}^Q \not\simeq k\mathbb{R}^{hQ}.$$

Proof: Consider the descent spectral sequence

$$E_2^{s,t} = H^{-s}(Q; ku_t) \Rightarrow (k\mathbb{R}^{hQ})_{s+t}.$$

Since $\tau(v) = -v$ by 2.2.10, we find the action of Q on ku_* is trivial in degrees $4n$ and by -1 in degrees $4n+2$, and so

$$E_2 \quad = \quad \mathbb{Z}[y,z]/(2y, 2z) \otimes \mathbb{Z}[v^2]$$

with $y \in E_2^{-2,0} = H^2(Q, \mathbb{Z})$, $v^2 \in E_2^{0,4} = H^0(Q, \mathbb{Z})$, and $z \in E_2^{-1,2} = H^1(Q, \mathbb{Z}(-1))$, where $\mathbb{Z}(-1)$ is $\mathbb{Z}$ with Q acting by negation.

The differentials are forced by comparison with the periodic case,

$$E_2^{s,t} = H^{-s}(Q; KU_t) \Rightarrow KO_{s+t}$$

for which

$$E_2 \quad = \quad \mathbb{Z}[y,z]/(2y,2z) \otimes \mathbb{Z}[v^2, v^{-2}].$$

Since $E_2 = E_3$ by degree, it suffices to show that $d_3(z) = 0$, $d_3(y) \neq 0$ and $d_3(v^2) \neq 0$, since there can then be no further differentials by sparsity. The multiplicative structure shows that it is enough to establish only one of the non-vanishing d_3s.

We have seen that $KO = K\mathbb{R}^Q \simeq K\mathbb{R}^{hQ}$, and for the present purposes, we take the coefficients KO_* to be known. The element z detects $\eta \in ko_1 = KO_1$, so that yv^2 detects η^2, and yzv^2 detects η^3. Since z is an infinite cycle, so is z^3; since $\eta^3 = 0$, z^3 must be a boundary, and $d_3(v^2) = yzv^2$. (Note that v^2 is indecomposable in $E_2 = E_3$.) We must also have $d_3(y) = y^2z$ so that the edge homomorphism $KO_0 \longrightarrow KU_0$ will be an isomorphism.

In the periodic case, this leaves E_4 concentrated in columns $s = 0, -1, -2$, and hence $E_4 = E_\infty$. Hence $E_4^{0,*} = E_\infty^{0,*} = \mathbb{Z}[2v^2, v^4]$ in the connective case and $E_4^{0,*} = E_\infty^{0,*} = \mathbb{Z}[2v^2, v^4, v^{-4}]$ in the periodic case. Similarly, $E_4^{1,*}$ is $z\ \mathbb{F}_2[v^4]$ and $z\ \mathbb{F}_2[v^4, v^{-4}]$, and Similarly, $E_4^{2,*}$ is $yv^2\ \mathbb{F}_2[v^4]$ and $yv^2\ \mathbb{F}_2[v^4, v^{-4}]$, respectively. In the connective case, we also have $y^{2i} \in E_\infty^{-4i,0}$ for $i > 0$, contributing $k\mathbb{R}^{hQ}_{-4i} = \mathbb{F}_2$, and showing that $ko = k\mathbb{R}^Q \longrightarrow k\mathbb{R}^{hQ}$ is not onto. □

Po Hu **[60]** has proved analogous statements in higher chromatic periodicity when $G = 1$.

REMARK 3.2.3. There is an alternative approach to deducing KO-theory from KU-theory: using the fact that $\eta^3 = 0$ we may obtain a Bockstein spectral sequence, which works just as well in the connective case. This is the subject of Chapter 4.

3.3. Statement of periodicity for equivariant connective real K-theory

In seeking periodicity in the connective case, we are guided by Atiyah's result that $K\mathbb{R}$ is $1 \oplus \xi$-periodic. We refer to this as $(1,1)$-periodicity, but with different grading conventions it is sometimes called $(2,1)$-periodicity. The $(1,1)$-periodicity is a special case of the Thom isomorphism for representations $\hat{U}$, as in Theorem 2.2.5, but the special case will cover our applications.

THEOREM 3.3.1. *(Atiyah) There are $\widehat{G}$-equivalences*

$$S^{1\oplus\xi} \wedge K\mathbb{R} \simeq K\mathbb{R}$$

and

$$S^8 \wedge K\mathbb{R} \simeq K\mathbb{R}. \quad \square$$

The first consequence of the periodicity results is

$$K\mathbb{R} \wedge S^{4\xi} \simeq K\mathbb{R} \wedge S^4,$$

so that 4ξ is $K\mathbb{R}$-orientable. On taking Q-fixed points, we also deduce

$$(K\mathbb{R} \wedge S^{-\xi})^Q \simeq \Sigma KO.$$

In that form, the results generalize to the connective case.

THEOREM 3.3.2. *(a) The representation 4ξ is orientable for Borel $k\mathbb{R}$-homology in the sense that there is an equivalence*

$$k\mathbb{R} \wedge EQ_+ \wedge S^{4\xi} \simeq k\mathbb{R} \wedge EQ_+ \wedge S^4$$

of $\widehat{G}$-spectra, where $\widehat{G} = G \times Q$.

(b) There are G-equivalences of ko-module spectra

$$(k\mathbb{R} \wedge S^{-\xi})^Q \simeq \Sigma ko,$$

$$(k\mathbb{R} \wedge S^{-2\xi})^Q \simeq \Sigma^{-6} ko\langle 4\rangle$$

and

$$(k\mathbb{R} \wedge S^{-3\xi})^Q \simeq \Sigma^{-5} ko\langle 2\rangle.$$

Proof: We argue here that it suffices to prove each of the statements in the case $G = 1$, and the rest of the chapter will be devoted to that case.

By way of example, we consider the result about $k\mathbb{R} \wedge S^{-\xi}$. Indeed, we consider the defining pullback square in Definition 2.4.1. We may construct a map of forks

$$\begin{array}{ccccc}
F(EG_+, \inf_1^G \Sigma ko) & \longrightarrow & F(EG_+, \inf_1^G \Sigma KO) & \longleftarrow & \Sigma KO \\
\downarrow & & \downarrow & & \downarrow \\
F(EG_+, \inf_Q^{\widehat{G}} k\mathbb{R} \wedge S^{-\xi})^Q & \longrightarrow & F(EG_+, \inf_Q^{\widehat{G}} K\mathbb{R} \wedge S^{-\xi})^Q & \longleftarrow & (K\mathbb{R} \wedge S^{-\xi})^Q
\end{array}$$

On the right, Atiyah's periodicity gives an equivalence, and the right hand square commutes since that is compatible with forgetting equivariance. The central vertical is an equivalence by the Q-equivariant version of Atiyah's result. The left hand square commutes since the Q-equivariant periodic and connective results are compatible. All checks of compatibility are straightforward since the vertical maps are determined by the image of the unit, since they are module maps. It remains to prove the Q-equivariant connective result. □

REMARK 3.3.3. (*Deviation from 4ξ-periodicity*) As stated in the theorem, the Borel theory is precisely 4ξ-periodic. The behaviour for connective real K-theory itself is a bit more complicated. We shall see in the rest of the chapter that in the case $G = 1$ the deviation is a single Eilenberg-MacLane spectrum, and the deviation from $4k\xi$-periodicity is a wedge of k Eilenberg-MacLane spectra.

To see what happens in the more general case, we insert this behaviour into the defining pullback square of Definition 2.4.1. We have exact periodicity on the right hand (periodic) side, so the deviation arises by applying $F(EG_+, \cdot)$ to the Q-equivariant deviation on $k\mathbb{R}$. Thus we find that the Eilenberg-MacLane spectrum $H\mathbb{F}_2$ (with homotopy $\mathbb{F}_2$ in a single degree) is replaced by the Borel-Eilenberg-MacLane spectrum $F(EG_+, H\mathbb{F}_2)$ (with homotopy $H^*(BG; \mathbb{F}_2)$). Furthermore, if the suspension by $1 \oplus \xi = \hat{\mathbb{R}}$ (of real dimension 2) is replaced by the more general suspension $\hat{U}$ (of real dimension $2n$) associated to a representation U over $\mathbb{R}$, we acquire local G-coefficients, so that the deviation from $\hat{U}$-periodicity is given in homotopy by k copies of $H^*(BG^U; \mathbb{F}_2)$.

3.4. Periodicity for connective real K-theory

For the rest of the chapter, we describe Q-equivariant results (i.e., $G = 1, \widehat{G} = Q$). In view of the calculations required to prove the result, the authors find the simplicity of the statements remarkable.

We have recorded the situation in the periodic case in Theorem 3.3.1 above.

The statements generalize to the connective case, using connective covers. For any Q-spectrum X, we write $X\langle k\rangle$ for the cover of X in which all homotopy below degree k has been killed. More precisely, it is characterized by the fact that there is a map $X\langle k\rangle \longrightarrow X$ which is an isomorphism of π_n^Q and π_n for $n \geq k$, and so that $\pi_n^Q(X\langle k\rangle) = \pi_n(X\langle k\rangle) = 0$ for $n < k$.

We divide the statement into twisting and periodicity.

THEOREM 3.4.1. *Real connective K-theory is 8-periodic in the sense that*

$$S^8 \wedge k\mathbb{R} \simeq k\mathbb{R}\langle 8\rangle.$$

The first three negative twists are shifts of covers of $k\mathbb{R}$: there are Q-equivalences

- $S^{-1-\xi} \wedge k\mathbb{R} \simeq K\mathbb{R}\langle -2\rangle \simeq \Sigma^{-8}k\mathbb{R}\langle 6\rangle$
- $S^{-2-2\xi} \wedge k\mathbb{R} \simeq K\mathbb{R}\langle -4\rangle \simeq \Sigma^{-8}k\mathbb{R}\langle 4\rangle$
- $S^{-3-3\xi} \wedge k\mathbb{R} \simeq K\mathbb{R}\langle -6\rangle \simeq \Sigma^{-8}k\mathbb{R}\langle 2\rangle$.

Taking Q-fixed points of the equivalences, and noting that $(X\langle k\rangle)^Q \simeq X^Q\langle k\rangle$, we obtain the following.

COROLLARY 3.4.2. *There are non-equivariant equivalences*

(i) $(S^{-1-\xi} \wedge k\mathbb{R})^Q \simeq KO\langle -2\rangle$
(ii) $(S^{-2-2\xi} \wedge k\mathbb{R})^Q \simeq KO\langle -4\rangle$
(iii) $(S^{-3-3\xi} \wedge k\mathbb{R})^Q \simeq KO\langle -6\rangle$

In particular, the first and second statements give

$$(S^{-\xi} \wedge k\mathbb{R})^Q \simeq \Sigma ko,$$

$$(S^{-2\xi} \wedge k\mathbb{R})^Q \simeq \Sigma^{-2}ksp \quad \square$$

When it comes to periodicity, the connective case is definitely more complicated.

THEOREM 3.4.3. *(a) Connective real Borel K-homology is 4ξ-periodic in the sense that there*

$$k\mathbb{R} \wedge EQ_+ \wedge S^{4\xi} \simeq k\mathbb{R} \wedge EQ_+ \wedge S^4.$$

(b) Connective real K-theory is almost 4ξ-periodic in the sense that there is a cofibre sequence

$$k\mathbb{R} \wedge S^4 \longrightarrow k\mathbb{R} \wedge S^{4\xi} \longrightarrow S^{\infty\xi} \wedge H\mathbb{F}_2,$$

and a cofibre sequence

$$k\mathbb{R} \wedge S^{-4\xi} \longrightarrow k\mathbb{R} \wedge S^{-4} \longrightarrow S^{\infty\xi} \wedge \Sigma^{-4}H\mathbb{F}_2.$$

More generally, there are cofibre sequences

$$k\mathbb{R} \wedge S^{4k} \longrightarrow k\mathbb{R} \wedge S^{4k\xi} \longrightarrow S^{\infty\xi} \wedge \bigvee_{i=1}^{k} \Sigma^{4i-4}H\mathbb{F}_2,$$

and a cofibre sequence

$$k\mathbb{R} \wedge S^{-4\xi} \longrightarrow k\mathbb{R} \wedge S^{-4} \longrightarrow S^{\infty\xi} \wedge \bigvee_{i=1}^{k} \Sigma^{-4i} H\mathbb{F}_2.$$

REMARK 3.4.4. The situation for $k\mathbb{R}$ itself is also visible in the $RO(Q)$-graded coefficient groups, which are recorded in Section 3.7.

We now describe the proof of the theorems. The equivalence $\Sigma^8 k\mathbb{R} \simeq k\mathbb{R}\langle 8\rangle$ follows immediately from the definition and the equivalence $\Sigma^8 K\mathbb{R} \simeq K\mathbb{R}$. We explain the strategy for the twisted periodicities in more detail: its implementation relies on calculations made in the rest of the chapter.

First we will calculate the homotopy groups of the spectra $S^{k\xi} \wedge k\mathbb{R}$. Non-equivariantly, the suspension just shifts degrees by 1, which is elementary, but equivariantly the twisting is quite subtle. The calculation of the equivariant part is in two stages: first we calculate the homotopy groups of $S^{k\xi} \wedge k\mathbb{R} \wedge EQ_+$, and then deduce the answer from the norm-restriction sequence. In fact we need to know the restriction and induction maps relating equivariant to non-equivariant homotopy: we need the homotopy groups $\underline{\pi}^Q_*$ as Mackey functors, with the ko_*-module structure of π^Q_* and the ku_*-module structure of π_*. We refer to such objects as $\underline{k\mathbb{R}}^Q_*$-modules.

It is well known that the homotopy groups of $k\mathbb{R}$ are as follows

	0	1	2	3	4	5	6	7
π^Q_k	$\mathbb{Z}$	2	2	0	$\mathbb{Z}'$	0	0	0
	$\downarrow 1$				$\downarrow 2$			
π_k	$\mathbb{Z}$	0	$\mathbb{Z}$	0	$\mathbb{Z}$	0	$\mathbb{Z}$	0

We have only displayed the restriction maps, since that is all we require. The induction maps are multiplication by 2 and 1 in degrees 0 and 4 respectively (as follows from the restriction and the Mackey axiom) and is surjective in degree 2 (as follows from the behaviour of complexification). The homotopy groups of the covers as $\underline{k\mathbb{R}}^Q_*$-modules are obtained by truncation. It will be seen that the calculations of the homotopy groups of the twists give the predicted values as $\underline{k\mathbb{R}}^Q_*$-modules, and we give the rest of the proof based on this. Observe from the above table that the four twists $k\mathbb{R}$, $k\mathbb{R} \wedge S^{-\xi}$, $k\mathbb{R} \wedge S^{-2\xi}$ and $k\mathbb{R} \wedge S^{-3\xi}$ are essentially different (i.e., they are not G-fixed suspensions of one another).

Since $k\mathbb{R}$ is a ring spectrum, we may then specify a $k\mathbb{R}$-module map $f : \Sigma^i k\mathbb{R} \longrightarrow M$ by choosing an element of $\pi^Q_i(M)$. In particular, the 4ξ-periodicity of the Borel theory follows since $\underline{\pi}^Q_*(S^{4\xi} \wedge k\mathbb{R} \wedge EQ_+)$ is a free $\underline{k\mathbb{R}}^Q_*$-module on one generator of degree 4. For $S^{4\xi} \wedge k\mathbb{R}$ itself we need more care. First, we pick a generator $q \in \pi^Q_4(k\mathbb{R} \wedge S^{4\xi}) \cong \mathbb{Z}$, and use the $\underline{k\mathbb{R}}^Q_*$-module structure to show its cofibre has homotopy in a single degree: this gives the cofibre sequence

$$k\mathbb{R} \wedge S^4 \xrightarrow{q} k\mathbb{R} \wedge S^{4\xi} \longrightarrow S^{\infty\xi} \wedge H\mathbb{F}_2.$$

Now desuspend by $4 \oplus 4\xi$ to obtain the cofibre sequence

$$S^{-4\xi} \wedge k\mathbb{R} \xrightarrow{p} S^{-4} \wedge k\mathbb{R} \longrightarrow S^{\infty\xi} \wedge \Sigma^{-4} H\mathbb{F}_2.$$

In particular, p is a non-equivariant equivalence, and a homotopy isomorphism above degree -4.

The cofibre sequences for the higher twists are constructed similarly. The splitting of the error term into Eilenberg-MacLane spectra follows from the splitting of $t(k\mathbb{R})$ (Proposition 3.6.1). This completes the outline of the proof of Theorem 3.4.3.

We turn now to the identifications of the twistings in Theorem 3.4.1. First, we construct a map $a : k\mathbb{R} \longrightarrow S^{-(1+\xi)} \wedge k\mathbb{R}$ by choosing a generator of $\pi_0^Q(S^{-(1+\xi)} \wedge k\mathbb{R}) \cong \mathbb{Z}$; since the restriction is an isomorphism in $\underline{\pi}_0^Q$, it follows that a is non-equivariantly multiplication by v. Iterating, we obtain maps

$$k\mathbb{R} \xrightarrow{a} S^{-(1+\xi)} \wedge k\mathbb{R} \xrightarrow{a} S^{-2(1+\xi)} \wedge k\mathbb{R} \xrightarrow{a} S^{-3(1+\xi)} \wedge k\mathbb{R} \xrightarrow{pa} K\mathbb{R}\langle -8 \rangle.$$

Now we may verify that the effect in homotopy is completely determined by the $\underline{k\mathbb{R}}_*^Q$-module structure from the evident fact that non-equivariantly, the sequence is simply

$$ku = KU\langle 0\rangle \xrightarrow{v} KU\langle -2\rangle \xrightarrow{v} KU\langle -4\rangle \xrightarrow{v} KU\langle -6\rangle \xrightarrow{v} KU\langle -8\rangle.$$

One observes from the homotopy groups as $\underline{k\mathbb{R}}_*^Q$-modules that the maps are all isomorphisms where the domain is non-zero, and it follows that the the spectra are the specified covers. Indeed, in all cases we are considering a map $a : M\langle k\rangle \longrightarrow M$, of $\underline{k\mathbb{R}}_*^Q$-modules, where M is itself the cover $\underline{k\mathbb{R}}_*^Q\langle k-2\rangle$. In the non-equivariant part, the map is simply v and hence an isomorphism in degrees $\geq k$. Since the restriction is an isomorphism in degrees divisible by 8 the image contains a power of the Bott class in the equivariant part. The map a therefore induces an epimorphism on equivariant parts in all high degrees. Since multiplication by the Bott class is an isomorphism on $M\langle k\rangle$ above degree k, we infer the map a is also an isomorphism in lower degrees above k.

The rest of the chapter will calculate the homotopy groups with Mackey and module structure, and show they take the specified values, thereby completing the proofs.

3.5. Twistings of $k\mathbb{R}$

We give a systematic account of the homotopy fixed points and homotopy orbits of the twistings $k\mathbb{R} \wedge S^{k\xi}$ of real connective K-theory. This completes the proof of the periodicity statement in Section 3.4, and hence also in Section 3.3. This enables us to give an equivariant proof of the results connecting real and complex K-theory in Chapter 4. Giving a systematic account makes things clearer, and also permits us to give an analysis of the homotopy fixed points and geometric fixed points.

The analysis is organized around the Tate spectral sequence. This is an upper half-plane spectral sequence constructed by applying π_*^Q to the $\mathbb{Z}$-graded filtration of $S^{\infty\xi} \wedge k\mathbb{R}$ with kth term $S^{k\xi} \wedge k\mathbb{R}$ for all $k \in \mathbb{Z}$. It is convenient to grade it homologically and work from the E^1-term.

Since the filtration quotients are

$$S^{k\xi}/S^{(k-1)\xi} \wedge k\mathbb{R} \simeq \Sigma^k Q_+ \wedge ku,$$

it is clear that $E^1_{s,t} = 0$ if t is odd or negative, and $E^1_{s,2n} = \mathbb{Z}$. The d^1 differential takes the form

$$\cdots \longleftarrow \mathbb{Z} \xleftarrow{0} \mathbb{Z} \xleftarrow{2} \mathbb{Z} \xleftarrow{0} \mathbb{Z} \xleftarrow{2} \mathbb{Z} \xleftarrow{0} \mathbb{Z} \xleftarrow{2} \cdots$$

on each non-zero row. The zero maps are from even columns on rows $4m$ and from odd columns on rows $4m+2$. The result is depicted in Figure 3.5.2.

FIGURE 3.5.2. The E^1-term of the Tate spectral sequence. Bullets represent copies of $\mathbb{Z}$ which will give $\mathbb{Z}/2$ at E^2, and open dots represent copies of $\mathbb{Z}$ which are mapped monomorphically by d^1. The only differentials are d^1 and d^3.

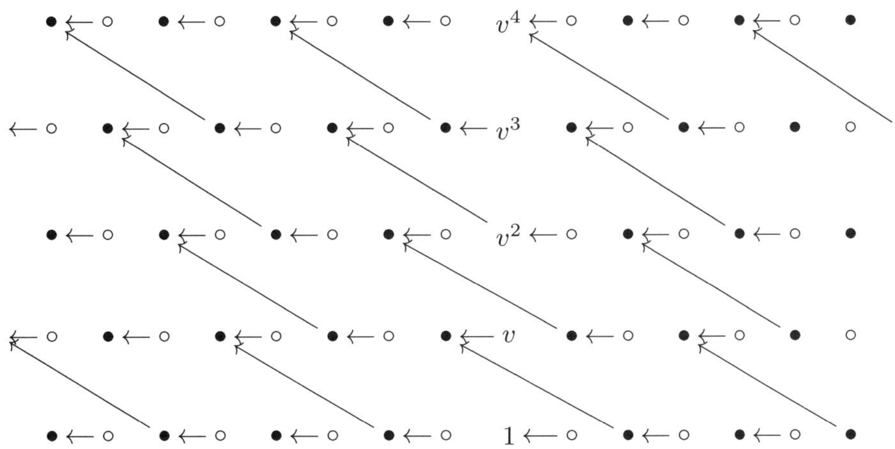

The spectral sequence converges conditionally to $t(k\mathbb{R})^Q_*$. To see this, we use an alternative view of the spectral sequence. Before, we took the obvious filtration of $S^{\infty\xi}$ and smashed with $k\mathbb{R}$, but now we smash it with $F(EQ_+, k\mathbb{R})$ instead. This has no effect on the spectral sequence since the map $k\mathbb{R} \longrightarrow F(EQ_+, k\mathbb{R})$ is a non-equivariant equivalence and the subquotients are free, but now conditional convergence to the homotopy groups of $t(k\mathbb{R}) = F(EQ_+, k\mathbb{R}) \wedge S^{\infty\xi}$ is straightforward to check by using the approximations in

$$t(K\mathbb{R}) = F(EQ_+, K\mathbb{R}) \wedge S^{\infty\xi} = \underrightarrow{\text{holim}}\, F(EQ_+, K\mathbb{R}) \wedge S^{k\xi} = \underrightarrow{\text{holim}}\, F(EQ^{-k\xi}, K\mathbb{R}).$$

(or see [**53**]).

Various truncations of the spectral sequence are also useful, and their behaviour is easy to understand, given that the pattern of differentials (suitably truncated) is the same in all of them; this is why we have displayed the spectral sequence from E^1. The Tate spectral sequence for the periodic theory $K\mathbb{R}$ is a whole-plane spectral sequence, and the upper half-plane is the Tate spectral sequence for $k\mathbb{R}$. The right quadrant gives the homotopy orbit spectral sequence for $k\mathbb{R}$. The left quadrant gives the homotopy fixed point spectral sequence for $k\mathbb{R}$.

First we consider the second quadrant truncations. If we include columns $s \leq 0$ we obtain the spectral sequence for $\pi_*(k\mathbb{R}^{hQ})$. More generally, if we include columns for $s \leq k$ we obtain the spectral sequence for

$$k\mathbb{R}^*_Q(S^{-k\xi} \wedge EQ_+) \cong \pi_*((S^{k\xi} \wedge k\mathbb{R})^{hQ}).$$

The calculations are summarized in tables below: the untwisted version is in Section 3.8, and the three twisted versions in Section 3.9.

PROPOSITION 3.5.1. *We have the following calculations for $k\mathbb{R}^*_Q(S^{k\xi} \wedge EQ_+)$. These depend only on k mod 4, and the value of*

$$\pi_{-k+s}(F(EQ_+, S^{-k\xi} \wedge k\mathbb{R})) = k\mathbb{R}_Q^{k-s}(S^{k\xi} \wedge EQ_+)$$

depends only on s mod 8 provided $s \geq 0$. The display is collected in groups of four successive homotopy groups. Roughly speaking, the pattern is that of repeating homotopy groups 2000 in sufficiently negative degrees and $\mathbb{Z}220\mathbb{Z}'000$ (i.e., the homotopy groups of ko_*, with $\mathbb{Z}'$ referring to the even integers) in sufficiently positive degrees. However the boundary between these two behaviours depends on k mod 4, and is indicated in the table with a vertical bar. To the left of the boundary, the pattern is constant along vertical lines, and to the right of the boundary it is constant along lines of slope -1 (reflecting $(1,1)$-periodicity).

The table below records

$$\pi_{-4i+t}(F(EQ_+, S^{-(4i+\epsilon)\xi} \wedge k\mathbb{R})) = k\mathbb{R}_Q^{4i-t}(S^{(4i+\epsilon)\xi} \wedge EQ_+)$$

The label above a group of four homotopy groups refers to the degree of the first of them (i.e., the left hand one).

t	-20	-16	-12	-8	-4	0	$+4$	$+8$	$+12$
$S^{4i\xi}$	2000	2000	2000	2000	$2000\|$	$\mathbb{Z}220$	$\mathbb{Z}'000$	$\mathbb{Z}220$	$\mathbb{Z}'000$
$S^{(4i+1)\xi}$	2000	2000	2000	2000	$200\|0$	$0\mathbb{Z}22$	$0\mathbb{Z}'00$	$0\mathbb{Z}22$	$0\mathbb{Z}'00$
$S^{(4i+2)\xi}$	2000	2000	2000	2000	$20\|\mathbb{Z}'0$	$00\mathbb{Z}2$	$20\mathbb{Z}'0$	$00\mathbb{Z}2$	$20\mathbb{Z}'0$
$S^{(4i+3)\xi}$	2000	2000	2000	2000	$2\|20\mathbb{Z}'$	$000\mathbb{Z}$	$220\mathbb{Z}'$	$000\mathbb{Z}$	$220\mathbb{Z}'$

There is a little more information in the E_∞-terms displayed in Examples 3.8.5, 3.9.1, 3.9.2 and 3.9.3 respectively.

Proof: The reader will find it useful to refer to the tables in Section 3.9 in working through the following argument.

Essentially, we just observe that truncating the pattern of differentials gives the illustrated E_∞-terms, and then give additional arguments to resolve extensions. However, to avoid duplicated work, and to see the effect of the inclusions $S^{k\xi} \longrightarrow S^{(k+1)\xi}$, we prefer to fill in the table row by row. The case $k = 0$ was discussed in detail in Section 3.2.

Assume that the spectral sequence for a particular value of k (i.e., the truncation for $s \leq -k$) has been analyzed and shown to give the stated result. To obtain the truncation for $s \leq -k-1$, one inspects the groups and differentials removed. This gives the values for $k = 1, 2, 3$ and 4 in turn. One observes that the spectral sequence for $S^{-4\xi}$ is a shifted version of that for S^0, and hence the table is correct for $k \geq 0$. Similarly, one shows that the answer for k implies that for $k-1$, giving values for $k = -1, -2, -3$ and -4 in turn. One observes that the spectral sequence for $S^{4\xi}$ is a shifted version of that for S^0, and hence the table is correct for $k \leq 0$. □

The first quadrant truncations are a little harder to understand. If we include columns for $s \geq 1$ we obtain the spectral sequence for $k\mathbb{R}^Q_*(\Sigma EQ_+) = \pi^Q_*(k\mathbb{R} \wedge \Sigma EQ_+)$. More generally, if we include columns for $s \geq k$ we obtain the spectral sequence for

$$k\mathbb{R}^Q_*(S^{k\xi} \wedge \Sigma EQ_+) \cong \pi^Q_*(S^{\infty\xi}/S^{k\xi} \wedge k\mathbb{R}).$$

From these spectral sequences it is easy to deduce the geometric fixed points of $k\mathbb{R}$. The calculations are summarized in tables below: the untwisted version is in Section 3.8, and the three twisted versions in Section 3.10.

PROPOSITION 3.5.2. *We have the following calculations for* $k\mathbb{R}^Q_*(S^{k\xi} \wedge EQ_+)$. *These depend only on* k *mod 4, and the value of* $k\mathbb{R}^Q_{k+s}(S^{k\xi} \wedge EQ_+)$ *depends only on* s *mod 8 provided* $s \geq 1$.

The following table shows the values

$$k\mathbb{R}^Q_{4i+t}(S^{(4k+\epsilon)\xi} \wedge EQ_+).$$

t	0	1	2	3	4	5	6	7	8	9	10	11	12
$S^{4i\xi}$	$\mathbb{Z}$	2	2	2	$\mathbb{Z}$	0	0	2	$\mathbb{Z}$	2	2	2	$\mathbb{Z}$
$S^{(4i+1)\xi}$		2	0	$\mathbb{Z} \oplus 2$	0	0	0	$\mathbb{Z} \oplus 2$	2	2	0	$\mathbb{Z} \oplus 2$	0
$S^{(4i+2)\xi}$			$\mathbb{Z}$	2	0	0	$\mathbb{Z}$	$2 \oplus 2$	2	0	$\mathbb{Z}$	2	0
$S^{(4i+3)\xi}$				2	0	$\mathbb{Z}$	2	$2 \oplus 2$	0	$\mathbb{Z}$	0	2	0

There is a little more information in the E_∞*-terms displayed in Examples 3.8.6, 3.10.3, 3.10.2 and 3.10.1 respectively.*

The conclusion is described in an alternative layout in Section 3.7.

Proof: The reader will find the tables in Section 3.10 helpful in following the argument.

The strategy is exactly the same as in the second quadrant case. Essentially, we just observe that truncating the pattern of differentials gives the illustrated E_∞-terms, and then give additional arguments to resolve extensions. However, to avoid duplicated work, and to see the effect of the inclusions $S^{k\xi} \longrightarrow S^{(k+1)\xi}$, we prefer to fill in the table row by row.

The behaviour for the truncation for $s \geq 1$ can be obtained directly from the spectral sequence, or from the $s \leq 0$ case and the Tate sequence. Now observe that the behaviour of the $s \geq k+1$ truncation follows from that of the $s \geq k$ truncation by noting the effect of removing groups and differentials. One observes that after 4 steps one obtains a shift of the original picture. Similarly, the behaviour of the $s \geq k-1$ truncation follows from that of the $s \geq k$ truncation by noting the effect of adding groups and differentials, and the same 4-fold periodicity appears.

It is straightforward to read off the following pattern, where 2 denotes the group of order 2 and where $A \cdot B$ denotes a group with subgroup A and quotient group B. The 2^* entries indicate groups coming from $t(k\mathbb{R})^Q_*$.

t	0	1	2	3	4	5	6	7	8	9	10	11	12
$S^{4i\xi}$	$\mathbb{Z}$	2	2	2^*	$\mathbb{Z}$	0	0	2^*	$\mathbb{Z} \cdot 2$	2	2	2^*	$\mathbb{Z}$
$S^{(4i+1)\xi}$		2	0	$\mathbb{Z} \cdot 2^*$	0	0	0	$\mathbb{Z} \cdot 2 \cdot 2^*$	2	2	0	$\mathbb{Z} \cdot 2^*$	0
$S^{(4i+2)\xi}$			$\mathbb{Z}$	2^*	0	0	$\mathbb{Z} \cdot 2$	$2 \cdot 2^*$	2	0	$\mathbb{Z}$	2^*	0
$S^{(4i+3)\xi}$				2^*	0	$\mathbb{Z} \cdot 2$	2	$2 \cdot 2^*$	0	$\mathbb{Z}$	0	2^*	0

It now remains to resolve the extension questions. This can be done by considering the long exact Tate sequence arising from the cofibre sequence

$$EQ_+ \wedge S^{k\xi} \longrightarrow F(EQ_+, k\mathbb{R}) \wedge S^{k\xi} \longrightarrow t(k\mathbb{R}).$$

It turns out that those involving $\mathbb{Z}$ can be resolved by looking at the sequence itself. Those of the form $2 \cdot 2$ split by using the action of η in $F(EQ_+, k\mathbb{R}) \wedge S^{k\xi}$. □

COROLLARY 3.5.3. *The geometric fixed points of $k\mathbb{R}$ have homotopy groups $\mathbb{Z}/2$ in degrees of the form $4n \geq 0$ and are zero otherwise.*

Proof: Since $\pi_*(\Phi^Q k\mathbb{R}) = \pi_*^Q(S^{\infty\xi} \wedge k\mathbb{R})$ the answer may be read off from $\pi_*^Q(k\mathbb{R}) = ko_*$ and the calculation of $k\mathbb{R}_*^Q(EQ_+)$, using the cofibre sequence $EQ_+ \longrightarrow S^0 \longrightarrow S^{\infty\xi}$. □

3.6. Splittings

We have seen that the homotopy fixed points, homotopy orbits and Tate spectra for $k\mathbb{R}$ have a regular and expected behaviour together with an error term whose homotopy consists of $\mathbb{F}_2$ in suitable degrees of form $4k$. In this section, we observe that this error is actually a wedge of Eilenberg-MacLane spectra.

PROPOSITION 3.6.1. *We have an equivalence*

$$t(k\mathbb{R})^Q \simeq \bigvee_{k \in \mathbb{Z}} \Sigma^{4k} H\mathbb{F}_2.$$

Proof: The homotopy groups we have calculated are consistent with this statement, and it remains to construct a map of spectra inducing this isomorphism.

The spectrum $k\mathbb{R}$ is connective, and the bottom homotopy Mackey functor is the constant functor $\mathbb{Z}$ (i.e., its restriction map is the identity and the induction map is multiplication by 2). Accordingly we have a map $k\mathbb{R} \longrightarrow H\mathbb{F}_2$ of Q-spectra, where $H\mathbb{F}_2$ is the Eilenberg-MacLane spectrum for the constant Mackey functor $\mathbb{F}_2$. Applying the Tate construction, we obtain a map

$$t(k\mathbb{R})^Q \longrightarrow t(H\mathbb{F}_2)^Q.$$

The Tate spectral sequence for $t(H\mathbb{F}_2)$ collapses at E_2 to show that $\pi_*^Q(t(H\mathbb{F}_2)) = \mathbb{F}_2[t, t^{-1}]$, where t is in degree 1, and the comparison map of Tate spectral sequences takes the bottom edge for $t(k\mathbb{R})$ to $\mathbb{F}_2[t^4, t^{-4}]$.

Finally, $t(H\mathbb{F}_2)$ is a module over $H\mathbb{F}_2$, so $t(H\mathbb{F}_2)^Q$ is a module over the nonequivariant ring spectrum $(H\mathbb{F}_2)^Q = H\mathbb{F}_2$. Thus

$$t(H\mathbb{F}_2) \simeq \bigvee_{k \in \mathbb{Z}} \Sigma^k H\mathbb{F}_2.$$

Projecting onto the factors in degrees which are multiples of 4, we obtain the required map

$$t(k\mathbb{R})^Q \longrightarrow t(H\mathbb{F}_2)^Q \simeq \bigvee_{k \in \mathbb{Z}} \Sigma^k H\mathbb{F}_2 \longrightarrow \bigvee_{k \in \mathbb{Z}} \Sigma^{4k} H\mathbb{F}_2.$$

□

The same proof gives the following result for homotopy orbits and homotopy fixed points.

COROLLARY 3.6.2. *There are cofibre sequences*

$$ko \longrightarrow k\mathbb{R}^{hQ} \longrightarrow \bigvee_{k \geq 1} \Sigma^{-4k} H\mathbb{F}_2,$$

and

$$k\mathbb{R}_{hQ} \longrightarrow ko \longrightarrow \bigvee_{k\geq 0} \Sigma^{4k} H\mathbb{F}_2.$$

More generally, note that suspensions by multiples of ξ do not affect the final term. We therefore get cofibre sequences

$$(k\mathbb{R} \wedge S^{k\xi})^Q \longrightarrow (k\mathbb{R} \wedge S^{k\xi})^{hQ} \longrightarrow \bigvee_{k\geq 1} \Sigma^{-4k} H\mathbb{F}_2,$$

and

$$(k\mathbb{R} \wedge S^{k\xi})_{hQ} \longrightarrow (k\mathbb{R} \wedge S^{k\xi})^Q \longrightarrow \bigvee_{k\geq 0} \Sigma^{4k} H\mathbb{F}_2.$$

The fact that the wedge of Eilenberg-MacLane spectra does not depend on k quantifies the fact that $k\mathbb{R}$ is not 4ξ periodic. We will give precise statements in the twisted cases in Sections 3.9 and 3.10 along with the appropriate tables. The full $RO(Q)$-graded homotopy groups are displayed in Section 3.7.

3.7. The $RO(Q)$-graded homotopy of $k\mathbb{R}$

It seems useful to record the homotopy groups of all the spectra $k\mathbb{R} \wedge S^{k\xi}$, and natural to do so in the form

$$[S^{b+a\xi}, k\mathbb{R}]^Q = [S^b, S^{-a\xi} \wedge k\mathbb{R}]^Q,$$

with the a-axis horizontal and the b-axis vertical. Thus the homotopy groups $\pi_*^Q(k\mathbb{R} \wedge S^{k\xi})$ occur in the vertical line $a = -k$.

3.7.A. Description of the $RO(Q)$-graded homotopy of $k\mathbb{R}$. The main content of this section is Figure 3.7.3 displaying

$$\pi^Q_{a\xi+b}(k\mathbb{R}) \cong \pi_b(k\mathbb{R} \wedge S^{-a\xi})$$

at the point (a, b). It is worthwhile highlighting some features.

The crudest feature of the display is that since $k\mathbb{R} \wedge S^{-a\xi}$ has no cells below $-a$, is that non-zero groups have $b \geq -a$. Next, the $a = 0$ column gives the familiar homotopy groups $\pi_*^Q(k\mathbb{R}) = ko_*$. One might hope to fill in the rest of the half-plane with $(1,1)$-periodicity and $(-4,4)$-periodicity. The deviation from this is fairly small and easily described: the required change is to add 2 throughout each of the rows $b = 0, 4, 8, \ldots$ strictly below the vanishing line $b = -a$, and to add 2 throughout each of the rows $b = -5, -9, -13, \ldots$ on or above the vanishing line $b = -a$. This is indicated in the diagram by a $+$. Properly speaking, all copies of $\mathbb{Z}$ on the line $b = -a$ with $b \leq 0$ should be $\mathbb{Z}'$.

3.7.B. Description of the $RO(Q)$-graded homotopy of $k\mathbb{R} \wedge EQ_+$. The homotopy of $k\mathbb{R}$ is calculated from that of $k\mathbb{R} \wedge EQ_+$. This was described in Proposition 3.5.2, though displayed in a different way. If displayed on the above chart with

$$\pi^Q_{a\xi+b}(k\mathbb{R} \wedge EQ_+) \cong \pi_b((k\mathbb{R} \wedge S^{-a\xi})_{hQ})$$

at (a, b), the chart is very similar to Figure 3.7.3. It too is formed by minor changes to the $(1,1)$- and $(-4,4)$-periodic chart. This time, $+$ modifications are placed in all rows with $b \equiv -1 \bmod 4$ at or above the vanishing line $b = -a$. Properly speaking, all copies of $\mathbb{Z}$ on the line $b = -a$ should be $\mathbb{Z}'$.

FIGURE 3.7.3. The $RO(G)$-graded homotopy groups $\pi^Q_{a\xi+b}(k\mathbb{R})$ of $k\mathbb{R}$. The ath column displays $\pi^Q_*(k\mathbb{R} \wedge S^{-a\xi})$. The symbol + indicates the addition of a group of order 2. Note the predominantly $(1,1)$- and $(-4,4)$-periodic pattern.

		-4				0				4				8			
	$\mathbb{Z}$	0	0	0	$\mathbb{Z}'$	0	2	2	$\mathbb{Z}$	0	0	0	$\mathbb{Z}'$	0	2	2	$\mathbb{Z}$
	0	0	0	$\mathbb{Z}'$	0	2	2	$\mathbb{Z}$	0	0	0	$\mathbb{Z}'$	0	2	2	$\mathbb{Z}$	0
	0	0	$\mathbb{Z}'$	0	2	2	$\mathbb{Z}$	0	0	0	$\mathbb{Z}'$	0	2	2	$\mathbb{Z}$	0	0
8	0	$\mathbb{Z}'$	0	2	2	$\mathbb{Z}$	0	0	0	$\mathbb{Z}'$	0	2	2	$\mathbb{Z}$	0	0	0
	$\mathbb{Z}'$	0	2	2	$\mathbb{Z}$	0	0	0	$\mathbb{Z}'$	0	2	2	$\mathbb{Z}$	0	0	0	$\mathbb{Z}'$
	0	2	2	$\mathbb{Z}$	0	0	0	$\mathbb{Z}'$	0	2	2	$\mathbb{Z}$	0	0	0	$\mathbb{Z}'$	0
	2	2	$\mathbb{Z}$	0	0	0	$\mathbb{Z}'$	0	2	2	$\mathbb{Z}$	0	0	0	$\mathbb{Z}'$	0	2
4	+	$\mathbb{Z}$	0	0	0	$\mathbb{Z}'$	0	2	2	$\mathbb{Z}$	0	0	0	$\mathbb{Z}'$	0	2	2
	.	.	0	0	$\mathbb{Z}'$	0	2	2	$\mathbb{Z}$	0	0	0	$\mathbb{Z}'$	0	2	2	$\mathbb{Z}$
	.	.	.	$\mathbb{Z}'$	0	2	2	$\mathbb{Z}$	0	0	0	$\mathbb{Z}'$	0	2	2	$\mathbb{Z}$	0
	.	.	.	.	2	2	$\mathbb{Z}$	0	0	0	$\mathbb{Z}'$	0	2	2	$\mathbb{Z}$	0	0
0	+	+	+	+	+	$\mathbb{Z}$	0	0	0	$\mathbb{Z}'$	0	2	2	$\mathbb{Z}$	0	0	0
	.	.	.	.	.	.	0	0	$\mathbb{Z}'$	0	2	2	$\mathbb{Z}$	0	0	0	$\mathbb{Z}'$
	.	.	.	.	.	.	.	$\mathbb{Z}'$	0	2	2	$\mathbb{Z}$	0	0	0	$\mathbb{Z}'$	0
	.	.	.	.	.	.	.	.	2	2	$\mathbb{Z}$	0	0	0	$\mathbb{Z}'$	0	2
-4	.	.	.	.	.	.	.	.	.	$\mathbb{Z}'$	0	0	0	$\mathbb{Z}'$	0	2	2
	.	.	.	.	.	.	.	.	.	.	0^+	0^+	$\mathbb{Z}'^+$	0^+	2^+	2^+	$\mathbb{Z}^+$
	.	.	.	.	.	.	.	.	.	.	.	$\mathbb{Z}'$	0	2	2	$\mathbb{Z}$	0
	.	.	.	.	.	.	.	.	.	.	.	.	2	2	$\mathbb{Z}$	0	0
-8	.	.	.	.	.	.	.	.	.	.	.	.	.	$\mathbb{Z}'$	0	0	0
	.	.	.	.	.	.	.	.	.	.	.	.	.	.	0^+	0^+	$\mathbb{Z}'^+$
	.	.	.	.	.	.	.	.	.	.	.	.	.	.	.	$\mathbb{Z}'$	0
	.	.	.	.	.	.	.	.	.	.	.	.	.	.	.	.	2
	.	.	.	.	.	.	.	.	.	.	.	.	.	.	.	.	.

In other words, all the displayed + modifications in rows $-5, -9, -13, \ldots$ are retained. In rows $b = 0, 4, 8, \ldots$ the + modifications below the vanishing line $b = -a$ are removed and instead + modifications are placed in the next row down at or above the vanishing line, and the bottom $\mathbb{Z}$ in each column $a = 0, -4, -8, \ldots$ should be replaced by a $\mathbb{Z}'$.

3.7.C. Calculation of the $RO(Q)$-graded homotopy of $k\mathbb{R}$. The calculation in the $-k$th column follows from the cofibre sequence

$$(k\mathbb{R} \wedge S^{k\xi})_{hQ} \longrightarrow (k\mathbb{R} \wedge S^{k\xi})^Q \longrightarrow \Phi^Q k\mathbb{R}$$

given the calculations we have made for the first and last term. In particular, the connecting map $\Phi^Q k\mathbb{R} \longrightarrow (k\mathbb{R} \wedge S^{k\xi})_{hQ}$ can be read off from the homotopy orbit spectral sequence, and the triviality of the extensions in negative rows $b = -5, -9, \ldots$ is part of the calculation of the homotopy orbit spectral sequence (see Proposition 3.5.2).

3.7.D. Formula for $RO(Q)$-graded homotopy of $k\mathbb{R}$. To avoid confusion, we restate the information in the chart formulaically.

- $\pi^Q_{a\xi+b}(k\mathbb{R}) = KO_{b-a}$ provided $a+b \geq 0$ and $b \notin \{-5,-9,-13,\ldots\}$, except when $(a,b) = (8n+4, -8n-4)$ for $n \geq 0$.
- $\pi^Q_{(8n+4)\xi-(8n+4)}(k\mathbb{R}) = 2.KO_0$ for $n \geq 0$
- $\pi^Q_{a\xi+b}(k\mathbb{R}) = KO_{b-a} \oplus 2$ provided $a+b \geq 0$ and $b \in \{-5,-9,-13,\ldots\}$.
- $\pi^Q_{a\xi+b}(k\mathbb{R}) = 2$ provided $a+b < 0$ and $b \in \{0,4,8,\ldots\}$.
- $\pi^Q_{a\xi+b}(k\mathbb{R}) = 0$ provided $a+b < 0$ and $b \notin \{0,4,8,\ldots\}$.

3.8. Six truncations of the periodic Tate spectral sequence

Continuing with $G = 1$, it may be helpful to summarize the calculations of Sections 3.2 and 3.5. We should start with the whole plane spectral sequence, obtained from the pattern of Figure 3.5.2, and summarize the truncations of that. For clarity, the indexing is taken from that of the Tate spectral sequence in all cases, even in the case of homotopy quotients, where other indexing might come to mind first.

EXAMPLE 3.8.1. *(The whole plane spectral sequence for $t(K\mathbb{R})$.)* We have seen that $0 = E_4 = E_\infty$. Equally, it is immediate from the fact that $\eta^3 = 0$ that $t(K\mathbb{R}) \simeq *$ [**37**]. It is not worth displaying the E_∞ term.

EXAMPLE 3.8.2. *(The left half-plane spectral sequence for $K\mathbb{R}^{hQ}$.)* We have seen that $E_4 = E_\infty$, and that the calculation gives $KO \simeq K\mathbb{R}^{hQ}$. It is worth displaying the E_∞ term.

.	.	.	.	.	2	.	$\mathbb{Z}'$	12
.	.	.	.	.	.	.	.	11
.	.	.	.	.	.	2	.	10
.	.	.	.	.	.	.	.	9
.	.	.	.	.	.	.	$\mathbb{Z}$	8
.	.	.	.	.	.	.	.	7
.	.	.	.	.	.	.	.	6
.	.	.	.	.	.	.	.	5
.	.	.	.	.	2	.	$\mathbb{Z}'$	4
.	.	.	.	.	.	.	.	3
.	.	.	.	.	.	2	.	2
.	.	.	.	.	.	.	.	1
.	.	.	.	.	.	.	$\mathbb{Z}$	0
.	.	.	.	.	.	.	.	-1
.	.	.	.	.	.	.	.	-2
.	.	.	.	.	.	.	.	-3
.	.	.	.	.	2	.	$\mathbb{Z}'$	-4

EXAMPLE 3.8.3. *(The right half-plane spectral sequence for $\Sigma K\mathbb{R}_{hQ}$.)* We have seen that $E_4 = E_\infty$. The facts that $t(K\mathbb{R}) \simeq *$ and $K\mathbb{R}^{hQ} \simeq KO$ show that $K\mathbb{R}_{hQ} \simeq KO$, and that the geometric fixed point spectrum $\Phi^Q K\mathbb{R}$ is contractible.

It is worth displaying the E_∞ term.

12	$\mathbb{Z}$	.	.	.	.	.	.
11	.	.	.	.	.	.	.
10	2	.	.	.	.	.	.
9	.	.	.	.	.	.	.
8	$\mathbb{Z}$	2	.	.	.	.	.
7	.	.	.	.	.	.	.
6	.	.	2	.	.	.	.
5	.	.	.	.	.	.	.
4	$\mathbb{Z}$	.	.	.	.	.	.
3	.	.	.	.	.	.	.
2	2	.	.	.	.	.	.
1	.	.	.	.	.	.	.
0	$\mathbb{Z}$	2	.	.	.	.	.
−1	.	.	.	.	.	.	.
−2	.	.	2	.	.	.	.
−3	.	.	.	.	.	.	.
−4	$\mathbb{Z}$	.	.	.	.	.	.

The interesting feature here is that the homotopy in degrees 0 mod 8 is given as an extension $0 \longrightarrow \mathbb{Z} \longrightarrow \mathbb{Z} \longrightarrow \mathbb{Z}/2 \longrightarrow 0$, which replaces the occurence of $\mathbb{Z}'$ in degrees 4 mod 8 in the homotopy fixed point spectal sequence for KO_4.

One can also see exactly how the norm map $K\mathbb{R}_{hQ} \longrightarrow K\mathbb{R}^{hQ}$ must behave in the spectral sequence, since its mapping cone is $t(K\mathbb{R})^Q \simeq *$.

When we come to the connective case, we indicate the persistent groups along the lower edge with an asterisk.

EXAMPLE 3.8.4. *(The upper half-plane spectral sequence for $t(k\mathbb{R})$.)* We have seen that $E_4 = E_\infty$, and that the homotopy is $\mathbb{Z}/2$ in degrees which are multiples of 4. It is worth displaying the E_∞ term for comparison with the following two examples.

.	.	.	.	.	.	.	.	.	.	.	.	.	.	.
.	.	.	.	.	.	.	.	.	.	.	.	.	.	.
.	.	.	.	.	.	.	.	.	.	.	.	.	.	.
.	.	.	.	.	.	.	.	.	.	.	.	.	.	.
.	.	.	.	.	.	.	.	.	.	.	.	.	.	.
.	.	.	.	.	.	.	.	.	.	.	.	.	.	.
.	.	2*	.	.	.	2*	.	.	.	2*	.	.	.	2*
−6	−5	−4	−3	−2	−1	0	1	2	3	4	5	6	7	8

This corresponds to the equivalence

$$t(k\mathbb{R}) \simeq \bigvee_{k \in \mathbb{Z}} \Sigma^{4k} H\mathbb{F}_2.$$

EXAMPLE 3.8.5. *(The second quadrant spectral sequence for $k\mathbb{R}^{hQ}$.)* We have seen that $E_4 = E_\infty$, and that the calculation shows $ko \not\simeq k\mathbb{R}^{hQ}$. It is worth displaying the E_∞ term.

−9	−8	−7	−6	−5	−4	−3	−2	−1	0	
.	.	.	.	.	.	.	2	.	$\mathbb{Z}'$	12
.	.	.	.	.	.	.	.	.	.	11
.	.	.	.	.	.	.	.	2	.	10
.	.	.	.	.	.	.	.	.	.	9
.	.	.	.	.	.	.	.	.	$\mathbb{Z}$	8
.	.	.	.	.	.	.	.	.	.	7
.	.	.	.	.	.	.	.	.	.	6
.	.	.	.	.	.	.	.	.	.	5
.	.	.	.	.	.	.	2	.	$\mathbb{Z}'$	4
.	.	.	.	.	.	.	.	.	.	3
.	.	.	.	.	.	.	.	2	.	2
.	.	.	.	.	.	.	.	.	.	1
.	2*	.	.	.	2*	.	.	.	$\mathbb{Z}$	0

The behaviour of the map $ko = k\mathbb{R}^Q \longrightarrow k\mathbb{R}^{hQ}$ on homotopy follows by naturality from the periodic case. The mapping cone has homotopy $\mathbb{Z}/2$ in each negative degree which is a multiple of 4, and 3.6.2 shows that this corresponds to a cofibre sequence

$$ko \longrightarrow k\mathbb{R}^{hQ} \longrightarrow \bigvee_{k \geq 1} \Sigma^{-4k} H\mathbb{F}_2.$$

EXAMPLE 3.8.6. *(The first quadrant spectral sequence for $\Sigma k\mathbb{R}_{hQ}$.)* We have seen that $E_4 = E_\infty$, and that the calculation shows $ko \not\simeq k\mathbb{R}_{hQ}$. It is worth displaying the E_∞ term.

	1	2	3	4	5	6	7	8	9	10
12	$\mathbb{Z}$	.	.	.	.	.	.	.	.	.
11	.	.	.	.	.	.	.	.	.	.
10	2	.	.	.	.	.	.	.	.	.
9	.	.	.	.	.	.	.	.	.	.
8	$\mathbb{Z}$	2	.	.	.	.	.	.	.	.
7	.	.	.	.	.	.	.	.	.	.
6	0	.	2	.	.	.	.	.	.	.
5	.	.	.	.	.	.	.	.	.	.
4	$\mathbb{Z}$	.	.	.	.	.	.	.	.	.
3	.	.	.	.	.	.	.	.	.	.
2	2	.	.	.	.	.	.	.	.	.
1	.	.	.	.	.	.	.	.	.	.
0	$\mathbb{Z}$	2	.	2*	.	.	.	2*	.	.

Once again we see that there is an extension $0 \longrightarrow \mathbb{Z} \longrightarrow \mathbb{Z} \longrightarrow \mathbb{Z}/2 \longrightarrow 0$ in positive degrees which are multiples of 8. However, note that the quotient 2 which was present in degree 0 for KO_{hQ} is not present here, so that the map $ko \longrightarrow KO \longrightarrow K\mathbb{R}_{hQ}$ does not lift to a map $ko \longrightarrow k\mathbb{R}_{hQ}$. By degree there are no additive extensions between groups along the bottom edge and the other groups.

From the homotopy groups of $t(k\mathbb{R})$ we may read off the effect of the norm map $k\mathbb{R}_{hQ} \longrightarrow k\mathbb{R}^{hQ}$ in homotopy.

The geometric fixed point spectrum $\Phi^Q k\mathbb{R}$, is the mapping cone of a map $k\mathbb{R}_{hQ} \longrightarrow ko$, and has homotopy $\mathbb{Z}/2$ in each degree of the form $4k \geq 0$. Indeed, there is a cofibre sequence

$$k\mathbb{R}_{hQ} \longrightarrow ko \longrightarrow \Phi^Q k\mathbb{R} \simeq \bigvee_{k\geq 0} H\mathbb{F}_2.$$

3.9. Three twists of the homotopy fixed point spectral sequence

In the previous section we recorded the principal truncations of the Tate spectral sequence. The homotopy fixed point spectral sequence lies in the second quadrant; by Theorem 3.4.1 the truncation for suspensions by multiples of 4ξ is obtained by a simple shift. It is convenient to record representatives for the other truncations too, corresponding to twists by ξ, 2ξ and 3ξ. By now it is clear that the patterns for the periodic case are obtained by deleting the lower edge and repeating the patterns from the connective case, so we just provide the charts in the connective case.

EXAMPLE 3.9.1. *(The second quadrant spectral sequence for $(k\mathbb{R} \wedge S^{-\xi})^{hQ}$.)* We have seen that $E_4 = E_\infty$. It is worth displaying the E_∞ term.

.	.	.	.	.	.	.	2	.	12
.	.	.	.	.	.	.	.	.	11
.	.	.	.	.	.	.	.	$\mathbb{Z}$	10
.	.	.	.	.	.	.	.	.	9
.	.	.	.	.	.	.	.	.	8
.	.	.	.	.	.	.	.	.	7
.	.	.	.	.	.	2	.	$\mathbb{Z}'$	6
.	.	.	.	.	.	.	.	.	5
.	.	.	.	.	.	.	2	.	4
.	.	.	.	.	.	.	.	.	3
.	.	.	.	.	.	.	.	$\mathbb{Z}$	2
.	.	.	.	.	.	.	.	.	1
.	2*	.	.	.	2*	.	.	.	0
−9	−8	−7	−6	−5	−4	−3	−2	−1	

The cofibre sequence discussed in Section 3.1 then becomes

$$\Sigma ko \simeq (k\mathbb{R} \wedge S^{-\xi})^Q \longrightarrow (k\mathbb{R} \wedge S^{-\xi})^{hQ} \longrightarrow \bigvee_{k\geq 1} \Sigma^{-4k} H\mathbb{F}_2.$$

EXAMPLE 3.9.2. *(The second quadrant spectral sequence for $(k\mathbb{R} \wedge S^{-2\xi})^{hQ}$.)* We have seen that $E_4 = E_\infty$. It is worth displaying the E_∞ term.

.	.	.	.	.	.	.	$\mathbb{Z}$	12
.	.	.	.	.	.	.	.	11
.	.	.	.	.	.	.	.	10
.	.	.	.	.	.	.	.	9
.	.	.	.	.	2	.	$\mathbb{Z}'$	8
.	.	.	.	.	.	.	.	7
.	.	.	.	.	.	2	.	6
.	.	.	.	.	.	.	.	5
.	.	.	.	.	.	.	$\mathbb{Z}$	4
.	.	.	.	.	.	.	.	3
.	.	.	.	.	.	.	.	2
.	.	.	.	.	.	.	.	1
.	2^*	.	.	.	2^*	.	$\mathbb{Z}'$	0
-9	-8	-7	-6	-5	-4	-3	-2	

The cofibre sequence discussed in Section 3.1 then becomes

$$\Sigma^{-6} ko\langle 4\rangle \simeq (k\mathbb{R} \wedge S^{-2\xi})^Q \longrightarrow (k\mathbb{R} \wedge S^{-2\xi})^{hQ} \longrightarrow \bigvee_{k\geq 1} \Sigma^{-4k} H\mathbb{F}_2.$$

EXAMPLE 3.9.3. *(The second quadrant spectral sequence for $(k\mathbb{R} \wedge S^{-3\xi})^{hQ}$.)* We have seen that $E_4 = E_\infty$. It is worth displaying the E_∞ term.

.	.	.	.	.	.	.	12
.	.	.	.	.	.	.	11
.	.	.	.	2	.	$\mathbb{Z}'$	10
.	.	.	.	.	.	.	9
.	.	.	.	.	2	.	8
.	.	.	.	.	.	.	7
.	.	.	.	.	.	$\mathbb{Z}$	6
.	.	.	.	.	.	.	5
.	.	.	.	.	.	.	4
.	.	.	.	.	.	.	3
.	.	.	.	2	.	$\mathbb{Z}'$	2
.	.	.	.	.	.	.	1
.	2^*	.	.	.	2^*	.	0
-9	-8	-7	-6	-5	-4	-3	

The cofibre sequence discussed in Section 3.1 then becomes

$$\Sigma^{-5} ko\langle 2\rangle \simeq (k\mathbb{R} \wedge S^{-3\xi})^Q \longrightarrow (k\mathbb{R} \wedge S^{-3\xi})^{hQ} \longrightarrow \bigvee_{k\geq 1} \Sigma^{-4k} H\mathbb{F}_2.$$

3.10. Three twists of the homotopy orbit spectral sequence

In the previous section we recorded the principal truncations of the Tate spectral sequence. The homotopy orbit point spectral sequence lies in the first quadrant, and the truncation for suspensions by multiples of 4ξ is obtained by a simple shift. It is convenient to record representatives for the other truncations too, corresponding to twists by ξ, 2ξ and 3ξ. By now it is clear that the patterns for the periodic case are obtained by deleting the lower edge and repeating the patterns from the connective case, so we just provide the charts in the connective case.

As before, we retain the indexing from the Tate spectral sequence, so that we calculate the simple suspension of the homotopy orbit spectra of interest.

EXAMPLE 3.10.1. *(The first quadrant spectral sequence for $\Sigma(k\mathbb{R} \wedge S^{-\xi})_{hQ}$.)* We have seen that $E_4 = E_\infty$. It is worth displaying the E_∞ term.

12	2	.	.	.	.	.	.	.	.	.	.
11	.	.	.	.	.	.	.	.	.	.	.
10	$\mathbb{Z}$	2	.	.	.	.	.	.	.	.	.
9	.	.	.	.	.	.	.	.	.	.	.
8	0	.	2	.	.	.	.	.	.	.	.
7	.	.	.	.	.	.	.	.	.	.	.
6	$\mathbb{Z}$	.	.	.	.	.	.	.	.	.	.
5	.	.	.	.	.	.	.	.	.	.	.
4	2	.	.	.	.	.	.	.	.	.	.
3	.	.	.	.	.	.	.	.	.	.	.
2	$\mathbb{Z}$	2	.	.	.	.	.	.	.	.	.
1	.	.	.	.	.	.	.	.	.	.	.
0	2^*	.	2	.	2^*	.	.	.	2^*	.	.
	0	1	2	3	4	5	6	7	8	9	10

Once again we see that there is an extension $0 \longrightarrow \mathbb{Z} \longrightarrow \mathbb{Z} \longrightarrow \mathbb{Z}/2 \longrightarrow 0$ in positive degrees which of the form $8k+1$. There are no other additive extensions between groups along the bottom edge and the other groups. The cofibre sequence discussed in Section 3.1 then becomes

$$(k\mathbb{R} \wedge S^{-\xi})_{hQ} \longrightarrow \Sigma ko \longrightarrow \bigvee_{k \geq 0} H\mathbb{F}_2.$$

EXAMPLE 3.10.2. *(The first quadrant spectral sequence for $\Sigma(k\mathbb{R} \wedge S^{-2\xi})_{hQ}$.)* We have seen that $E_4 = E_\infty$. It is worth displaying the E_∞ term.

12	$\mathbb{Z}$	2	.	.	.	.	.	.	.	.	.	.
11	.	.	.	.	.	.	.	.	.	.	.	.
10	0	.	2	.	.	.	.	.	.	.	.	.
9	.	.	.	.	.	.	.	.	.	.	.	.
8	$\mathbb{Z}$	.	.	.	.	.	.	.	.	.	.	.
7	.	.	.	.	.	.	.	.	.	.	.	.
6	2	.	.	.	.	.	.	.	.	.	.	.
5	.	.	.	.	.	.	.	.	.	.	.	.
4	$\mathbb{Z}$	2	.	.	.	.	.	.	.	.	.	.
3	.	.	.	.	.	.	.	.	.	.	.	.
2	0	.	2	.	.	.	.	.	.	.	.	.
1	.	.	.	.	.	.	.	.	.	.	.	.
0	$\mathbb{Z}$	2^*	.	.	.	2^*	.	.	.	2^*	.	.
	-1	0	1	2	3	4	5	6	7	8	9	10

Once again we see that there is an extension $0 \longrightarrow \mathbb{Z} \longrightarrow \mathbb{Z} \longrightarrow \mathbb{Z}/2 \longrightarrow 0$ in positive degrees which of the form $8k+2$. There are no additive extensions between groups along the bottom edge and the other groups.

The cofibre sequence discussed in Section 3.1 then becomes

$$(k\mathbb{R} \wedge S^{-2\xi})_{hQ} \longrightarrow \Sigma^{-6}ko\langle 4\rangle \longrightarrow \bigvee_{k\geq 0} H\mathbb{F}_2.$$

EXAMPLE 3.10.3. *(The first quadrant spectral sequence for $\Sigma(k\mathbb{R} \wedge S^{-3\xi})_{hQ}$.)* We have seen that $E_4 = E_\infty$. It is worth displaying the E_∞ term.

	−2	−1	0	1	2	3	4	5	6	7	8	9	10
12	0	.	2	.	.	.	.	.	.	.	.	.	.
11	.	.	.	.	.	.	.	.	.	.	.	.	.
10	$\mathbb{Z}$	.	.	.	.	.	.	.	.	.	.	.	.
9	.	.	.	.	.	.	.	.	.	.	.	.	.
8	2	.	.	.	.	.	.	.	.	.	.	.	.
7	.	.	.	.	.	.	.	.	.	.	.	.	.
6	$\mathbb{Z}$	2	.	.	.	.	.	.	.	.	.	.	.
5	.	.	.	.	.	.	.	.	.	.	.	.	.
4	0	.	2	.	.	.	.	.	.	.	.	.	.
3	.	.	.	.	.	.	.	.	.	.	.	.	.
2	$\mathbb{Z}$	.	.	.	.	.	.	.	.	.	.	.	.
1	.	.	.	.	.	.	.	.	.	.	.	.	.
0	2	.	2*	.	.	.	2*	.	.	.	2*	.	.

Once again we see that there is an extension $0 \longrightarrow \mathbb{Z} \longrightarrow \mathbb{Z} \longrightarrow \mathbb{Z}/2 \longrightarrow 0$ in positive degrees which of the form $8k+3$. There are no additive extensions between groups along the bottom edge and the other groups.

There is a cofibre sequence

$$(k\mathbb{R} \wedge S^{-3\xi})_{hQ} \longrightarrow \Sigma^{-5}ko\langle 2\rangle \longrightarrow \bigvee_{k\geq 0} H\mathbb{F}_2.$$

CHAPTER 4

The Bockstein Spectral Sequence

In this section we describe our basic tool for deducing real K-theory from complex K-theory. As we explained in Chapter 3, the descent spectral sequence does not converge to the right group in the connective case, so we use a Bockstein spectral sequence based on η instead. This Bockstein spectral sequence collapses at the E^4-page, and is very close to being algebraic. This makes it a delight to use and a remarkably effective calculation tool.

We will give two constructions, one based on classical non-equivariant homotopy theory, and one based on K-theory with reality. These are very well known in the periodic case, and we describe the relationships between the various approaches.

4.1. Constructions

4.1.A. Bockstein spectral sequences in general. We describe the general form of a Bockstein spectral sequence before specializing to the cases in hand. Let E be a ring spectrum and let $a \in E_d$. Define E/a by the cofibre sequence

$$\Sigma^d E \xrightarrow{a} E \longrightarrow E/a.$$

We may form an inverse system

$$\begin{array}{ccccccc} E & \longleftarrow & \Sigma^d E & \longleftarrow & \Sigma^{2d} E & \longleftarrow & \cdots \\ \downarrow & & \downarrow & & \downarrow & & \\ E/a & & \Sigma^d E/a & & \Sigma^{2d} E/a & & \end{array}$$

The Bockstein spectral sequence is the spectral sequence obtained by applying $[X, \cdot]_*$ to the diagram. Thus

$$E^1_{s,t} = [X, E/a]_{t-sd},$$

and we always display the Bockstein spectral sequence on the (s,t)-plane. It is convenient to introduce a formal variable $\tilde{a}$ of bidegree $(1,d)$, so that

$$E^1 = [X, E/a]_* \otimes \mathbb{Z}[\tilde{a}].$$

The differential

$$d^r : E^r_{s,t} \longrightarrow E^r_{s+r,t-1}$$

takes a representative element x of $[X, E/a]_*$, with image $\overline{x} \in [X, E]_*$, divides by a^{r-1} to find $y \in [X, E]_*$ with $a^{r-1} y = \overline{x}$ and then $d^r x$ is the image of y in $[X, E/a]_*$; note that this is independent of s. Thus at the E^r-page, multiplication by $\tilde{a}$ is an isomorphism at the $(r-1)$st row and above. The spectral sequence converges to $E^*(X)/(a\text{-divisible elements})$. With this layout the Bockstein spectral sequence is like the Adams spectral sequence, so that the entries in the t-th column all contribute to $[X, E]_t$, and the differential maps from the t-th to the $(t-1)$-st column.

4.1.B. The classical η-Bockstein spectral sequence. The Bockstein spectral sequence we shall use derives from the cofibre sequence

$$\Sigma ko \xrightarrow{\eta} ko \xrightarrow{c} ku \xrightarrow{R} \Sigma^2 ko$$

REMARK 4.1.1. The existence of such a sequence was noted by Bott in his 1963 lectures ([**20**] but note the typo in the sequence at the top of page 75). He observed that Bott periodicity tells us in particular that $\Omega(U/O) = BO \times \mathbb{Z}$. Thus, U/O is the first space in the KO spectrum, and the infinite loop fibration

$$U/O \longrightarrow BO \times \mathbb{Z} \xrightarrow{c} BU \times \mathbb{Z}$$

provides a cofibre sequence $\Sigma KO \longrightarrow KO \xrightarrow{c} KU \longrightarrow \Sigma^2 KO$. It also implies the existence of a cofibre sequence $\Sigma ko \longrightarrow ko \xrightarrow{c} ku \longrightarrow \Sigma^2 ko$. This left open the identification of the two maps other than complexification. The next step was taken by Atiyah [**12**] who derived the cofibre sequence

$$\Sigma KO \xrightarrow{\eta} KO \xrightarrow{c} KU \xrightarrow{R} \Sigma^2 KO$$

from his Q-equivariant real K-theory by means of the coefficient sequence

$$Q_+ \longrightarrow S^0 \longrightarrow S^\xi$$

together with the isomorphisms $\widetilde{K\mathbb{R}}^q_Q(X \wedge Q_+) \cong \widetilde{KU}^q(X)$, $\widetilde{K\mathbb{R}}^q_Q(X \wedge S^\xi) \cong \widetilde{K\mathbb{R}}^{q+1}(X)$, and when X has trivial involution, $K\mathbb{R}^q_Q(X) \cong KO^q(X)$.

However, he left it to the reader to determine that R is an extension of realification over complex Bott periodicity. This point is obscured by the fact that multiplication by the Bott map is an isomorphism in the periodic theory. The only place we know in the literature which carefully identifies this map is Mimura and Toda's book [**85**, Lemma 5.11]. □

4.1.C. Four contexts. There are four contexts where we are interested in a spectral sequence going from complex to real K-theory, depending on whether the theories are periodic or connective and whether they are complete theories or fully equivariant. These spectral sequences are compatible in the sense that inverting the Bott element on the connective ones gives the periodic ones, and the equivariant ones applied to suitable G-free spaces give the Borel theory, which agrees with the completed construction by the completion theorem.

We apply the construction of the Bockstein spectral sequence as in Subsection 4.1.A in all four cases with $a = \eta$ (i.e., the product of the Hopf map and the unit) in degree $d = 1$. Furthermore, in these cases $\eta^3 = 0$ in E_*, so that there are no η-divisible elements, and $d^r = 0$ for $r \geq 4$. Furthermore, at $E^4 = E^\infty$ the only non-zero entries have $s = 0, 1$ or 2.

We also consider two equivariant constructions of a spectral sequence in appropriate cases. The first is the descent spectral sequence considered in Chapter 3. This is of a similar form to the Bockstein spectral sequence of Subsection 4.1.A but the degree d is not an integer. Using the twisting results of Chapter 3 this agrees with the η-Bockstein spectral sequence in the periodic case, but does not converge to ko^G_* in the connective case. Finally, we describe a second equivariant construction that applies to both the connective and periodic cases. In each of the four contexts, all the available constructions agree with each other.

Case 1 *(Periodic and complete).* This is Atiyah's classical non-equivariant η-Bockstein spectral sequence. We take $E = KO$, and $a = \eta$. It is well known that $E/a = KO \wedge C\eta = KU$, where $C\eta = \Sigma^{-2}\mathbb{C}P^2$ is the cofibre of η, ([**12**] or invert the Bott map in the connective case proved below). So, if we take $X = BG$, the spectral sequence reads

$$E^1_{s,t} = KU^{s-t}(BG) \cdot \widetilde{\eta}^s \Rightarrow KO^{-t}(BG).$$

Case 2 *(Periodic and equivariant).* We take the G-equivariant spectrum $E = KO$, and $a = \eta$. It is again well known that $E/a = KO \wedge C\eta = KU$ ([**12**] or invert the Bott map in the connective case proved below). So, if we take $X = S^0$, and use equivariant maps the spectral sequence reads

$$E^1_{s,t} = KU_G^{s-t} \cdot \widetilde{\eta}^s \Rightarrow KO_G^{-t}.$$

or equivalently

$$E^1_{s,t} = KU^G_{t-s} \cdot \widetilde{\eta}^s \Rightarrow KO^G_t.$$

In this case the descent spectral sequence gives an alternative construction; this is Atiyah's construction via K-theory with reality. We consider the product group $\widehat{G} = G \times Q$, and the cofibre sequence

$$S^\xi \longleftarrow S^0 \longleftarrow Q_+.$$

Applying $F(\cdot, K\mathbb{R})$ where $K\mathbb{R}$ is the $\widehat{G}$-equivariant spectrum representing equivariant K-theory with reality, we obtain the cofibre sequence

$$K\mathbb{R} \wedge S^{-\xi} \longrightarrow K\mathbb{R} \longrightarrow K\mathbb{R} \wedge Q_+$$

of $\widehat{G}$-spectra. Taking Q-fixed points, and using Lemma 3.3.1, we obtain the cofibre sequence

$$\Sigma KO \longrightarrow KO \longrightarrow KU$$

from which the Bockstein spectral sequence was constructed. The map $\Sigma KO \longrightarrow KO$ is multiplication by the Euler class of ξ, which Atiyah shows to be η [**12**].

Case 3 *(Connective and complete).* Here we take $E = ko$, and $a = \eta$. It is perhaps less well known that $E/a = ko \wedge C\eta = ku$ (see Lemma 4.1.2 below), so that if we take $X = BG$, the spectral sequence reads

$$E^1_{s,t} = ku^{s-t}(BG) \cdot \widetilde{\eta}^s \Rightarrow ko^{-t}(BG).$$

LEMMA 4.1.2. $ko \wedge C\eta = ku$

Proof: The map $\eta : \Sigma ku \longrightarrow ku$ is trivial, so that the composite $c\eta : \Sigma ko \longrightarrow ku$ is trivial. Hence complexification, $c : ko \longrightarrow ku$, extends to a map $ko \wedge C\eta \longrightarrow ku$, which we claim is an equivalence.

$$\begin{array}{ccccccc}
\Sigma ko & \xrightarrow{\eta} & ko & \longrightarrow & ko \wedge C\eta & \longrightarrow & \Sigma^2 ko \\
 & & & \searrow^{c} & \downarrow{\simeq} & \nearrow_{R} & \\
 & & & & ku & &
\end{array}$$

To see this, we show that it induces an isomorphism in mod p homology for each prime p. In mod 2 homology, complexification $c : ko \longrightarrow ku$ induces the inclusion

$$\mathbb{F}_2[\overline{\xi}_1^4, \overline{\xi}_2^2, \overline{\xi}_i \mid i \geq 3] \longrightarrow \mathbb{F}_2[\overline{\xi}_1^2, \overline{\xi}_2^2, \overline{\xi}_i \mid i \geq 3]$$

while $S \wedge C\eta \longrightarrow ko \wedge C\eta \longrightarrow ku$ induces

$$\mathbb{F}_2[\overline{\xi}_1^2]/(\overline{\xi}_1^4) \longrightarrow \mathbb{F}_2[\overline{\xi}_1^2, \overline{\xi}_2^2, \overline{\xi}_i \mid i \geq 3].$$

Tensoring the domains produces an isomorphism, and hence $ko \wedge C\eta \longrightarrow ku$ is an equivalence at 2. Away from two, η is trivial, so $ko \wedge C\eta = ko \vee \Sigma^2 ko$, and this is simply the grouping into even and odd terms in the Adams splitting $ku \simeq l \vee \Sigma^2 l \vee \cdots \Sigma^{2(p-2)} l$. □

Case 4 *(Connective and equivariant).* There are two ways we might attempt to make an equivariant construction. The first is to take the G-equivariant spectrum $E = ko$, and $a = \eta$. The fact that we have a G-equviariant equivalence $ko/a = ko \wedge C\eta = ku$ follows from the previous cases together with the definitions of ko and ku by pullback squares. Indeed, the non-equivariant map $ko \wedge C\eta \longrightarrow ku$ and the periodic G-equivariant map $KO \wedge C\eta \longrightarrow KU$ are compatible, and hence induce a map of forks which is an equivalence on all three terms. Now, if we take $X = S^0$, and use equivariant maps the spectral sequence reads

$$E^1_{s,t} = ku_G^{s-t} \cdot \widetilde{\eta}^s \Rightarrow ko_G^{-t}.$$

or equivalently

$$E^1_{s,t} = ku^G_{t-s} \cdot \widetilde{\eta}^s \Rightarrow ko^G_t.$$

The second attempt to make an equivariant construction would be as in Case 2, inserting suspensions S^ξ. As we saw in Chapter 3 this does not calculate ko^G_*. However it is useful to be able to compare the descent and η-Bockstein spectral sequences, so we describe a variant of the construction from Case 2. We again consider the product group $\widehat{G} = G \times Q$, and the cofibre sequence

$$S^\xi \longleftarrow S^0 \longleftarrow Q_+.$$

Applying $F(\cdot, k\mathbb{R})$ where $k\mathbb{R}$ is the $\widehat{G}$-equivariant connective equivariant K-theory with reality, we obtain the cofibre sequence

$$k\mathbb{R} \wedge S^{-\xi} \longrightarrow k\mathbb{R} \longrightarrow k\mathbb{R} \wedge Q_+$$

of $\widehat{G}$-spectra. Taking Q-fixed points, and using Lemma 3.3.2, we obtain the cofibre sequence

$$\Sigma ko \longrightarrow ko \longrightarrow ku$$

of G-spectra from which the Bockstein spectral sequence was constructed. The calculation that proved $(k\mathbb{R} \wedge S^{-\xi})^Q \simeq \Sigma ko$ (Theorem 3.4.1) showed that the map $\Sigma ko \longrightarrow ko$ is multiplication by η. So far the two equivariant constructions agree.

Now however, we cannot just use the Bockstein spectral sequence type construction with twists by ξ as in the periodic case, since this gives the descent spectral sequence that we know does not converge to ko^G_*. Instead of considering the diagram for the descent spectral sequence we consider the variant. The descent spectral sequence comes from the diagram

$$\begin{array}{ccccccccccc}
S^{-\xi}\wedge k\mathbb{R} & \longrightarrow & S^0\wedge k\mathbb{R} & \longrightarrow & Q_+\wedge k\mathbb{R} & \longrightarrow & S^{1-\xi}\wedge k\mathbb{R} & \longrightarrow & S^1\wedge k\mathbb{R} & \longrightarrow & \\
 & & & & & & \uparrow{=} & & & & \\
 & & & \longrightarrow & S^{1-2\xi}\wedge k\mathbb{R} & \longrightarrow & S^{1-\xi}\wedge k\mathbb{R} & \longrightarrow & S^{1-\xi}\wedge Q_+\wedge k\mathbb{R} & \longrightarrow & \\
 & & & & \uparrow{=} & & & & & & \\
 & \longrightarrow & S^{1-3\xi}\wedge k\mathbb{R} & \longrightarrow & S^{1-2\xi}\wedge k\mathbb{R} & \longrightarrow & S^{1-2\xi}\wedge Q_+\wedge k\mathbb{R} & \longrightarrow & & & \\
 & & \uparrow{=} & & & & & & & & \\
S^{1-4\xi}\wedge k\mathbb{R} & \longrightarrow & S^{1-3\xi}\wedge k\mathbb{R} & \longrightarrow & S^{1-3\xi}\wedge Q_+\wedge k\mathbb{R} & \longrightarrow & & & & &
\end{array}$$

in which each row is a cofibre sequence. Going one step down and one step left we obtain the ξth desuspension.

The variant comes from the diagram

$$\begin{array}{ccccccccccc}
S^{-\xi}\wedge k\mathbb{R} & \longrightarrow & S^0\wedge k\mathbb{R} & \longrightarrow & Q_+\wedge k\mathbb{R} & \longrightarrow & S^{1-\xi}\wedge k\mathbb{R} & \longrightarrow & S^1\wedge k\mathbb{R} & \longrightarrow & \\
 & & & & & & \uparrow & & & & \\
 & & & \longrightarrow & S^{2-\xi}\wedge k\mathbb{R} & \longrightarrow & S^2\wedge k\mathbb{R} & \longrightarrow & S^2\wedge Q_+\wedge k\mathbb{R} & \longrightarrow & \\
 & & & & \uparrow & & & & & & \\
 & \longrightarrow & S^{3-\xi}\wedge k\mathbb{R} & \longrightarrow & S^3\wedge k\mathbb{R} & \longrightarrow & S^3\wedge Q_+\wedge k\mathbb{R} & \longrightarrow & & & \\
 & & \uparrow & & & & & & & & \\
S^{4-\xi}\wedge k\mathbb{R} & \longrightarrow & S^4\wedge k\mathbb{R} & \longrightarrow & S^4\wedge Q_+\wedge k\mathbb{R} & \longrightarrow & & & & &
\end{array}$$

in which row is a cofibre sequence. Going one step down and one step left we obtain a $\widehat{G}$-fixed desuspension. Instead of having equivariant $\widehat{G}$-equivalences between terms in successive rows, we just have a map $S^1\wedge k\mathbb{R}\longrightarrow S^{-\xi}\wedge k\mathbb{R}$. This is the map obtained from $\eta : S^1\longrightarrow S^{-\xi}\wedge k\mathbb{R}$ by extending to a $k\mathbb{R}$-module map. We need to know that it is non-equivariantly $v : S^1\wedge ku\longrightarrow S^{-1}\wedge ku$ and in Q-fixed points it is the equivalence $S^1\wedge ko\longrightarrow (S^{-\xi}\wedge k\mathbb{R})^Q$ of Chapter 3.

This shows that when we take Q-fixed points of the whole diagram, the maps relating the rows are G-equivalences, and we obtain the integer graded equivariant construction described above. As before, we therefore obtain an exact couple when we apply $[X,\cdot]^{\widehat{G}}$ for a $\widehat{G}$-space X inflated from a G-space.

REMARK 4.1.3. There is a map from the first to the second displayed diagram, and hence a map from the descent spectral sequence to the Bockstein spectral sequence. Note that if we arrange both spectral sequences on the page following the Bockstein spectral sequence conventions the map at E^1 is an isomorphism in the first octant, but not in the second, third or fourth, because the descent spectral sequence is not constant along the n-diagonal when n is negative. At E^2, an η-step along the n-diagonal goes from $H^i(Q;ku^G_j)$ to $H^{i+1}(Q;ku^G_{j+2})$ where $n=j-2i$; in particular, in the range where ku^*_G is 2-periodic, remembering the action of Q on v and the 2-periodicity of $H^*(Q)$, these are isomorphic.

4.1.D. Comparison. Both in the periodic and in the connective case, there are two comparisons to be made. Since the periodic form is obtained from the connective by inverting the Bott map, it suffices to deal with the connective cases.

The first comparison arises from two constructions of the spectral sequence for

$$ko^*(BG)=ko^*_G(EG)=(ko^*_G)^\wedge_J,$$

where the first equality is elementary and the second is the completion theorem. For the two equivariant theories we start from the basic construction described in Case 4, and either apply $F(EG_+, \cdot)$ for the central expression or J-adic completion to obtain the right hand expression. These agree by the completion theorem. After taking Q-fixed points in the construction of the Bockstein spectral sequence for ko^*_G in Case 4 we obtain

$$\Sigma ko \simeq (\Sigma^{-\xi} \wedge k\mathbb{R})^Q \longrightarrow (k\mathbb{R})^Q \longrightarrow ku,$$

where the first map is multiplication by η. Similarly, the nonequivariant construction is obtained from $\Sigma ko \longrightarrow ko \longrightarrow ku$, where the first map is multiplication by η. These two constructions are the same by using the equivalence

$$F(BG_+, ko) \simeq F(EG_+, \mathrm{inf}(ko))^G \simeq F(EG_+, ko)^G.$$

The second comparison is parallel, but with the completion theorem replaced by the local cohomology theorem. Because of the nature of the local cohomology theorem, an equality is replaced by a spectral sequence. For simplicity we describe the argument for a finite group G; precisely similar comparison applies to compact Lie groups, but a suspension by the adjoint representation must be inserted in the usual way. We have two constructions of the Bockstein spectral sequence corresponding to

$$ko_*(BG) = ko^G_*(EG) \Leftarrow H^*_J(ko^*_G)$$

where the first equality is the Adams isomorphism (proved by a transfer argument) and the second is the local cohomology theorem. For the two equivariant theories we start from the construction described in Case 4, and either apply $EG_+ \wedge (\cdot)$ for the central expression or local cohomology for the right hand expression. The fact that these agree is best expressed at the level of G-spectra by the equivalence

$$EG_+ \wedge k\mathbb{R} \simeq \Gamma_J(k\mathbb{R}).$$

This is natural for smashing with a spectrum (such as $S^{k\xi}$), so can be applied to the whole equivariant construction in Case 4. The two resulting Bockstein spectral sequences agree by the local cohomology theorem.

Having established this equivalence, we may focus attention on the $\Gamma_J k\mathbb{R}$ construction of the Bockstein spectral sequence. Taking Q-fixed points commutes with Γ_J, so we can equally say that the spectral sequence is obtained by applying Γ_J to the cofibre sequence

$$\Sigma ko \simeq (\Sigma^{-\xi} \wedge k\mathbb{R})^Q \longrightarrow (k\mathbb{R})^Q \longrightarrow ku,$$

where the first map is multiplication by η. Similarly the nonequivariant construction is obtained from $\Sigma ko \longrightarrow ko \longrightarrow ku$, where the first map is multiplication by η by smashing with BG_+. These two constructions are compatible same by using the equivalence

$$ko \wedge BG_+ \simeq (EG_+ \wedge \mathrm{inf}(ko))^G \simeq (EG_+ \wedge ko)^G.$$

4.1.E. General behaviour. Since Cases 1 and 3 are simply completions of Cases 2 and 4, it is most economical to describe the equivariant case. Anxious readers can recover Cases 1 and 3 by completion. The periodic case is entirely representation theory, but since it coincides with the connective case in the region of the plane $t \geq s$, it is useful to record it separately. In any case, the authors find this a convenient way to calculate and remember the 8 different groups of KO^G_i.

We have noted that each differential d^r is independent of s. In particular d^1 is induced by $E/a \xrightarrow{R} \Sigma^{d+1}E \xrightarrow{c} \Sigma^{d+1}E/a$, and multiplication by the element $\tilde{a}$ of bidegree $(1, d)$ is an isomorphism. For us $d = 1$, and we will usually just record the d^1-sequence

$$\cdots \longleftarrow [X, E/a]_{i-2} \xleftarrow{(cR)^*} [X, E/a]_i \xleftarrow{(cR)^*} [X, E/a]_{i+2} \xleftarrow{(cR)^*} [X, E/a]_{i+4} \longleftarrow \cdots$$

for i even and i odd. The entry on the zero line is then the relevant kernel

$$E^2_{0,i} = \ker([X, E/a]_i \longrightarrow [X, E/a]_{i-2})$$

and the entry for $s \geq 1$ is the relevant homology group

$$E^2_{s,i} = \frac{\ker([X, E/a]_{i-s} \longrightarrow [X, E/a]_{i-2-s})}{\operatorname{im}([X, E/a]_{i+2-s} \longrightarrow [X, E/a]_{i-s})}.$$

In many cases $[X, E/a]$ is entirely in even degrees.

4.2. The periodic case

The periodic case is completely described by representation theory, and it will be useful to make this explicit.

4.2.A. Conjugation. The complex K-theory KU^G_* is entirely in even degrees with $KU^G_{2i} = RU(G)$. Using the equivariant construction of the spectral sequence we see that the d^1-sequence is

$$\cdots \longleftarrow RU(G) \xleftarrow{1+\tau} RU(G) \xleftarrow{1-\tau} RU(G) \xleftarrow{1+\tau} \cdots,$$

where τ is the conjugation action, and where $1 - \tau$ maps from degrees $\equiv 0 \bmod 4$ and $1 + \tau$ maps from degrees $\equiv 2 \bmod 4$. It is convenient to write

$$RU_{sc} := \ker(1 - \tau)$$

for the self-conjugate virtual representations,

$$RU_{asc} := \ker(1 + \tau)$$

for the anti-self-conjugate virtual representations and

$$H := RU_{sc}/(1 + \tau)RU(G).$$

It is clear that $2H = 0$. Next, note that $RU_{asc}/(1 - \tau)RU = 0$, since this is the odd homology of the group $Q = \{e, \tau\}$ with coefficients in the permutation representation given by the simple modules.

This allows us to write down the E^2-page.

4.2.B. The E^2-term. In discussing the Bockstein spectral sequence, we refer to the line of slope 1 meeting the 0-line at n as the *n-diagonal*; at E^1 all the groups on the n-diagonal are the same.

Turning to the E^2-page, along the 0-line we have

$$E^2_{0,4i} = RU_{sc} \text{ and } E^2_{0,4i+2} = RU_{asc}.$$

Above the 0-line, everything is concentrated on the n-diagonals with n divisible by 4, where we have

$$E^2_{s,4i+s} = H \text{ for } s \geq 1.$$

All other entries are 0.

We may display the result as follows.

$s=3$		·		H		·		H		·		H
$s=2$	·		H		·		H		·		H	
$s=1$		H		·		H		·		H		·
$s=0$	RU_{sc}		RU_{asc}		RU_{sc}		RU_{asc}		RU_{sc}		RU_{asc}	
E^2	0		2		4		6		8		10	

We will see many examples of this in special cases later.

4.2.C. The differentials. Since all entries are in even $(t-s)$-degree, $d^2 = 0$, so that $E^2 = E^3$ and it only remains to describe d^3. Off the 0-line there are two cases. Since the whole spectral sequence is 8-periodic, there are two families of differentials $d^3 : H \longrightarrow H$. We write $\delta_0 : H \longrightarrow H$ for the d^3s leaving an n-diagonal with $n \equiv 0 \bmod 8$; these have kernel consisting of *real* representations, and cokernel of the same rank. We write $\delta_4 : H \longrightarrow H$ for the d^3s leaving an n-diagonal with $n \equiv 4 \bmod 8$; these have kernel consisting of *quaternionic* representations, and cokernel of the same rank.

Since the E^∞ term of the Bockstein spectral sequence is zero above the 2-line, the sequence

$$\cdots \longrightarrow H \xrightarrow{\delta_0} H \xrightarrow{\delta_4} H \xrightarrow{\delta_0} H \xrightarrow{\delta_4} \cdots$$

is exact. The differential from the 0-line is the composite

$$RU_{sc} \longrightarrow RU_{sc}/(1+\tau) = H \xrightarrow{\delta_i} H$$

from the appropriate diagonal.

Writing

$$H' := \ker(\delta_0 : H \longrightarrow H),\ H'' := \ker(\delta_4 : H \longrightarrow H)$$

and similarly for RU'_{sc}, RU''_{sc}, we may display the result as follows.

$s=3$		·		·		·		·		·		·
$s=2$	·		H'		·		H''		·		H'	
$s=1$		H'		·		H''		·		H'		·
$s=0$	RU'_{sc}		RU_{asc}		RU''_{sc}		RU_{asc}		RU'_{sc}		RU_{asc}	
E^∞	0		2		4		6		8		10	

We will see many examples of this in special cases later.

4.2.D. Relationship with representation theory. Finally, we may see KO^G_* (2.2.12) at the E^∞ stage. For this, we recall the familiar sequence of coefficient groups of KO^G_*. As in Chapter 2, we have maps

$$RO \xrightarrow{c} RU \xrightarrow{q} RSp$$

and

$$RSp \xrightarrow{\tilde{c}} RU \xrightarrow{r} RO,$$

and we informally write their cokernels as quotients, so that the sequence of homotopy groups is RO, RO/RU, RU/RSp, 0, RSp, RSp/RU, RU/RO, 0 for $i = 0, 1, 2, 3, 4, 5, 6, 7 \bmod 8$.

Along the 0-line we have $RO(G)$ in degrees $\equiv 0 \bmod 8$, and $RSp(G)$ in degrees $\equiv 4 \bmod 8$, expressed as finite index subgroups of the group RU_{sc} of self-conjugate representations. In degrees $\equiv 2 \bmod 8$ we have the torsion free quotient of $RU(G)/RSp(G)$ expressed as RU_{asc}; similarly, in degrees $\equiv 6 \bmod 8$ we have the torsion free quotient of $RU(G)/RO(G)$ expressed as RU_{asc}. The torsion (all of $RO(G)/RU(G)$ and $RSp(G)/RU(G)$, and only part of $RU(G)/RSp(G)$ and $RU(G)/RO(G)$) occurs on the 1 and 2 line as an appropriate subgroup of H. We summarize this as a lemma.

LEMMA 4.2.1. (i) *We have an isomorphism* $RU'_{sc} \cong RO$, *which is to say there is an exact sequence*

$$0 \longrightarrow RO \longrightarrow RU_{sc} \xrightarrow{\delta_0} H \xrightarrow{\delta_4} H$$

(ii) *We have an isomorphism*

$$RO/RU \cong H' = \ker(\delta_0)$$

(iii) *There is an exact sequence*

$$0 \longrightarrow H' \longrightarrow RU/RSp \longrightarrow RU_{asc} \longrightarrow 0$$

where $H' = \ker(\delta_0)$.

(iv) *We have an isomorphism* $RU''_{sc} \cong RSp$, *which is to say there is an exact sequence*

$$0 \longrightarrow RSp \longrightarrow RU_{sc} \xrightarrow{\delta_4} H \xrightarrow{\delta_0} H$$

(v) *We have an isomorphism*

$$RSp/RU \cong H'' = \ker(\delta_4)$$

(vi) *There is an exact sequence*

$$0 \longrightarrow H'' \longrightarrow RU/RO \longrightarrow RU_{asc} \longrightarrow 0,$$

where $H'' = \ker(\delta_4)$.

□

4.2.E. The easy life away from 2. If 2 is inverted, $(1-\tau)/2$ and $(1+\tau)/2$ are orthogonal idempotents. Thus away from 2, we find that in degrees divisible by 4 we have

$$KO[1/2]_{4k} = KU[1/2]^{sc}_{4k} = RU_{sc}$$

and in degrees congruent to 2 mod 4 we have

$$KO[1/2]_{4k+2} = KU[1/2]^{sc}_{4k+2} = RU_{asc}.$$

4.3. The connective case

We may now describe our standard procedure for calculating with the Bockstein spectral sequence to deduce ko^G_* from ku^G_*.

4.3.A. Complex connective K-theory. In the connective case the E^1-term involves ku^G_*, and this is best understood using the map

$$ku^G_* \longrightarrow ku^G_*[1/v] = KU^G_* = RU(G)[v, v^{-1}].$$

We let QU denote the image, and consider the natural exact sequence

$$0 \longrightarrow TU \longrightarrow ku^G_* \longrightarrow QU \longrightarrow 0,$$

where TU consists of the v-power torsion. We showed in Theorem 2.9.3 that TU is concentrated in degrees ≤ -6. In any case, above degree -6, ku^G_* is in even degrees and

$$ku^G_{2n} = \begin{cases} RU(G) & \text{if } n \geq 0 \\ JU_1(G) & \text{if } n = -1 \\ JU_2(G) & \text{if } n = -2, \end{cases}$$

where $JU_1(G) = \ker(RU(G) \longrightarrow \mathbb{Z})$ is generated by the first Chern classes $c_1(V)$ of complex representations V, and $JU_2(G)$ is generated by $JU_1(G)^2$ and second Chern classes $c_2(V)$ of complex representations V.

4.3.B. Comparison of d^1. In this subsection we apply the general comparisons of Subsection 4.1.D to the first differential.

The non-equivariant Bockstein spectral sequence is based on the cofibre sequence

$$\Sigma ko \xrightarrow{\eta} ko \longrightarrow ku.$$

Indeed, it is obtained from the cofibre sequence

$$S^1 \xrightarrow{\eta} S^0 \longrightarrow C(\eta) = \Sigma^{-2}\mathbb{C}P^2$$

by smashing with ko. Furthermore d^1 is the composite

$$ku \xrightarrow{R} \Sigma^2 ko \xrightarrow{c} \Sigma^2 ku$$

which is obtained by smashing

$$\mathbb{C}P^2 \longrightarrow S^4 \longrightarrow \Sigma^2\mathbb{C}P^2$$

with $\Sigma^{-2}ko$.

Accordingly it is straightforward to compare the spectral sequence with the corresponding sequence for other theories of interest to us. In fact we may smash the sequence with the diagram

$$KO \longleftarrow ko \longrightarrow H\mathbb{Z} \longrightarrow H\mathbb{F}_2$$

to obtain comparisons of Bockstein spectral sequences. In particular we find the image of the d^1 differential in periodic K-homology and cohomology and in mod 2 homology and cohomology.

Picking out the behaviour on the n-diagonal, we obtain the following.

LEMMA 4.3.1. (i) *The image of $d^1 : ku_G^{-n} \longrightarrow ku_G^{-n+2}$ in periodic K-cohomology is $1 - \tau$ when $n \equiv 0$ mod 4 and $1 + \tau$ when $n \equiv 2$ mod 4.*

(ii) *The image of $d^1 : ku^{-n}(BG) \longrightarrow ku^{-n+2}(BG)$ in periodic K-cohomology is $1 - \tau$ when $n \equiv 0$ mod 4 and $1 + \tau$ when $n \equiv 2$ mod 4.*

(iii) *The image of $d^1 : ku_n(BG) \longrightarrow ku_{-n-2}(BG)$ in periodic K-cohomology is $1 - \tau$ when $n \equiv 3$ mod 4 and $1 + \tau$ when $n \equiv 1$ mod 4.*

(iv) *The image of $d^1 : ku^{-n}(X) \longrightarrow ku^{-n+2}(X)$ in mod 2 cohomology is $Sq^2 : H^{-n}(X : \mathbb{F}_2) \longrightarrow H^{-n+2}(X : \mathbb{F}_2)$*

(v) *The image of* $d^1 : ku_n(X) \longrightarrow ku_{n-2}(X)$ *in mod 2 homology is* $(Sq^2)^\vee : H_n(X : \mathbb{F}_2) \longrightarrow H_{n-2}(X : \mathbb{F}_2)$

Proof: For Part (i) we use the equivariant construction, noting that both spectral sequences are obtained from the same diagram by smashing with KO or KU and related by inverting the Bott element. We have discussed the differential d^1 in periodic equivariant K-theory already.

Part (ii) follows by applying $F(EG_+, \cdot)$, and using the completion theorem as described in Subsection 4.1.D. Since the the equivariant and non-equivariant constructions of the Bockstein spectral sequences agree, it does not matter which we use.

Similarly Part (iii) follows by applying $EG_+ \wedge (\cdot)$, and using the local cohomology theorem. Since the the equivariant and non-equivariant constructions of the Bockstein spectral sequences agree, it does not matter which we use.

For Parts (iii) and (iv) we need only observe that $H\mathbb{F}_2 \wedge \mathbb{C}P^2$ is a wedge of two copies of $H\mathbb{F}_2$, and that the image from connective complex K-theory gives the bottom one. Now d^1 is obtained from the composite

$$\mathbb{C}P^2 \longrightarrow S^4 \longrightarrow \Sigma^2\mathbb{C}P^2$$

by smashing with $H\mathbb{F}_2$ and picking out the relevant factors to obtain

$$\Sigma^2 H\mathbb{F}_2 \longrightarrow \mathbb{C}P^2 \wedge H\mathbb{F}_2 \longrightarrow S^4 \wedge H\mathbb{F}_2 \longrightarrow \Sigma^2\mathbb{C}P^2 \wedge H\mathbb{F}_2 \longrightarrow \Sigma^4 H\mathbb{F}_2.$$

Since $Sq^2 : H^2(\mathbb{C}P^2; \mathbb{F}_2) \longrightarrow H^4(\mathbb{C}P^2; \mathbb{F}_2)$ is non-trivial, we conclude the composite is $Sq^2 : H\mathbb{F}_2 \longrightarrow \Sigma^2 H\mathbb{F}_2$ as required. □

4.3.C. The E^2-term. Since the construction of the Bockstein spectral sequence in the connective and periodic cases is compatible, we may deduce the behaviour of differentials on QU from the behaviour in the periodic case: it is $1-\tau$ when the domain is in degrees 0 mod 4 and $1+\tau$ when the domain is in degrees 2 mod 4. From the fact that η is detected by Sq^2 we see that on the part of TU detected in mod 2 cohomology, the differential is Sq^2. In any case, the d^1 sequence may be displayed in even degrees as

$$\begin{array}{ccccccccccccccc}
\cdots & \xleftarrow{Sq^2} & TU_{-8} & \xleftarrow{Sq^2} & TU_{-6} & \longleftarrow & 0 & \longleftarrow & 0 & \longleftarrow & 0 & \longleftarrow & 0 & \longleftarrow & \cdots \\
 & & \downarrow & & \downarrow & & \downarrow & & \downarrow & & \downarrow & & \downarrow & & \\
\cdots & \longleftarrow & ku^G_{-8} & \longleftarrow & ku^G_{-6} & \longleftarrow & ku^G_{-4} & \longleftarrow & ku^G_{-2} & \longleftarrow & ku^G_0 & \longleftarrow & ku^G_2 & \longleftarrow & \cdots \\
 & & \downarrow & & \downarrow & & \downarrow & & \downarrow & & \downarrow & & \downarrow & & \\
\cdots & \xleftarrow{1-\tau} & QU_{-8} & \xleftarrow{1+\tau} & QU_{-6} & \xleftarrow{1-\tau} & JU_2 & \xleftarrow{1+\tau} & JU_1 & \xleftarrow{1-\tau} & RU & \xleftarrow{1+\tau} & RU & \xleftarrow{1-\tau} & \cdots
\end{array}$$

Indeed, we may regard

$$0 \longrightarrow TU \longrightarrow ku^G_* \longrightarrow QU \longrightarrow 0.$$

as a short exact sequence of chain complexes, inducing a long exact sequence in homology

$$\cdots \longrightarrow H(TU)_n \longrightarrow H(ku^G_*)_n \longrightarrow H(QU)_n \longrightarrow H(TU)_{n-2} \longrightarrow \cdots,$$

remembering that the differential d^1 reduces degree by 2. Similarly, there is a sequence

$$0 \longrightarrow Z(TU)_n \longrightarrow Z(ku_*^G)_n \longrightarrow Z(QU)_n \longrightarrow H(TU)_{n-2} \longrightarrow \cdots$$

for cycles. Note that in odd degrees, $QU = 0$ so that the d^1 sequence in odd degrees only involves TU.

4.3.D. The $E^{1\frac{1}{2}}$-term. It is often convenient to display the homology of QU and that of TU on a Bockstein spectral sequence chart, and then treat the connecting homomorphism of the short exact sequence $0 \longrightarrow TU \longrightarrow ku_G^* \longrightarrow QU \longrightarrow 0$ as a late d^1. We refer to this display as the $E^{1\frac{1}{2}}$-term.

4.3.E. The differentials. We have described a systematic method for calculating the E^2-term. The nature of the Bockstein spectral sequence shows that we only have to find d^2 and d^3. In positive degrees, the spectral sequence agrees with the periodic ones, so we understand all differentials. Using the fact that the E^4-term is zero above the 2-line, one can usually infer the behaviour of the remaining differentials by working downwards from degree 0. We will give numerous examples of this process in subsequent chapters.

4.3.F. The easy life away from 2. If 2 is inverted, $(1-\tau)/2$ and $(1+\tau)/2$ are orthogonal idempotents. Thus away from 2 we find that in degrees divisible by 4 we have

$$QO[1/2]_{\equiv 0(4)} = QU[1/2]^{sc}_{\equiv 0(4)}$$

and in degrees congruent to 2 mod 4 we have

$$QO[1/2]_{\equiv 2(4)} = QU[1/2]^{asc}_{\equiv 2(4)}.$$

Things are even simpler for the torsion part, where we find the torsion is

$$Z(TU)[1/2] = TU[1/2].$$

CHAPTER 5

Characteristic Classes

Although our primary focus is on finite groups, the values of ko^*_G for the groups $O(n)$, $U(n)$ and $Sp(n)$ serve as universal examples for real, complex and quaternionic representations. The Yoneda Lemma tells us that the value of $ko^*_{G(n)}$ or $ku^*_{G(n)}$ gives all characteristic classes in ko^* or ku^* for $G(n)$ representations.

The symplectic groups give Pontrjagin classes, and these are the best behaved. The unitary groups are well behaved for ku, but have somewhat complicated relations and torsion for ko. The orthogonal groups are quite complicated. We need a few low dimensional cases which we compute completely, but we state some results more generally when it does not represent too large a digression.

To summarize, the groups we will consider are

- $Sp(n)$
- $U(n)$
- $SU(n)$
- $O(n)$
- $SO(n)$
- $T(n) = T^n$
- $V(n) = C_2^n$

The special groups $SO(n)$ and $SU(n)$ are included because they are universal for representations of determinant 1. Of course, the 2-tori $V(n)$ are finite, and are worked out in detail in Chapters 9-12.

5.1. Preliminaries

We refer to [**22**] for representation rings and to [**85**] for cohomology rings. The induced maps in representation theory and cohomology can also be found in these sources. We start each section by reminding the reader of these results. All the representation rings we consider, except those of tori and 2-tori, come with a defining representation whose exterior powers generate the representation ring (except for $SO(2n)$ which requires one additional generator). We will denote these by $\lambda_i(k^n)$, with $k = \mathbb{R}$, $\mathbb{C}$, or $\mathbb{H}$ if it is necessary to be completely precise, abbreviating this to λ_i^k or simply λ_i when possible without ambiguity. In particular, the defining representation is λ_1.

The following simple observation will prove useful.

PROPOSITION 5.1.1. *If $H^*(X,\mathbb{Z})$ is concentrated in even degrees then ku^*X is also, and $v : ku^nX \longrightarrow ku^{n-2}X$ is a monomorphism.*

Proof: The Atiyah-Hirzebruch spectral sequence $H^*(X, ku^*) \Longrightarrow ku^*X$ is concentrated in even degrees, so ku^*X is also. The boundary map $\overline{Q}_1 : H^n(X,\mathbb{Z}) \longrightarrow ku^{n+3}X$ is therefore 0, and hence v acts monomorphically. □

5.2. Tori

The representation ring is

$$RU(T^n) = \mathbb{Z}[t_1^{\pm 1}, \ldots, t_n^{\pm 1}]$$

where t_i is the one dimensional representation obtained by projecting onto the i-th factor. All simple representations but the trivial one are complex. The integral cohomology ring is

$$H^*BT^n = \mathbb{Z}[y_1, \ldots, y_n]$$

with $y_i = c_1(t_i)$. Similarly, we have

THEOREM 5.2.1.

$$ku^*BT^n = ku^*[[y_1, \ldots, y_n]]$$

with $y_i = c_1^{ku}(t_i)$ *and*

$$ku^*_{T^n} = \mathrm{ModRees}(RU(T^n)) = ku^*[y_1, \overline{y}_1 \ldots, y_n, \overline{y}_n]/(vy_i\overline{y}_i = y_i + \overline{y}_i)$$

with $\overline{y}_i = c_1^{ku}(t_i^{-1})$.

Proof: The completed statement follows from the $n = 1$ case by the completed Künneth Theorem, since ku^*BT^1 is flat. Alternatively, the Atiyah-Hirzebruch spectral sequence collapses. The equivariant case then follows by the defining pullback square ([**46**], as in Theorem 2.6.1). For this, we must calculate $ku^*BT^n \cap KU^*_{T^n}$ in $KU^*(BT^n)$. Now

$$\begin{aligned} \overline{y}_i &= [-1](y_i) = \frac{1 - t_i^{-1}}{v} = \frac{1 - (1 - vy_i)^{-1}}{v} \\ &= -y_i(1 + vy_i + v^2y_i^2 + \cdots) \end{aligned}$$

in $KU^*_{T^n}$ and ku^*BT^n, respectively. The relation $vy_i\overline{y}_i = y_i + \overline{y}_i$ then follows either by calculation or by noting that the formal sum of y_i and $\overline{y}_i$ is zero. The intersection

$$ku^*[[y_1, \ldots, y_n]] \cap KU^*[t_1^{\pm 1}, \ldots, t_n^{\pm 1}]$$

can therefore equally be written

$$ku^*[[y_1, \overline{y}_1, \ldots, y_n, \overline{y}_n]] \cap KU^*[y_1, \overline{y}_1, \ldots, y_n, \overline{y}_n]$$

which is evidently $ku^*[y_1, \overline{y}_1 \ldots, y_n, \overline{y}_n]/(vy_i\overline{y}_i = y_i + \overline{y}_i)$. It is immediate that this is $\mathrm{ModRees}(RU(T^n))$: all irreducible representations are one dimensional, so $JU_n = (JU_1)^n$. □

This is proved in [**47**], where $ku^*_{T(n)}$ is shown to be the representing ring for T^n-equivariant multiplicative formal group laws.

To calculate ko^*BT^n, we first observe that the reduced mod 2 cohomology has a special property which dramatically simplifies the Adams spectral sequence converging to $ko^*BT(n)$.

LEMMA 5.2.2. *If the mod 2 cohomology of* X *is concentrated in even degrees and is free over* $E[Sq^2]$, *then* η *acts trivially on* ko^*X, *and hence complexification* $c : ko^*X \longrightarrow ku^*X$ *embeds* ko^*X *into* ku^*X *as the kernel of* $R : ku^*X \longrightarrow ko^{*+2}X$.

Proof: We may suppose without loss of generality that X is 2-complete. The hypothesis implies that as an $\mathcal{A}(1)$-module H^*X is the sum of copies of even suspensions of $H^*C\eta$. The mod 2 Adams spectral sequence

$$\mathrm{Ext}_{\mathcal{A}(1)}(\mathbb{F}_2, H^*X) \Longrightarrow ko^*X$$

is therefore concentrated in even degrees. It follows that η must act trivially upon ko^*X and hence that the ηcR sequence is short exact:

$$0 \longrightarrow ko^*X \xrightarrow{c} ku^*X \xrightarrow{R} ko^{*+2}X \longrightarrow 0.$$

□

This applies to $BT(n) = (\mathbb{C}P^\infty)^n$ by the Künneth Theorem and the fact that it applies to $\mathbb{C}P^\infty$.

We can be more explicit by use of the Bockstein spectral sequence. The representation ring is given by $RU(T^n) = \mathbb{Z}[t_1^{\pm 1}, \ldots, t_n^{\pm 1}]$, and (with the exception of the trivial representation) there are no simple symplectic or real representations. When $n = 1$ this gives us

$$RO/RU = \mathbb{F}_2\{1\}, \qquad \text{and} \qquad RSp/RU = 0$$

and

$$RU/RO = \mathbb{Z}\{t^n \mid n \geq 1\} \qquad \text{and} \qquad RU/RSp = RU/RO \oplus RO/RU$$

5.3. Symplectic groups

In this section we first compute the connective real and complex K-theory of $Sp(1)^n$ in Theorem 5.3.1. We then define Pontrjagin classes in representation theory in Definition 5.3.3. These may be divided by appropriate powers of the Bott maps to give classes in connective K-theory and these classes generate the connective K-theory of $Sp(n)$ (Theorem 5.3.5). This allows us to define Pontrjagin classes in connective K-theory for symplectic representations in Definition 5.3.6. Next, we determine the relation between the Pontrjagin classes and Chern classes in Theorem 5.3.9. Finally, we note in Theorem 5.3.10 that there is a *symplectic splitting principle,* entirely analogous to the complex case, which splits symplectic bundles into sums of symplectic line bundles.

We have chosen to break with tradition in our sign convention for Pontrjagin classes because we prefer that $p_\bullet(qV)$ be related to $c_\bullet(V)c_\bullet(\overline{V})$ rather than a version of this twisted by signs. In ordinary cohomology this amounts to replacing the symmetric polynomial $\sigma_i(z_1^2, \ldots, z_n^2)$ by $\sigma_i(-z_1^2, \ldots, -z_n^2)$ since $c_1(\overline{V}) = -c_1(V)$. The difference is much greater in K-theory, and determines our choice. A number of formulas simplify as a consequence.

The representation ring is

$$RU(Sp(n)) = \mathbb{Z}[\lambda_1, \ldots, \lambda_n]$$

The λ_{2i} are real and the λ_{2i+1} are quaternionic. In particular, they are all self conjugate. Note that $\lambda_1 = \mathbb{H}^n = \mathbb{C}^{2n}$, which is $2n$ dimensional, but its higher exterior powers $\lambda_{n+1}, \ldots \lambda_{2n}$ can be expressed in terms of the first n. The integral cohomology is

$$H^*BSp(n) = \mathbb{Z}[p_1, \ldots, p_n]$$

with $|p_i| = 4i$. Restriction along $Sp(1)^n \longrightarrow Sp(n)$ will play much the same role for $Sp(n)$ as restriction along $T(n) = U(1)^n \longrightarrow U(n)$ plays for $U(n)$, so we start by considering $Sp(1)^n$. Its representation ring is

$$RU(Sp(1)^n) = \mathbb{Z}[s_1, \ldots, s_n]$$

where s_i is the one dimensional symplectic representation obtained by projecting onto the i-th factor. Complexification factors as $Sp(1) \cong SU(2) \subset U(2)$. As a result we have $c_1^{ku}(s_i) = vc_2^{ku}(s_i)$ (see Section 5.4) and

$$v^2 c_2^{ku}(s_i) = vc_1^{ku}(s_i) = c_2^R(s_i) = c_1^R(s_i) = 2 - s_i.$$

Accordingly, we have classes $z_i = c_2^{ku}(s_i) \in ku^4(BSp(1)^n)$ which satisfy

$$v^2 z_i = 2 - s_i.$$

We will see in a moment that z_i comes from ko^*. The integral cohomology ring is

$$H^* BSp(1)^n = \mathbb{Z}[z_1, \ldots, z_n]$$

with $z_i = p_1(s_i)$, the first Pontrjagin class of s_i in cohomology.

THEOREM 5.3.1. *We have compatible generators z_i so that*

- $ku^* BSp(1)^n = ku^*[[z_1, \ldots, z_n]]$
- $ko^* BSp(1)^n = ko^*[[z_1, \ldots, z_n]]$
- $ku^*_{Sp(1)^n} = ku^*[z_1, \ldots, z_n] = \mathrm{ModRees}(RU(Sp(1)^n))$
- $ko^*_{Sp(1)^n} = ko^*[z_1, \ldots, z_n]$

and both complexification and completion send the generators to generators of the same name. In particular, $z_i^{ku} \in ku^4_{Sp(1)^n}$ and $z_i^{ko} \in ko^4_{Sp(1)^n}$ satisfy

- $v^2 z_i^{ku} = 2 - s_i \in ku^0_{Sp(1)^n} = RU(Sp(1)^n)$,
- $\alpha z_i^{ko} = 2(2 - s_i) \in ko^0_{Sp(1)^n} = RO(Sp(1)^n)$ *and*
- $\beta z_i^{ko} = 2 - s_i \in ko^{-4}_{Sp(1)^n} = RSp(Sp(1)^n)$.

Proof: The proof goes just as for T^n except that the calculations are so much simpler that we can do the ko calculation at the same time. Computing the completed rings is trivial because the cohomology of $Sp(1)^n$ is a trivial $\mathcal{A}(1)$-module concentrated in degrees congruent to 0 mod 4. Thus, the Adams spectral sequence collapses at $E_2^{*,*} = H^* BSp(1)^n \otimes \mathrm{Ext}^{*,*}_{\mathcal{A}(1)}(\mathbb{F}_2, \mathbb{F}_2)$ for ko^* and similarly for $E(1)$ and ku^*. Hence we can choose compatible ring generators for ko^* and ku^* which map to the generators of the same name in integral cohomology. That the result is a power series ring rather than a polynomial ring is evident from the lim-lim^1 sequence based on the skeleta.

The equivariant cases then follow by the defining pullback squares ([**46**] and Theorem 2.6.1). For ku we have

$$\begin{array}{ccc} ku^*_{Sp(1)^n} & \longrightarrow & KU^*[z_1, \ldots, z_n] \\ \downarrow & & \downarrow \\ ku^*[[z_1, \ldots, z_n]] & \longrightarrow & KU^*[[z_1, \ldots, z_n]] \end{array}$$

where the periodic groups are as claimed because we can change generators from the s_i to the $z_i = (2 - s_i)/v^2$. It is immediate that this is $\mathrm{ModRees}(RU(Sp(1)^n))$: all irreducible representations are two dimensional, so $JU_{2n} = JU_{2n-1} = (JU_2)^n$.

For ko, it is evident that $KO^*BSp(1)^n = KO^*[[z_1,\ldots,z_n]]$ because this is $ko^*BSp(1)^n[\beta^{-1}]$. The periodic equivariant real K-theory is a bit more delicate. Precisely, $RO(Sp(1)^n)$ is spanned by the monomials of even total degree (1, z_1^2, z_1z_2, etc.) and twice the monomials of odd total degree (z_1,$z_1z_2^2$, etc.), while $RSp(Sp(1)^n)$ is spanned by the odd powers and twice the even powers. The various quotients (RO/RU, etc.) are then seen to give exactly $KO^*[z_1,\ldots,z_n]$. The pullback square then works exactly as for the complex case.

That $v^2z_i^{ku} = 2 - s_i$ follows from the fact that we may define $z_i^{ku} = c_2^{ku}(s_i)$. The relations for αz_i^{ko} and βz_i^{ko} then follow because we may choose the z_i^{ko} to make this true, using the fact that, when $G = Sp(1)^n$, the realification maps $ko^{4i}_{Sp(1)^n} \longrightarrow ku^{4i}_{Sp(1)^n}$ are monomorphisms. □

We can use this to give representation theoretic descriptions of all the equivariant connective real K-theory of $Sp(1)^n$, not just the part near the edge of periodicity.

First, we establish some convenient notation. Recall that we have chosen to name elements of RO and RSp by their image in RU, so that the complexification maps $c : RO \longrightarrow RU$ and $\widetilde{c} : RSp \longrightarrow RU$ are the evident inclusions.

Let us write $z^{\langle i\rangle} = z_1^{i_1}\cdots z_n^{i_n}$ for any monomial in the $z_1,\ldots,z_n$ with exponent sum $i_1+\cdots+i_n = i$. Thus $z^{\langle 2i\rangle}$ denotes a monomial with $i_1+\cdots+i_n$ even, and $z^{\langle 2i+1\rangle}$ a monomial with $i_1+\cdots+i_n$ odd. With this notation,

$$\begin{array}{lcl}
RO & = & \mathbb{Z}\langle z^{\langle 2i\rangle}, 2z^{\langle 2i+1\rangle} \mid i \geq 0\rangle \\
JSp_{2k} & = & \mathbb{Z}\langle z^{\langle 2i\rangle}, 2z^{\langle 2i+1\rangle} \mid i \geq k\rangle \\
JO\cdot JSp_{2k} & = & \mathbb{Z}\langle 2z^{\langle 2i+1\rangle}, z^{\langle 2i+2\rangle} \mid i \geq k\rangle \\
RSp & = & \mathbb{Z}\langle 2z^{\langle 2i\rangle}, z^{\langle 2i+1\rangle} \mid i \geq 0\rangle \\
JSp_{2k-1} & = & \mathbb{Z}\langle z^{\langle 2i-1\rangle}, 2z^{\langle 2i\rangle} \mid i \geq k\rangle \\
JO\cdot JSp_{2k-1} & = & \mathbb{Z}\langle 2z^{\langle 2i\rangle}, z^{\langle 2i+1\rangle} \mid i \geq k\rangle
\end{array}$$

and

$$JU_{2k} = JU_{2k-1} = JU^k = \mathbb{Z}\langle z^{\langle i\rangle} \mid i \geq k\rangle$$

Thus, there are four ways to write each of the quotients occuring in $ko_*^{Sp(1)^n}$: the two listed below and the two obtained from them by the fact that $JU_{2i-1} = JU_{2i}$. In the Corollary, we have chosen the version that is consistent with Theorem 2.9.5.

$$\begin{array}{lclclcl}
\dfrac{RU}{RO} & \supset & \dfrac{JU_{4k-4}}{JSp_{2k-2}} & = & \dfrac{JU_{4k-2}}{JO\cdot JSp_{2k-2}} & = & \mathbb{F}_2\langle z^{\langle 2i-1\rangle} \mid i \geq k\rangle \\[2ex]
\dfrac{RSp}{RU} & \supset & \dfrac{JSp_{2k-1}}{JU_{4k-2}} & = & \dfrac{JO\cdot JSp_{2k-3}}{JU_{4k-4}} & = & \mathbb{F}_2\langle z^{\langle 2i-1\rangle} \mid i \geq k\rangle \\[2ex]
\dfrac{RU}{RSp} & \supset & \dfrac{JU_{4k-2}}{JSp_{2k-1}} & = & \dfrac{JU_{4k}}{JO\cdot JSp_{2k-1}} & = & \mathbb{F}_2\langle z^{\langle 2i\rangle} \mid i \geq k\rangle \\[2ex]
\dfrac{RO}{RU} & \supset & \dfrac{JSp_{2k}}{JU_{4k}} & = & \dfrac{JO\cdot JSp_{2k-2}}{JU_{4k-2}} & = & \mathbb{F}_2\langle z^{\langle 2i\rangle} \mid i \geq k\rangle
\end{array}$$

COROLLARY 5.3.2. *The coefficients of $Sp(1)^n$-equivariant connective real K-theory are*

$$ko_i^{Sp(1)^n} = \begin{cases} 0 & i = 8k+7 \geq 0 \\ RU/RO & i = 8k+6 \geq 0 \\ RSp/RU & i = 8k+5 \geq 0 \\ RSp & i = 8k+4 \geq 0 \\ 0 & i = 8k+3 \geq 0 \\ RU/RSp & i = 8k+2 \geq 0 \\ RO/RU & i = 8k+1 \geq 0 \\ RO & i = 8k \geq 0 \\ 0 & i = -8k+7 < 0 \\ JU_{4k-3}/JO \cdot JSp_{2k-2} & i = -8k+6 \\ JSp_{2k-1}/JU_{4k-3} & i = -8k+5 \\ JSp_{2k-1} & i = -8k+4 \\ 0 & i = -8k+3 < 0 \\ JU_{4k-2}/JSp_{2k-1} & i = -8k+2 \\ JSp_{2k}/JU_{4k} & i = -8k+1 \\ JSp_{2k} & i = -8k \end{cases}$$

Proof: As noted in the proof of Theorem 5.3.1, the collapse of the Adams spectral sequence tells us that each $z^{\langle i \rangle} = z_1^{i_1} \cdots z_n^{i_n}$ contributes a copy of ko_* shifted to start in $ko_{-4i}^{Sp(1)^n}$. Observe that $\alpha z^{\langle i \rangle} \in ko_{-8i+4}^{Sp(1)^n}$ will map to $2z^{\langle i \rangle} \in RU(Sp(1)^n)$ since $\alpha \mapsto 2v^2$ and that similar factors of 2 appear when mapping RU to RO or RSp here. It follows that

$$\begin{aligned} ko_{-8k+6} &= \mathbb{F}_2\langle z^{\langle 2i-1 \rangle} \mid i \geq k \rangle \\ ko_{-8k+5} &= \mathbb{F}_2\langle z^{\langle 2i-1 \rangle} \mid i \geq k \rangle \\ ko_{-8k+4} &= \mathbb{Z}\langle z^{\langle 2i-1 \rangle}, 2z^{\langle 2i \rangle} \mid i \geq k \rangle \\ ko_{-8k+2} &= \mathbb{F}_2\langle z^{\langle 2i \rangle} \mid i \geq k \rangle \\ ko_{-8k+1} &= \mathbb{F}_2\langle z^{\langle 2i \rangle} \mid i \geq k \rangle \\ ko_{-8k} &= \mathbb{Z}\langle z^{\langle 2i \rangle}, 2z^{\langle 2i+1 \rangle} \mid i \geq k \rangle \end{aligned}$$

Multiplication by β^k embeds these into the periodic part of $ko_*^{Sp(1)^n}$, determining, for example, that we will write $ko_{-8k+6}^{Sp(1)^n}$ as the quotient of something contained in RU by something in RSp. □

Now we turn to the definition of Pontrjagin classes. Restriction along the inclusion $Sp(1)^n \longrightarrow Sp(n)$ embeds $RU(Sp(n))$ into $RU(Sp(1)^n)$ as the symmetric polynomials. As in the complex case ([**28**, Def 1.3.5]), it is natural to define the *first* Pontrjagin class as the virtual representation $p_1^R(V) = |V| - V$, and to define Pontrjagin classes in general by requiring them to restrict to the symmetric polynomials in the first Pontrjagin classes of the line bundles which generate $RU(Sp(1)^n)$. The only difference from the complex case is the factor of 2 which appears because a trivial n-dimensional quaternionic bundle is a trivial $2n$-dimensional complex bundle.

DEFINITION 5.3.3. The kth representation theoretic *Pontrjagin class* of an n-dimensional symplectic representation $V : G \longrightarrow Sp(n)$ is

$$p_k^R(V) = \sum_{j=0}^{k} (-1)^j 2^{k-j} \binom{n-j}{n-k} \Lambda^j(V)$$

As usual, it is equivalent to simply define elements $p_k^R = p_k^R(\lambda_1)$ in the representation ring of $Sp(n)$, and obtain all others by naturality.

PROPOSITION 5.3.4. *The restriction $RU(Sp(n)) \longrightarrow RU(Sp(1)^n)$ sends p_k^R to $\sigma_k(2-s_1, \dots, 2-s_n)$. The representation p_k^R is real if k is even, and quaternionic if k is odd.*

Accordingly, we shall generally consider p_{2i}^R as an element of $RO(G)$ and p_{2i+1}^R as an element of $RSp(G)$. Note, however, that representations which are not irreducible can be both real and quaternionic.

Proof: We shall work on the universal case $\lambda_1 \in RU(Sp(n))$. Restriction to $Sp(1)^n$ sends λ_i to $\sigma_i(s_1, \dots, s_n)$. Let us write p_k for the element which restricts to the symmetric polynomial $\sigma_k(2-s_1, \dots, 2-s_n)$. Then we may compute:

$$\begin{aligned}
\sum_{k=0}^{n} p_k x^{n-k} &= \prod_{i=1}^{n} (x + (2 - s_i)) \\
&= \prod_{i=1}^{n} ((x+2) - s_i)) \\
&= \sum_{j=0}^{n} (-1)^j \lambda_j (x+2)^{n-j} \\
&= \sum_{j=0}^{n} (-1)^j \lambda_j \sum_{i=0}^{n-j} \binom{n-j}{i} 2^{n-i} x^i
\end{aligned}$$

Comparing coefficients of x^{n-k} we see that $p_k = p_k^R(\lambda_1)$. The terms λ_{2i+1} are quaternionic, the terms λ_{2i} are real, and $2\lambda_i$ is both real and quaternionic. Thus, the nature of p_k^R is determined by that of λ_k. □

To carry this out in the connective K-theory of $Sp(n)$, we must first note that it too is embedded as the symmetric polynomials in the connective K-theory of $Sp(1)^n$.

THEOREM 5.3.5. *We have*

- $ku^* BSp(n) = ku^*[[p_1, \dots, p_n]]$
- $ko^* BSp(n) = ko^*[[p_1, \dots, p_n]]$
- $ku^*_{Sp(n)} = ku^*[p_1, \dots, p_n]$
- $ko^*_{Sp(n)} = ko^*[p_1, \dots, p_n]$.

In each case, p_k restricts to $\sigma_k(z_1, \dots, z_n)$. In the complex case, we have $v^{2k} p_k^{ku} = p_k^R \in ku^0_{Sp(n)} = RU(Sp(n))$, while, in the real case, we have $\beta^k p_{2k}^{ko} = p_{2k}^R \in ko^0_{Sp(n)} = RO(Sp(n))$ and $\beta^k p_{2k+1}^{ko} = p_{2k+1}^R \in ko^4_{Sp(n)} = JSp(Sp(n))$.

Proof: Since $H^*BSp(n)$ is a trivial $\mathcal{A}(1)$-module, the E_2 term of the Adams spectral sequence is simply $\mathrm{Ext}_{\mathcal{A}(1)}(\mathbb{F}_2, \mathbb{F}_2)[p_1, \ldots, p_n]$, and there are no nonzero differentials because $\mathrm{Ext}_{\mathcal{A}(1)}(\mathbb{F}_2, \mathbb{F}_2)$ is 0 in degrees congruent to -1 mod 4. Thus the Adams spectral sequence for $ko^*BSp(n)$ must collapse at E_2. Similar arguments apply to ku^* (and $H\mathbb{Z}^*$), Thus we can choose generators $p_i = p_i^{ko} \in ko^*BSp(n)$ and $p_i = p_i^{ku} \in ku^*BSp(n)$ such that $p_i^{ko} \mapsto p_i^{ku} \mapsto p_i^H$ and each maps to the symmetric polynomial in the corresponding z_i. The relation to representation theory then follows, first in the complex case by noting that $v^{2k}\sigma_k(z_1, \ldots, z_n) = \sigma_k(2 - s_1, \ldots, 2 - s_n)$, then in the real case by comparison with the complex case.

The equivariant statements then follow by the defining pullback square as in Theorem 2.6.1 and the succeeding remark. □

DEFINITION 5.3.6. Let $V : G \longrightarrow Sp(n)$ be a symplectic representation. For $E = RU$, ko, KO, ku, KU or H, we define the *Pontrjagin class* $p_i^E(V) \in E_G^{4i}$ *to be* $V^*(p_i)$. It is convenient to collect these into the *total Pontrjagin class*

$$p_\bullet^E(V) = 1 + p_1^E(V) + p_2^E(V) + \cdots + p_n^E(V)$$

and to let $p_i^E(V) = 0$ if $i > n$.

COROLLARY 5.3.7.

$$p_\bullet^E(V \oplus W) = p_\bullet^E(V) p_\bullet^E(W)$$

Proof: This is a simple matter of invariant theory:

$$\sigma_n(z_1, \ldots, z_{k+j}) = \sum_{i=0}^{i=n} \sigma_i(z_1, \ldots, z_k)\sigma_{n-i}(z_1, \ldots, z_j)$$

□

It is worth noting the special case ko_G^*. For a symplectic representation $V \in RSp(G)$, the classes $p_i^{ko}(V) \in ko_G^{4i}$ satisfy

$$\beta^k p_{2k}^{ko}(V) = p_{2k}^R(V) \in ko_G^0 = RO(G)$$

and

$$\beta^k p_{2k+1}^{ko}(V) = p_{2k+1}^R(V) \in ko_G^4 = JSp(G).$$

Finally, we consider the relation to the Chern classes induced by the natural inclusions $U(n) \xrightarrow{q} Sp(n) \xrightarrow{\tilde{c}} U(2n)$. The latter is the natural representation λ_1, so this calculates the Chern classes $c_i^{ku}(\lambda_1)$. First consider the maximal torus.

LEMMA 5.3.8. *The restriction* $ku^*_{Sp(1)^n} \longrightarrow ku^*_{T(n)}$ *maps* z_i *to* $y_i\overline{y}_i$.

Proof: This follows from the fact that the inclusions $T(1) \longrightarrow Sp(1) \longrightarrow U(2)$ define the representation $t + t^{-1} \in \mathbb{Z}[t^{\pm 1}] = RU(T(1))$. Thus $z_i = c_2^{ku}(s_i) \mapsto c_2^{ku}(t_i + t_i^{-1}) = c_1(t_i)c_1(t_i^{-1})$. □

Let us write $\overline{c}_i(V) = c_i(\overline{V})$ for the Chern classes of the complex conjugate of a representation.

THEOREM 5.3.9. *The restriction maps* $ku^*_{U(2n)} \xrightarrow{q^*} ku^*_{Sp(n)} \xrightarrow{\widetilde{c}^*} ku^*_{U(n)}$ *obey*

$$c_k \mapsto \sum_{0 \le 2i \le k} \binom{k-i}{i} v^{k-2i} p_{k-i} \mapsto \sum_{i+j=k} c_i \overline{c}_j.$$

Proof: We shall do the calculation in $ku^*_{T(n)}$. First, consider the commutative square

$$\begin{array}{ccc} T(n) & \longrightarrow & U(n) \\ \downarrow & & \downarrow \\ Sp(1)^n & \longrightarrow & Sp(n) \end{array}$$

By Lemma 5.3.8, $p^{ku}_k \mapsto \sigma_k(z_1, \dots, z_n) \mapsto \sigma_k(y_1\overline{y}_1, \dots, y_n\overline{y}_n)$. Recall that $vy_i\overline{y}_i = y_i + \overline{y}_i$ (5.2.1). Using the generating functions $c_\bullet = \sum c_i t^i$ and $\overline{c}_\bullet = \sum \overline{c}_i t^i$ we have

$$c_\bullet \overline{c}_\bullet = \sum_n \left(\sum_{i+j=n} c_i \overline{c}_j \right) t^n$$

but also

$$\begin{aligned} c_\bullet \overline{c}_\bullet &= \prod (1 + y_i t) \prod (1 + \overline{y}_i t) \\ &= \prod (1 + (y_i + \overline{y}_i) t + y_i \overline{y}_i t^2) \\ &= \prod (1 + y_i \overline{y}_i (vt + t^2)) \\ &= \sum_i p_i \sum_{j=0}^{i} \binom{i}{j} v^j t^{2i-j} \end{aligned}$$

(omitting the induced maps from the square above to avoid cluttering the notation). Comparing coefficients of t^n we see that

$$\sum_{i+j=n} c_i \overline{c}_j = \sum_i \binom{i}{2i-n} v^{2i-n} p_i = \sum_{0 \le 2i \le n} \binom{n-i}{i} v^{n-2i} p_{n-i}.$$

This tells us that $q^* : ku^*_{Sp(n)} \longrightarrow ku^*_{U(n)}$ is as stated.

Now consider the commutative diagram of maximal tori:

$$\begin{array}{ccccc} U(n) & \longrightarrow & Sp(n) & \longrightarrow & U(2n) \\ \uparrow & & & & \uparrow \\ T^n & & \longrightarrow & & T^{2n} \end{array}$$

We may arrange this so that the map of tori induces $y_i \mapsto y_i$ and $y_{n+i} \mapsto \overline{y}_i$. Then the total Chern class of $U(2n)$ maps by

$$\sum_{i=0}^{2n} c_i t^i \mapsto \prod_1^{2n} (1 + y_i t) \mapsto \prod_1^{n} (1 + y_i t)(1 + \overline{y}_i t)$$

which is the image of $c_\bullet \overline{c}_\bullet$. □

Note that if we specialize to ordinary cohomology by setting $v = 0$ we obtain the usual relations up to sign ([**85**, Ch. III, Thm. 5.8]):

$$q^*(p_n^H) = \sum_{i+j=2n} c_i^H \overline{c}_j^H = \sum_{i+j=2n} (-1)^j c_i^H c_j^H,$$

$$\widetilde{c}^*(c_{2i-1}) = 0, \qquad \text{and} \qquad \widetilde{c}^*(c_{2i}) = p_i.$$

The sign is simpler here than classically because of our decision that the Pontrjagin classes should restrict to the symmetric polynomials in the z_i in $Sp(1)^n$, and hence to the symmetric polynomials in the $y_i\overline{y}_i$ in the maximal torus.

This theorem gives, for $Sp(4)$, for example,

$$\begin{array}{lllllll}
c_1 \mapsto & vp_1 & & & & \mapsto & c_1 + \overline{c}_1 \\
c_2 \mapsto & p_1 & +v^2p_2 & & & \mapsto & c_2 + c_1\overline{c}_1 + \overline{c}_2 \\
c_3 \mapsto & & 2vp_2 & +v^3p_3 & & \mapsto & c_3 + c_2\overline{c}_1 + c_1\overline{c}_2 + \overline{c}_3 \\
c_4 \mapsto & & p_2 & +3v^2p_3 & +v^4p_4 & \mapsto & c_4 + c_3\overline{c}_1 + c_2\overline{c}_2 + c_1\overline{c}_3 + \overline{c}_4 \\
c_5 \mapsto & & & 3vp_3 & +4v^3p_4 & \mapsto & c_4\overline{c}_1 + c_3\overline{c}_2 + c_2\overline{c}_3 + c_1\overline{c}_4 \\
c_6 \mapsto & & & p_3 & +6v^2p_4 & \mapsto & c_4\overline{c}_2 + c_3\overline{c}_3 + c_2\overline{c}_4 \\
c_7 \mapsto & & & & 4vp_4 & \mapsto & c_4\overline{c}_3 + c_3\overline{c}_4 \\
c_8 \mapsto & & & & p_4 & \mapsto & c_4\overline{c}_4
\end{array}$$

It is more difficult to get good expressions for the images of the individual p_i. However, for $i = 1$, using the fact that v acts monomorphically on $ku^*_{U(n)}$ (Proposition 5.1.1) together with Proposition 5.5.5 we have

$$p_1 \mapsto \frac{c_1 + \overline{c}_1}{v} = \sum_{k=2}^{n} (-v)^{k-2} \sum_{i+j=k} c_i\overline{c}_j = c_2 + c_1\overline{c}_1 + \overline{c}_2 - v\sum_{k=3}^{n}(-v)^{k-3} \sum_{i+j=k} c_i\overline{c}_j.$$

In cohomology, where $v = 0$ and $\overline{c}_i = (-1)^i c_i$, we have $p_1 \mapsto c_2 + c_1\overline{c}_1 + \overline{c}_2 = 2c_2 - c_1^2$ with our normalization of the p_i. Thus, if $c_1 = 0$, then $c_2 = p_1/2$.

Finally, we provide the following *symplectic splitting principle.*

THEOREM 5.3.10. *Let ξ be an $Sp(n)$ bundle over X. Then there exists a map $f : Y \longrightarrow X$ such that $f^*\xi$ is a sum of symplectic line bundles and $f^* : H^*X \longrightarrow H^*Y$ is a monomorphism.*

Proof: Let Y be the pullback

$$\begin{array}{ccc}
Y & \longrightarrow & BSp(1)^n \\
\Big\downarrow{\scriptstyle f} & & \Big\downarrow \\
X & \longrightarrow & BSp(n)
\end{array}$$

along the classifying map of the bundle ξ. The universal bundle over $BSp(n)$ splits as a sum of line bundles over $BSp(1)^n$, so $f^*\xi$ also splits in this manner.

The Serre spectral sequence for the bundle

$$Sp(n) \longrightarrow Sp(n)/Sp(1)^n \longrightarrow BSp(1)^n$$

shows that $H^*Sp(n)/Sp(1)^n$ is the quotient $\mathbb{Z}[z_1, \ldots, z_n]/(\sigma_1, \ldots, \sigma_n)$ of the polynomial ring $H^*Sp(1)^n$ by the ideal generated by the symmetric polynomials.

Now, the Serre spectral sequence

$$H^*(BSp(n), H^*(Sp(n)/Sp(1)^n)) \Longrightarrow H^*BSp(1)^n$$

collapses, and the local coefficients are trivial, so the same is true for the pullback, showing that $f^* : H^*X \longrightarrow H^*Y$ is a monomorphism, since the Serre spectral sequence collapses. □

Note that we can compute $ko_*BSp(n)$ by the collapsed local cohomology spectral sequence because the adjoint representation is ko oriented, being a *Spin* bundle, since w_1 and w_2 are trivially 0, so that

$$H^n_J(ko^*(BSp(n)) = ko_*(BSp(n)^{ad}) = \Sigma^{\dim(Sp(n))}ko_*(BSp(n)).$$

5.4. Special unitary groups

The representation ring is

$$RU(SU(n)) = \mathbb{Z}[\lambda_1, \dots, \lambda_{n-1}].$$

We have $\overline{\lambda_i} = \lambda_{n-i}$, so the λ_i are all complex unless $n = 2m$, when λ_m is real if m is even and quaternionic if m is odd. The integral cohomology is

$$H^*BSU(n) = \mathbb{Z}[c_2, \dots, c_n]$$

with $c_i = c_i(\lambda_1)$. The connective complex K-theory is easy to compute.

THEOREM 5.4.1. *$ku^*BSU(n) = ku^*[[c_2, \dots, c_n]]$ and*

$$ku^*_{SU(n)} = \mathrm{ModRees}(RU(SU(n))) = ku^*[c_2, \dots, c_n].$$

Proof: Since $H^*BSU(n)$ is concentrated in even degrees, the Atiyah-Hirzebruch spectral sequence shows that $ku^*BSU(n)$ must be the complete ku^* algebra freely generated by $c_2, \dots, c_n$. In $KU^*_{SU(n)}$, we have

$$\lambda_i = \sum_{j=0}^{i}(-1)^j\binom{n-j}{n-i}c_j^R = \sum_{j=0}^{i}(-1)^j\binom{n-j}{n-i}v^jc_j^{ku}.$$

and $\lambda_n = 1$. Hence, the $c_j = c_j^{ku}$ generate, and since $c_1 - vc_2 + \cdots + (-v)^{n-1}c_n = 0$, $KU^*_{SU(n)}$ is polynomial on any $n-1$ of $c_1, \dots, c_n$. In particular, $KU^*_{SU(n)} = KU^*[c_2, \dots, c_n]$. Then the pullback square ([**46**])

$$\begin{array}{ccc} ku^*_{SU(n)} & \longrightarrow & KU^*[c_2, \dots, c_n] \\ \downarrow & & \downarrow \\ ku^*[[c_2, \dots, c_n]] & \longrightarrow & KU^*[[c_2, \dots, c_n]] \end{array}$$

shows that $ku^*_{SU(n)} = ku^*[c_2, \dots, c_n]$. □

5.5. Unitary groups

The representation ring is

$$RU(U(n)) = \mathbb{Z}[\lambda_1, \dots, \lambda_n, \lambda_n^{-1}]$$

The representation $\lambda_i\lambda_{n-i}\lambda_n^{-1}$ is real for $1 \le i \le n/2$ and $\overline{\lambda_i} = \lambda_{n-i}\lambda_n^{-1}$.

The integral cohomology is

$$H^*BU(n) = \mathbb{Z}[c_1, \dots, c_n]$$

where $c_i = c_i(\lambda_1)$. Again, the complex connective K-theory follows immediately.

THEOREM 5.5.1. *$ku^*BU(n) = ku^*[[c_1, \ldots, c_n]]$ and*

$$ku^*_{U(n)} = \mathrm{ModRees}(RU(U(n))) = ku^*[c_1, \ldots, c_n, \Delta^{-1}]$$

where $\Delta = \lambda_n = 1 - vc_1 + v^2c_2 - \cdots + (-v)^n c_n$.

Proof: The argument is nearly the same as for $SU(n)$, except that $KU^*_{U(n)}$ is not polynomial, but is instead $KU^*[c_1, \ldots, c_n, \Delta^{-1}]$. □

In cohomology, restriction along the inclusion $SU(n) \longrightarrow U(n)$ is the quotient which sends c_1 to 0. The proper way to think of this is that it is taking the quotient by the Chern class of the determinant of the defining representation. In K-theory, the map this induces is more interesting.

THEOREM 5.5.2. *The restriction homomorphism $ku^*_{U(n)} \longrightarrow ku^*_{SU(n)}$ is the quotient $ku^*[c_1, \ldots, c_n, \Delta^{-1}] \longrightarrow ku^*[c_2, \ldots, c_n]$ which sends Δ to 1 and c_1 to $vc_2 - v^2c_3 + \cdots - (-v)^{n-1}c_n$.*

Proof: $SU(n)$ is the kernel of the determinant $U(n) \longrightarrow U(1)$. The determinant sends $y = (1 - \lambda_1)/v \in ku^2_{U(1)}$ to $(1 - \lambda_n)/v = c_1 - vc_2 + v^2c_3 - \cdots$, so this must go to zero in $ku^*_{SU(n)}$. After dividing by this, we have an isomorphism, by Theorem 5.4.1. □

COROLLARY 5.5.3. *If $c_1(V) = vx \in ku^*_G$ for some x then $c_1(V) = v(c_2(V) - vc_3(V) + \cdots + (-v)^{n-2}c_n(V))$, where $n = \dim(V)$.*

Proof: $c_1^{ku}(V) = vx$ implies $c_1^H(V) = 0$, which implies that the determinant representation $\det(V)$ is trivial, so that V must factor through $SU(n)$, forcing the relation to hold. □

It is sometimes useful to consider the conjugate Chern classes $\overline{c}_i(V) = c_i(\tau(V))$.

COROLLARY 5.5.4. *The restriction homomorphism $ku^*_{U(n)} \longrightarrow ku^*_{T(n)}$ sends $\overline{c}_i$ to $\sigma_i(\overline{y}_1, \ldots, \overline{y}_n)$. The $\overline{c}_i$ also satisfy*

- $\Delta v^i \overline{c}_i = \sum_{j=0}^{i} (-1)^j \binom{n-j}{n-i} \lambda_{n-j}$
- $\Delta \overline{c}_i = \sum_{k=i}^{n} (-1)^k \binom{k}{i} v^{k-i} c_k$

Proof: The first statement is immediate from the fact that conjugation on the maximal torus sends t_i to t_i^{-1} and hence $y_i = (1 - t_i)/v$ to $\overline{y}_i = (1 - t_i^{-1})/v$. Using the fact that $\overline{\lambda}_i = \lambda_{n-i}\Delta^{-1}$ and the definition of c_i ([**28**, Def. 1.3.5 and Lemma

1.3.6]) we find

$$\begin{aligned}\Delta v^i\bar{c}_i = &= \sum_{j=0}^{i}(-1)^j\binom{n-j}{n-i}\lambda_{n-j}\\ &= \sum_{j=0}^{i}(-1)^j\binom{n-j}{n-i}\sum_{k=0}^{n-j}(-1)^k\binom{n-k}{j}v^kc_k\\ &= \sum_{k=0}^{n}(-1)^kv^kc_k\sum_{j=0}^{\min(i,n-k)}(-1)^j\binom{n-j}{n-i}\binom{n-k}{j}\\ &= \sum_{k=i}^{n}(-1)^k\binom{k}{i}v^{k-i}c_k\end{aligned}$$

Since $ku^*BU(n)$ and $ku^*_{U(n)}$ have no v-torsion, the result follows. □

The conjugate of the determinant, $\overline{\Delta} = \overline{\lambda}_n = 1 - v\bar{c}_1 + v^2\bar{c}_2 - \cdots \pm v^n\bar{c}_n$ satisfies $\Delta\overline{\Delta} = 1$. Collecting terms and using the fact that v acts monomorphically on $ku^*_{U(n)}$ we find

PROPOSITION 5.5.5. *In* $ku^*_{U(n)}$,

$$c_1 + \bar{c}_1 = -\sum_{k=2}^{2n}(-v)^{k-1}\sum_{i+j=k}c_i\bar{c}_j$$

□

5.6. Orthogonal groups

The representation ring is

$$RU(O(n)) = \mathbb{Z}[\lambda_1, \ldots, \lambda_n]/(\lambda_n^2 - 1, \lambda_i\lambda_n - \lambda_{n-i}).$$

These representations are all real, so that complexification and quaternionification are isomorphisms

$$RO(O(n)) \xrightarrow{\cong} RU(O(n)) \xrightarrow{\cong} RSp(O(n)).$$

The integral cohomology is complicated. The best approach is to give the mod 2 cohomology, and if integral issues matter, the cohomology localized away from 2. We have

$$H\mathbb{F}_2^*BO(n) = \mathbb{F}_2[w_1, \ldots, w_n]$$

where $w_i = w_i(\lambda_1)$.

To describe the equivariant periodic complex K-theory in a way which relates well to the connective K-theory, we rewrite the representation ring in terms of Chern classes, as usual. Let $c_i = c_i^K(\lambda_1) \in K^{2i}_{O(n)}$, so that

$$\lambda_i = \sum_{j=0}^{i}(-1)^j\binom{n-j}{n-i}v^jc_j,$$

where $c_0 = 1$ ([**28**, Section 1.3]). In addition, rather than replace λ_n by the top Chern class, c_n, we prefer to use the first Chern class of the determinant representation, $c = c_1^K(\lambda_n) \in K^2_{O(n)}$. This satisfies $vc = 1 - \lambda_n$, which is much more convenient than $v^n c_n = 1 - \lambda_1 + \cdots + (-1)^n \lambda_n$.

PROPOSITION 5.6.1.

$$KU^*_{O(2n+1)} = KU^*[c_1, \ldots, c_n, c]/(vc^2 - 2c)$$

and

$$KU^*_{O(2n)} = KU^*[c_1, \ldots, c_n, c]/(vc^2 - 2c, c\sum_{i=0}^{n} \binom{2n-i}{n} (-v)^i c_i)$$

Proof: Since λ_n and c_n differ by a unit, modulo $(c_1, \ldots, c_{n-1})$, we may choose either the λ_i or the c_i as generators for the representation ring. The relations $\lambda_i \lambda_n = \lambda_{n-i}$ imply that the λ_i with $n/2 < i < n$ are redundant, as are the corresponding c_i. We then find

$$\begin{aligned} KU^*_{O(2n+1)} &= KU^*[\lambda_1, \ldots, \lambda_n, \lambda_{2n+1}]/(\lambda^2_{2n+1} - 1) \\ &= KU^*[c_1, \ldots, c_n, c]/(vc^2 - 2c) \end{aligned}$$

and

$$\begin{aligned} KU^*_{O(2n)} &= KU^*[\lambda_1, \ldots, \lambda_n, \lambda_{2n+1}]/(\lambda^2_{2n+1} - 1, \lambda_n(1 - \lambda_{2n})) \\ &= KU^*[c_1, \ldots, c_n, c]/(vc^2 - 2c, c\sum_{i=0}^{n} \binom{2n-i}{n} (-v)^i c_i) \end{aligned}$$

□

To compute $ku^*BO(n)$, we need to determine the $E(1)$-module structure of $H\mathbb{F}_2^*BO(n)$. We start with a general result identifying its stable type. Let ϵ be 0 or 1. First, note that the submodule

$$\mathbb{F}_2[w_2^2, w_4^2, \ldots, w_{2n}^2] \longrightarrow H^*BO(2n + \epsilon)$$

is a trivial $E(1)$-submodule. Second, the reduced homology of $BO(1)$ is the ideal (w_1) in $\mathbb{F}_2[w_1]$, and as an $E(1)$-submodule,

$$(w_1) \otimes \mathbb{F}_2[w_2^2, w_4^2, \ldots, w_{2n-2}^2] \longrightarrow H^*BO(2n - \epsilon)$$

is a direct sums of suspensions of (w_1). The sum of these two submodules exhausts the 'interesting' part of $H^*BO(n)$, in the sense that the complementary summand is $E(1)$-free.

THEOREM 5.6.2. *The inclusions*

$$\mathbb{F}_2[w_2^2, w_4^2, \ldots, w_{2n}^2] \oplus (w_1) \otimes \mathbb{F}_2[w_2^2, w_4^2, \ldots, w_{2n-2}^2] \longrightarrow H^*BO(2n)$$

and

$$\mathbb{F}_2[w_2^2, w_4^2, \ldots, w_{2n}^2] \oplus (w_1) \otimes \mathbb{F}_2[w_2^2, w_4^2, \ldots, w_{2n}^2] \longrightarrow H^*BO(2n+1)$$

induce isomorphisms in Q_0 and Q_1 homology.

COROLLARY 5.6.3. *As an $E[Q_0, Q_1]$-module, $H^*BO(n)$ is the sum of trivial modules, suspensions of $H^*BO(1)$, and free modules.*

Proof: The corollary follows by the result of [**5**], which tells us that Q_0 and Q_1 homology detects the stable isomorphism type of the module. □

In principle, this determines the structure of $H^*BO(n)$ as an $E(1)$-module but finding a good parametrization of the complementary $E(1)$-free submodule is non-trivial. The $\mathcal{A}(1)$-module structure is not as simple, as Sq^2 does not annihilate all squares.

The w_{2i}^2 detect Pontrjagin classes p_i of the defining representation and w_1^2 in the (w_1) summand detects the first Chern class of the determinant representation.

COROLLARY 5.6.4. *The Adams spectral sequence converging to $ku^*BO(n)$ collapses at E_2, and the natural homomorphism*

$$ku^*BO(n) \longrightarrow H^*BO(n) \oplus KU^*BO(n)$$

is a monomorphism.

Proof: The trivial summands and the summands isomorphic to (w_1) have Ext modules on which the Bott map is a regular element. Thus, there can be no differentials from the summands generated by the free submodules of $H^*BO(n)$ since these are annihilated by the Bott element. Since the generators of the trivial submodules and the submodules isomorphic to (w_1) lie in degrees congruent to 0 mod 4, and Ext of any such module is zero in degrees congruent to -1 mod 4, they do not support differentials either.

It follows that the homomorphism to the sum of homology and periodic K-theory is a monomorphism at E_∞, and hence is a monomorphism. □

Having given some consequences of 5.6.2, we now prove it.

Proof of Theorem 5.6.2: The Q_0 homology is straightforward and we start with it. Since (w_1) is Q_0-acyclic, the Q_0 homology of the submodule is $\mathbb{F}_2[w_2^2, \ldots, w_{2n}^2]$. It suffices to show that the Q_0 homology of $H^*BO(2n+\epsilon)$ is the same.

First consider $H^*BO(2n+1)$. For $i > 0$, let $u_{2i+1} = Q_0(w_{2i}) = w_{2i+1} + w_1 w_{2i}$. If $R = \mathbb{F}_2[w_1^2, w_2^2, u_3, \ldots, w_{2n}^2, u_{2n+1}]$, then, as an $E[Q_0]$-module, $H^*BO(2n+1)$ is the exterior algebra over R on w_1 and the w_{2i}. Thus, $H^*BO(2n+1)$, with differential Q_0, is the Koszul complex for the regular sequence $(w_1^2, u_3, u_5, \ldots, u_{2n+1})$. Accordingly, its homology is $R/(w_1^2, u_3, u_5, \ldots, u_{2n+1})$, as claimed.

For $H^*BO(2n)$ a more ad hoc argument is needed. As a module over $E[Q_0]$,

$$H^*BO(2n) = \mathbb{F}_2[w_1, w_{2n}] \otimes \bigotimes_{i=1}^{n-1} \mathbb{F}_2[w_{2i}, u_{2i+1}]$$

with $Q_0(w_{2n}) = w_1 w_{2n}$ and $Q_0(w_{2i}) = u_{2i+1}$. By the Künneth theorem, the Q_0 homology is then $\mathbb{F}_2[w_{2n}^2] \otimes \mathbb{F}_2[w_2^2, \ldots, w_{2n-2}^2]$ as desired.

For the Q_1 homology, we wish to show that the inclusions

$$\mathbb{F}_2[w_1, w_2^2, w_4^2, \ldots w_{2n}^2] \longrightarrow H^*BO(2n+1)$$

and

$$\mathbb{F}_2[w_2^2, w_4^2, \ldots w_{2n}^2] \oplus (w_1) \otimes \mathbb{F}_2[w_2^2, w_4^2, \ldots w_{2n-2}^2] \longrightarrow H^*BO(2n)$$

induce Q_1 homology isomorphisms. Since the Q_1 homology of $\mathbb{F}_2[w_1]$ is the exterior algebra on w_1^2, this is equivalent to showing that

$$H(H^*BO(2n+1), Q_1) = E[w_1^2] \otimes \mathbb{F}_2[w_2^2, \ldots, w_{2n}^2]$$

and

$$H(H^*BO(2n), Q_1) = E[w_1^2] \otimes \mathbb{F}_2[w_2^2, \ldots, w_{2n}^2]/(w_1^2 w_{2n}^2)$$

In the limit as n goes to infinity this is easy, so we do this first, then modify the proof to handle the finite cases. Recall that

$$\begin{aligned} Q_1(w_{2i}) &= w_{2i+3} + w_1 w_{2i+2} + w_1^2 w_{2i+1} + (w_1^3 + w_1 w_2 + w_3) w_{2i} \\ Q_1(w_{2i-1}) &= w_1 w_{2i+1} + w_1^2 w_{2i} + (w_1^3 + w_1 w_2 + w_3) w_{2i-1}. \end{aligned}$$

For w_1 this simplifies to $Q_1(w_1) = w_1^4$. We make the change of variables $u_1 = w_1$, $u_{2i} = w_{2i}$, $u_3 = w_3 + w_1 w_2 + w_1^3$ and, for $i > 1$,

$$u_{2i+1} = w_{2i+1} + w_1 w_{2i} + w_1^2 w_{2i-1} + u_3 w_{2i-2}.$$

The Q_1 action then becomes $Q_1(u_1) = u_1^4$, $Q_1(u_3) = u_3^2$ and $Q_1(u_{2i}) = u_{2i+3}$, so that H^*BO decomposes as

$$H^*BO = \mathbb{F}_2[u_1] \otimes \mathbb{F}_2[u_3] \otimes \bigotimes_{i=1}^{\infty} \mathbb{F}_2[u_{2i}, u_{2i+3}]$$

as a module over $E[Q_1]$. By the Kunneth theorem, the Q_1 homology is therefore

$$E[u_1^2] \otimes \mathbb{F}_2 \otimes \mathbb{F}_2[u_{2i}^2 \mid i \geq 1] = E[w_1^2] \otimes \mathbb{F}_2 \otimes \mathbb{F}_2[w_{2i}^2 \mid i \geq 1].$$

Now consider $H^*BO(2n+1) = \mathbb{F}_2[u_1, \ldots, u_{2n+1}]$. We still have $Q_1(u_{2n-2}) = u_{2n+1}$ but now $Q_1(u_{2n}) = Q_1(w_{2n}) = u_1^2 w_{2n+1} + u_3 u_{2n}$. The precise expression for w_{2n+1} in terms of the u_i is of no consequence. All that matters is that w_{2n+1} is u_{2n+1} modulo decomposables. Let

$$R = \mathbb{F}_2[u_1, u_2, \ldots, u_{2n-1}, u_{2n+1}],$$

omitting only u_{2n}, and let S be the subalgebra of R generated by the u_{2i}^2, u_1^2 and u_3^2:

$$S = \mathbb{F}_2[u_1^2, u_2^2, u_3^2, u_4^2, u_5, \ldots, u_{2n-2}^2, u_{2n-1}, u_{2n+1}].$$

Then, as an $E[Q_1]$-module, R is the exterior algebra over S on u_1, u_3, and the u_{2i} for $1 \leq i \leq n-1$. The values of Q_1 on these exterior generators are u_1^4, u_3^2, and u_{2i+3}. These form a regular sequence, so the homology of R with respect to Q_1 is $E[u_1^2] \otimes \mathbb{F}_2[u_2^2, u_4^2, \ldots, u_{2n-2}^2]$.

We now filter $H^*BO(2n+1)$ by $F_i = R + Ru_{2n} + \cdots + Ru_{2n}^i$. In the associated graded, half the terms are suspensions of R, whose Q_1 homology we have just calculated. The other half are suspensions of $(R + Ru_{2n})/R$ with differential

$$Q_1(xu_{2n}) = (u_3 x + Q_1(x)) u_{2n}.$$

Writing R as an exterior algebra over S as above, it is simple to check that the kernel and image of Q_1 are both $u_3 S$, so that these terms are exact. It follows that the Q_1 homology of $H^*BO(2n+1)$ is $H(R, Q_1) \otimes \mathbb{F}_2[u_{2n}^2] = E[w_1^2] \otimes \mathbb{F}_2[w_2^2, \ldots, w_{2n}^2]$ as claimed.

Now, for $H^*BO(2n)$, we *redefine* $u_{2n} = w_{2n} + w_1 w_{2n-1}$. The Q_1 action is as before, except for $Q_1(u_{2n-2}) = u_1 u_{2n} + u_3 u_{2n-2}$ and $Q_1(u_{2n}) = (u_3 + u_1^3) u_{2n}$. To simplify the notation, let us write $u = u_{2n-2}$ and $v = u_{2n}$. We then have a decomposition

$$H^*BO(2n) = \mathbb{F}_2[u_1, u_3, u, v] \otimes \bigotimes_{i=1}^{n-2} \mathbb{F}_2[u_{2i}, u_{2i+3}]$$

FIGURE 5.6.1. The matrix representing $Q_1 : R \longrightarrow R$ as an S-linear transformation

	1	u_1	u_3	u	v	u_1u_3	u_1u	u_1v	u_3u	u_3v	uv	u_1u_3u	u_1u_3v	u_1uv	u_3uv	u_1u_3uv
1		u_1^4	u_3^2											$u_1^2v^2$		
u_1						u_3^2					v^2					
u_3						u_1^4										$u_1^2v^2$
u							u_1^4									
v							u_1^2									
u_1u_3															v^2	
u_1u																
u_1v				1	u_1^2											
u_3u				1								u_1^4				
u_3v					1							u_1^2				
uv															u_3^2	
u_1u_3u							1									
u_1u_3v								1	1	u_1^2						
u_1uv											u_1^2					u_3^2
u_3uv																
u_1u_3uv															u_1^2	

FIGURE 5.6.2. Row reduced version of the matrix representing $Q_1 : R \longrightarrow R$ as an S-linear transformation

	1	u_1	u_3	u	v	u_1u_3	u_1u	u_1v	u_3u	u_3v	uv	u_1u_3u	u_1u_3v	u_1uv	u_3uv	u_1u_3uv
1.		u_1^4	u_3^2											$u_1^2v^2$		
2.						u_3^2					v^2					
3.						u_1^4										$u_1^2v^2$
4.											u_1^2					u_3^2
5.				1									u_1^4			
6.					1								u_1^2			
7.							1									
8.								1	1	u_1^2						
9.															1	

FIGURE 5.6.3. The Koszul complex for the regular sequence (u_1^2, u_3^2, v^2)

$$0 \longleftarrow S \xleftarrow{\begin{bmatrix} u_1^2 & u_3^2 & v^2 \end{bmatrix}} S^3 \xleftarrow{\begin{bmatrix} u_3^2 & v^2 & 0 \\ u_1^2 & 0 & v^2 \\ 0 & u_1^2 & u_3^2 \end{bmatrix}} S^3 \xleftarrow{\begin{bmatrix} v^2 \\ u_3^2 \\ u_1^2 \end{bmatrix}} S \longleftarrow 0$$

as $E(1)$-modules. The homology of $\mathbb{F}_2[u_{2i}, u_{2i+3}]$ is $\mathbb{F}_2[u_{2i}^2]$. The first factor is a bit more complicated, but we may again write $R = \mathbb{F}_2[u_1, u_3, u, v]$ as the exterior algebra over $S = \mathbb{F}_2[u_1^2, u_3^2, u^2, v^2]$ using the S-linearity of Q_1. The matrix representing Q_1 is displayed in Figure 5.6.1 and its row reduced version in Figure 5.6.2. It is elementary to check that the S-module $\mathrm{im}(Q_1)$ has the following generators:

(i) u_1^4, u_3^2, $u_1^2v^2$
(ii) $u_1v + u_3u$
(iii) $u_1^3v + u_3v$
(iv) $u_1^4u + u_1^2v + u_1u_3u$
(v) u_1u_3v
(vi) $u_3^2u_1 + u_1^4u_3$
(vii) $v^2u_1 + u_1^2u_1uv$
(viii) $u_1^2v^2u_3 + u_3^2u_1uv$
(ix) $v^2u_1u_3 + u_3^2uv + u_1^2u_1u_3uv$

We will see that the kernel has the same generators, except for item (i), which should be replaced by the single element 1. The kernel, $\ker(Q_1)$, is defined by equations (1)-(9) of the row reduced matrix (Figure 5.6.2). If we write a_x for the coefficient of the monomial x in the exterior algebra on u_1, u_3, u, and v, then the row reduced form shows that a_1, a_{u_3u}, a_{u_3v}, $a_{u_1u_3u}$ and $a_{u_1u_3v}$ can be chosen freely, and determine a_u, a_v, a_{u_1u}, a_{u_1v} and a_{u_3uv} by equations (5)-(9). This gives the generator 1, together with the generators listed in items (ii)-(iv) above.

Equation (1) says that

$$u_1^4a_{u_1} + u_3^2a_{u_3} + u_1^2v^2a_{u_1uv} = 0$$

which implies that a_{u_3} is divisible by u_1^2. If we write $a_{u_3} = u_1^2a'_{u_3}$, then

$$u_1^2a_{u_1} + u_3^2a'_{u_3} + v^2a_{u_1uv} = 0.$$

Similarly, equation (3) contains a factor of u_1^2 which we can cancel. Having done so, equation (1) says that $(a_{u_1}, a'_{u_3}, a_{u_1uv})$ lies in the kernel at degree 1 of the Koszul complex in Figure 5.6.3, while $(a_{u_1u_3}, a_{uv}, a_{u_1u_3uv})$ lies in the kernel at degree 2 in this complex. Since (u_1^2, u_3^2, v^2) is a regular sequence, the complex is exact, and these kernels are the images of the preceding maps. This gives generators (v)-(ix) for $\ker(Q_1)$ listed above.

It follows that the homology of $\mathbb{F}_2[u_1, u_3, u, v]$ with respect to Q_1 is

$$\mathbb{F}_2[w_1^2, u^2, v^2]/(w_1^4, w_1^2v^2) = \mathbb{F}_2[w_1^2, w_{2n-2}^2, v^2]/(w_1^4, w_1^2v^2).$$

Since $v^2 = w_{2n}^2 + w_1^2w_{2n-1}^2$, we need only observe that

$$w_1^2w_{2n-1}^2 = Q_1(w_{2n-2}w_{2n-1} + w_{2n}\sum_{i=0}^{n-2} w_1^{2i+1}w_{2(n-i)-4})$$

to see that we may replace v by w_{2n} so that the Q_1 homology of $H^*BO(2n)$ is as claimed. □

REMARK 5.6.5. Note that $\mathbb{F}_2[u_1, u_3, u, v]$ is a regraded version of $H^*BO(4)$: $u = u_2$ and $v = u_4$ when $2n = 4$. Thus, the preceding argument shows that all the complexity in $H^*BO(2n)$ is already present in $H^*BO(4)$.

We now turn to the special cases in which we have more information.

5.6.A. The group $O(1)$. Notational note: the generator $c = c_1(\det)$ for the orthogonal group $O(1)$ is the same as the generator $y = c_1(\lambda_1)$ for the elementary abelian group $O(1)$. We have chosen to make our notation consistent with $O(n)$ in this section.

Since this is the rank 1 elementary abelian 2-group, we have many ways to do this calculation. The simplest is probably to cite [**28**, Section 2.2] for $ku^*(BO(1)) = ku^*[[c]]/(vc^2 - 2c)$ and then observe that the pullback square

$$\begin{array}{ccc} ku^*_{O(1)} & \longrightarrow & KU^*_{O(1)} = KU^*[c]/(vc^2 - 2c) \\ \downarrow & & \downarrow \\ ku^*BO(1) = ku^*[[c]]/(vc^2 - 2c) & \longrightarrow & KU^*BO(1) = KU^*[[c]]/(vc^2 - 2c) \end{array}$$

says that $ku^*_{O(1)} = ku^*[c]/(vc^2 - 2c)$. Here $c = c_1^{ku}(\lambda_1) \in ku^2_{O(1)}$ and $RU(O(1)) = \mathbb{Z}[c^R]/((c^R)^2 - 2c^R)$ with $c^R = vc = 1 - \lambda_1$.

We also have multiple ways to get the real connective K-theory, of which the Bockstein spectral sequence is the most direct.

THEOREM 5.6.6. *There are unique elements $p_0 \in ko^0_{O(1)}$ and $p_1 \in ko^4_{O(1)}$ with complexifications $c(p_0) = vc$ and $c(p_1) = c^2$. The ring*

$$ko^*_{O(1)} = ko^*[p_0, p_1]/(\eta p_1, \alpha p_1 - 4p_0, \beta p_1 - \alpha p_0, p_0 p_1 - 2p_1, p_0^2 - 2p_0)$$

Proof: The Bockstein spectral sequence $ku^*_{O(1)}[\widetilde{\eta}] \Longrightarrow ko^*_{O(1)}$ is entirely determined by representation theory since $ku^*_{O(1)}$ has no v-torsion. Writing c for c^R for brevity, the d^1 differential along each diagonal of slope -1 is a truncation at the right of the sequence

$$\cdots \xleftarrow{0} (c^4) \xleftarrow{\cong} (c^3) \xleftarrow{0} (c^2) \xleftarrow{\cong} (c) \xleftarrow{0} RU \xleftarrow{2} RU \xleftarrow{0} RU \xleftarrow{2} \cdots$$

since we have $1 + \tau = 2$, $1 - \tau = 0$, and $2c^i = c^{i+1}$. The second quadrant is then $E^2_{0,4i} = E^\infty_{0,4i} = (c^{2i})$ and zero elsewhere. In the first quadrant the $E^2 = E^3$ term is

$$\begin{array}{ccccccccc} \cdot & \cdot & \cdot & \cdot & RU/2 & \cdot & \cdot & \cdot & RU/2 \\ \cdot & \cdot & \cdot & RU/2 & \cdot & \cdot & \cdot & RU/2 & \cdot \\ \cdot & \cdot & RU/2 & \cdot & \cdot & \cdot & RU/2 & \cdot & \cdot \\ \cdot & RU/2 & \cdot & \cdot & \cdot & RU/2 & \cdot & \cdot & \cdot \\ RU & \cdot & \cdot & \cdot & RU & \cdot & \cdot & \cdot & RU \\ \hline 0 & & 2 & & 4 & & 6 & & 8\cdots \end{array}$$

As in Sections 4.2 and 4.3 we write $d^3 = \delta_0$ and $d^3 = \delta_4$ for the differentials leaving the 0-diagonal and 4-diagonal, respectively. Since all the representations are real for $O(1)$, $\delta_0 = 0$, while $d^3 = \delta_4 : RU \twoheadrightarrow RU/2$ from the 0-line and $d^3 = \delta_4 : RU/2 \cong RU/2$ above it. Thus, in the second quadrant, the E^∞ term is

$$\begin{array}{ccccccccccc} \cdot & \cdot & \cdot & \cdot & \cdot & \cdot & \cdot & \cdot & \cdot & \cdot & \cdot \\ \cdot & \cdot & \cdot & \cdot & \cdot & \cdot & \cdot & \cdot & \cdot & \cdot & \cdot \\ \cdot & \cdot & RO/2 & \cdot & \cdot & \cdot & \cdot & \cdot & \cdot & \cdot & RO/2 \\ \cdot & RO/2 & \cdot & \cdot & \cdot & \cdot & \cdot & \cdot & \cdot & RO/2 & \cdot \\ RO & \cdot & \cdot & \cdot & RSp & \cdot & \cdot & \cdot & RO & \cdot & \cdot \\ \hline 0 & & 2 & & 4 & & 6 & & 8 & & 10\cdots \end{array}$$

Then $p_0 = 1 - \lambda_1 \in ko^0_{O(1)}$ maps to $vc = 1 - \lambda_1 \in ku^0_{O(1)}$ and $p_1 = 2(1 - \lambda_1) \in JSp = ko^4_{O(1)}$ maps to $c^2 \in ku^4_{O(1)}$ and these generate, as a ring. The relations are easily checked by representation theory and easily seen to be complete. □

In terms of representation theory, this can be written as follows.

COROLLARY 5.6.7. *$O(1)$-equivariant connective real K-theory has coefficient ring*

$$ko_i^{O(1)} = \begin{cases} RSp & i = 8k+4 \geq 0 \\ RO/2 & i = 8k+2 \geq 0 \\ RO/2 & i = 8k+1 \geq 0 \\ RO & i = 8k \geq 0 \\ JSp_k = JSp^k & i = -4k \\ 0 & \text{otherwise} \end{cases}$$

Finally, we justify the notation p_i by computing the restriction from $Sp(1)$.

THEOREM 5.6.8. *The restriction $ku^*_{Sp(1)} \longrightarrow ku^*_{O(1)}$ is:*

$$z = p_1(\lambda_1) \mapsto p_1, \qquad \alpha z \mapsto 4p_0, \qquad \text{and} \qquad \beta z \mapsto \alpha p_0,$$

Proof: That z maps to p_1 is evident by comparison with ku^*. The rest follows by the relations in $ko^*_{O(1)}$. □

Thus, p_1 really is the first Pontrjagin class of the quaternionic representation induced up from the defining representation of $O(1)$, while p_0 is a genuinely real class. We call it p_0 because of the relations which tie it to p_1.

5.6.B. The group $O(2)$. Specializing to $n = 2$, Proposition 5.6.1 can be rewritten

COROLLARY 5.6.9. $KU^*_{O(2)} = KU^*[c, c_2]/(vc^2 - 2c, cc_2)$

Proof: Proposition 5.6.1 says that $KU^*_{O(2)} = KU^*[c, c_1]/(vc^2 - 2c, c(2 - vc_1))$. Now

$$\begin{aligned} v^2c_2 &= 1 - \lambda_1 + \lambda_2 \\ &= 1 - (2 - vc_1) + (1 - vc) \\ &= v(c_1 - c) \end{aligned}$$

so that $c_1 = c + vc_2$. Then the relation $0 = c(2 - vc_1)$ becomes $v^2cc_2 = 0$ since $c(2 - vc) = 0$. □

The connective K-theory is similar but somewhat larger.

THEOREM 5.6.10. $ku^*_{O(2)} = ku^*[c, c_2]/(vc^2 - 2c, 2cc_2, vcc_2)$

We will prove this shortly, but we first describe the representation theory in a way which exhibits $O(2)$ as the 'continuous dihedral group' and makes clear its relation with the finite dihedral groups as in Chapter 8. The description also facilitates calculation of the JU_i.

The orthogonal group $O(2)$ consists of the reflections s_θ in a line making an angle θ with the x axis, and the rotations r_θ through an angle θ. The reflections fall into a single conjugacy class. The rotations r_0 and r_π are central, and the other conjugacy classes are the pairs $\{r_\theta, r_{-\theta}\}$. The subgroup $SO(2)$ of rotations is normal and of index 2.

The only one dimensional representations of $O(2)$ are the trivial representation ϵ and the determinant representation $\delta = \lambda_2$. The complexification $\sigma_1 = \lambda_1$ of the natural representation on $\mathbb{R}^2$ is simple, and the other two dimensional simple representations are σ_n given by pulling back σ_1 from the quotients $O(2)/C_n$ for $n = 1, 2, \ldots$. The character of σ_n is the function a_n vanishing on reflections and given by

$$\chi_n(r_\theta) = 2\cos(n\theta).$$

This gives all simple representations. It is easy to see that $\delta\sigma_n = \sigma_n$, generalizing $\delta\lambda_1 = \lambda_1$. The formula

$$\sigma_k \sigma_l = \sigma_{k+l} + \sigma_{k-l}$$

is easy to check by expanding $(e^{ik\theta} + e^{-ik\theta})(e^{il\theta} + e^{-il\theta})$. Here $\sigma_0 = 1 + \delta$. It follows that the Chern filtration is entirely determined by its first two terms.

Let us write elements of the augmentation ideal $JU_1 \subset RU(O(2))$ as pairs whose first coordinate is the character value on r_θ and whose second coordinate is the character value on s_θ. If we let $z = e^{i\theta}$ then $\chi_n(r_\theta) = z^n + z^{-n}$. Writing $q_n = 2 - z^n - z^{-n}$, we have

$$\begin{array}{rclcl} c_1^R(\delta) & = & 1-\delta & = & (0,2) \\ c_1^R(\sigma_n) & = & 2-\sigma_n & = & (q_n,2) \\ c_2^R(\sigma_n) & = & 1+\delta-\sigma_n & = & (q_n,0) \end{array}$$

PROPOSITION 5.6.11. *JU_n is the ideal generated by $c_1^R(\delta)^n = (0, 2^n)$ and $c_2^R(\sigma_1)^i = (q_1^i, 0)$, where i is the integer part of $(n+1)/2$.*

Proof: JU_1 is generated by $(0,2)$ and the $(q_n, 0)$ for $n \geq 1$. But q_n is a polynomial in q_1 with constant term 0, so JU_1 is the ideal generated by $(0,2)$ and $(q_1, 0)$. Since JU_2 is generated by JU_1^2 and second Chern classes, the Proposition is true for JU_2 as well. The remaining cases follow by induction from $JU_n = \sum JU_i JU_{n-i}$. □

Theorem 5.6.2 gives the submodule $(w_1) \oplus \mathbb{F}_2[w_2^2] \subset H^*BO(2)$. Calculating the complementary free submodule gives

LEMMA 5.6.12. *As an $E(1)$-module, $H^*BO(2)$ is the direct sum of*

(i) *$H^*BO(1) = (w_1)$*
(ii) *a trivial $E(1)$-module $F_2[w_2^2]$*
(iii) *a free $E(1)$-module $\mathbb{F}_2[w_1^2, w_2^2]\langle w_2\rangle$*

Proof: It is a simple matter to verify that the sum of these modules gives all of $H^*BO(2)$ and a check of Poincare series shows that the sum must be direct. Note that $\langle w_2\rangle = \{w_2, w_1w_2, w_1^3w_2 + w_1w_2^2, w_1^2w_2^2\}$. □

We can now prove that $ku^*_{O(2)}$ is as stated in Theorem 5.6.10.

Proof: First consider the associated graded of the completion given by the Adams spectral sequence. By Corollary 5.6.4 there is a monomorphism

$$ku^*BO(2) \longrightarrow KU^*BO(2) \oplus H\mathbb{F}_2^*BO(2)$$

which acts as follows:

$$\begin{array}{ccc} c & \mapsto & (0,2,w_1^2) \\ c_1 & \mapsto & (q_1,2,w_1^2) \\ c_2 & \mapsto & (q_1,0,w_2^2) \end{array}$$

Hence $c_1 = c + vc_2$ in connective, as well as periodic, complex K-theory. Now, c_2 and c map to $c_2^H = w_2^2$ and $c_1^H = w_1^2$, respectively. Hence, the decomposition in the lemma shows that the Adams spectral sequence converging to $ku^*BO(2)$ has E_2 term

$$\mathrm{Ext}^{*,*}_{E(1)}(\mathbb{F}_2, H^*BO(2)) = \mathbb{F}_2[h_0,v][c]/(vc^2 - h_0c) \oplus \mathbb{F}_2[h_0,v][c_2] \oplus \mathbb{F}_2[c,c_2] < cc_2 >$$

Here we are using the fact that the submodule $\langle w_2 \rangle$ has socle generated by $w_1^2w_2^2$, detecting cc_2. This decomposition shows that the $E_2 = E_\infty$ term, and hence $ku^*BO(2)$, are generated by c and c_2 as $\mathrm{Ext}^{*,*}_{\mathcal{A}(1)}(\mathbb{F}_2,\mathbb{F}_2)$ and ku^* algebras, respectively.

The relations $vc^2 = 2c$, $2cc_2 = 0$ and $vcc_2 = 0$, are immediate from the monomorphism into mod 2 cohomology plus periodic K-theory. These are the only relations since the Adams spectral sequence shows that, additively, $ku^{4i+2}BO(2)$ is spanned by c^{2i+1}, $c^{2i-1}c_2$, $\ldots$, cc_2^i and vc_2^{i+1}, $v^3c_2^{i+2}$, $\ldots$, while $ku^{4i}BO(2)$ is spanned by c^{2i}, $c^{2i-2}c_2$, $\ldots$, c_2^i, $v^2c_2^{i+1}$, $\ldots$.

Now, the pullback square assumes the form

$$\begin{array}{ccc} ku^*_{O(2)} & \longrightarrow & KU^*[c,c_2]/(vc^2-2c, cc_2) \\ \downarrow & & \downarrow \\ ku^*[[c,c_2]](vc^2-2c, vcc_2, 2cc_2) & \longrightarrow & KU^*[[c,c_2]](vc^2-2c, cc_2) \end{array}$$

from which the theorem is immediate. □

COROLLARY 5.6.13. *The image $v^n ku^{2n}_{O(2)} \subseteq ku^0_{O(2)}$ is $JU_n \subseteq RU(O(2))$. That is, $QU = \mathrm{im}(ku^*_{O(2))} \longrightarrow K^*_{O(2))})$ is the modified Rees ring of $RU(O(2))$ with respect to the Chern filtration.*

Now we turn to the real case. The Bockstein spectral sequence has a filtration shift which leaves some extensions in doubt, so we shall combine the Bockstein spectral sequence with the Adams spectral sequence to resolve these.

First, note that when all representations are real, the Bockstein spectral sequence for periodic K-theory behaves just as it does for $O(1)$. The d^1 is 0 or 2 on the n-diagonals for $n \equiv 0$ or 2 modulo 4, respectively. Then $\delta_0 = 0$ and δ_4 is an isomorphism above the 0-line and the natural epimorphism $RU \twoheadrightarrow RU/2$ from the 0-line. (See the proof of Theorem 5.6.6 for charts of this.) The complications arise in negative degrees.

LEMMA 5.6.14. *$d^1 = cR : ku^{s-t}_{O(2)} \longrightarrow ku^{s-t+1}_{O(2)} \cdot \widetilde{\eta}$ is determined by*

(i) $d^1(v) = 2\widetilde{\eta}$
(ii) $d^1(c) = c^2\widetilde{\eta}$
(iii) $d^1(c_2) = cc_2\widetilde{\eta}$

Proof: The first is generic (Section 4.2). The second follows by the comparison with periodic K-theory as noted above using $v^2c^2 = 2vc$. In the third case, $d^1 = 1-\tau = 0$

in periodic K-theory. Hence $d^1(c_2)$ lies in the kernel of $ku^4_{O(2)} \longrightarrow K^4_{O(2)}$, which is spanned by cc_2. Since c_2 maps to $w_2^2 \in H^4BO(2)$ and $Sq^2(w_2^2) = w_1^2w_2^2$, Lemma 4.3.1 implies that $d^1(c_2) = cc_2$. $\square$

It follows that the d^1 differential is diagonal with respect to the decomposition of $ku^*_{O(2)}$ by powers of c_2, according to the pattern:

$$ku^{16} \longleftarrow ku^{14} \longleftarrow ku^{12} \longleftarrow ku^{10} \longleftarrow ku^{8} \longleftarrow ku^{6} \longleftarrow ku^{4} \longleftarrow ku^{2}$$

$$\begin{array}{cccc}
c^8 \xleftarrow{\cong} c^7 & c^6 \xleftarrow{\cong} c^5 & c^4 \xleftarrow{\cong} c^3 & c^2 \xleftarrow{\cong} c \\
c^6c_2 \quad c^5c_2 \xleftarrow{\cong} c^4c_2 & c^3c_2 \xleftarrow{\cong} c^2c_2 & cc_2 \twoheadleftarrow c_2 & \xleftarrow{2} vc_2 \\
c^4c_2^2 \xleftarrow{\cong} c^3c_2^2 & c^2c_2^2 \xleftarrow{\cong} cc_2^2 & c_2^2 \xleftarrow{2} vc_2^2 & v^2c_2^2 \xleftarrow{2} v^3c_2^2 \\
c^2c_2^3 \quad cc_2^3 \twoheadleftarrow c_2^3 & \xleftarrow{2} vc_2^3 & v^2c_2^3 \xleftarrow{2} v^3c_2^3 & v^4c_2^3 \xleftarrow{2} v^5c_2^3
\end{array}$$

The resulting E^2 is equal to E^3 for dimensional reasons.

LEMMA 5.6.15. *The E^3 term of the Bockstein spectral sequence for $ko^*_{O(2)}$ is as folllows. In non-negative degrees it is nonzero only along diagonals congruent to 0 mod 4, where it is $R = RU(O(2))$ on row 0 and $R/2$ in higher rows. In negative degrees congruent to 0 mod 4, row 0 is the subalgebra generated by c^2, c_2^2, and $2c_2$, together with all $v^{2i}c_2^{2i+j}$, $j \geq 0$ and $i > 0$. In negative degrees congruent to 2 mod 4, row 0 is $\mathbb{F}_2[c^2, c_2^2] \cdot cc_2$. Above row 0 in negative degrees it is the $\mathbb{F}_2$ vector space spannned by c_2^4 and the $v^{2i}c_2^{2i+j}$.*

Proof: Inspection of the pattern of differentials displayed before the lemma is sufficient to show this. $\square$

In non-negative degrees, the d^3 differential is nonzero only on diagonals congruent to 4 mod 8 where it is the quotient map $R \longrightarrow R/2$ from row 0 and is the isomorphism $R/2 \longrightarrow R/2$ above row 0. Since multiplication by the Bott class is a monomorphism in degrees congruent to 0 mod 4, this determines all the d^3 differentials. (The classes in degrees 2 mod 4 have nothing to hit.) This gives the E^∞ term of the Bockstein spectral sequence. We write $RO = RO(O(2))$, $RSp = RSp(O(2))$, and note that $RO = R = RU(O(2))$ here.

PROPOSITION 5.6.16. *The non-zero values of the E^∞-term of the Bockstein spectral sequence converging to $ko^*_{O(2)}$ are as follows.*

- *If $n \geq 0$ then $E^\infty_{0,8n} = RO$ and $E^\infty_{s,8n+s} = RO/2$ for $s = 1, 2$.*
- *If $n \geq 0$ then $E^\infty_{0,8n+4} = 2R = RSp$.*
- *If $n < 0$ then $E^\infty_{0,8n}$ is*
  $\mathbb{Z}\langle c^{4n}, v^{2i}c_2^{2n+i} \mid i \geq 0\rangle \oplus \mathbb{F}_2\langle c^{4i}c_2^{2j} \mid i+j = n; i,j > 0\rangle$.
- *If $n < 0$ then $E^\infty_{s,8n+s} = \mathbb{F}_2\langle v^{2i}c_2^{2n+i} \mid i \geq 0\rangle$ for $s = 1, 2$.*

- *If $n < 0$ then $E^{\infty}_{0,8n+4}$ is* $\mathbb{Z}\langle c^{4n+2}, 2v^{2i}c_2^{2n+i+1} \mid i \geq 0\rangle \oplus \mathbb{F}_2\langle c^{4i+2}c_2^{2j} \mid i+j=n; i \geq 0, j > 0\rangle$.
- *In negative degrees congruent to 2 mod 4, $E^{\infty}_{0,*}$ is $\mathbb{F}_2[c^2, c_2^2] \cdot cc_2$ concentrated in row 0.*

In degrees not congruent to 1 or 2 mod 8, $ko^*_{O(2)} \longrightarrow ku^*_{O(2)}$ is monomorphism.

Proof: The values of E^{∞} are immediate from the d^3 differential, which is determined by representation theory. The result is concentrated along row 0 except in degrees congruent to 1 or 2 mod 8, which implies that $ko^*_{O(2)} \longrightarrow ku^*_{O(2)}$ is a monomorphism except in these degrees. □

COROLLARY 5.6.17. *$ko^*_{O(2)}$ is generated as a ko^*-algebra by elements p_0, p_1, p_2, r_0, r_1 and s, with complexifications $p_0 \mapsto vc$, $p_1 \mapsto c^2$, $p_2 \mapsto c_2^2$, $r_0 \mapsto v^2c_2$, $r_1 \mapsto 2c_2$, and $s \mapsto cc_2$, respectively.*

Proof: The images of the six elements listed generate the E^{∞} term of the Bockstein spectral sequence. □

By the preceding corollary, their complexifications characterize all these generators except s. In the proof of the following theorem we will see that there is a unique choice of s satisfying $\beta s = \eta^2 r_0$.

THEOREM 5.6.18. $KO^*_{O(2)} = KO^*[p_0, r_0]/(p_0^2 - 2p_0, p_0r_0)$ *and*

$$ko^*_{O(2)} = ko^*[p_0, p_1, p_2, r_0, r_1, s]/I$$

where I is the ideal generated by the relations given in Figure 5.6.4. Complexification $KO^*_{O(2)} \longrightarrow KU^*_{O(2)}$ sends p_0 to vc and r_0 to v^2c_2.

Proof: First, we observe that the Bockstein spectral sequence implies that $ko^*_{O(2)}$ is $ko^*[p_0, r_0]/(p_0^2 - 2p_0, p_0r_0)$ in positive degrees, while in negative degrees it is as follows:

- $ko^{8n} = \mathbb{Z}\langle p_1^{2n}, r_0^i p_2^n\rangle \oplus \mathbb{F}_2\langle p_1^{2i}p_2^j \mid i > 0, j > 0, i+j=n\rangle$
- $ko^{8n+4} = \mathbb{Z}\langle p_1^{2n+1}, r_0^i r_1 p_2^n\rangle \oplus \mathbb{F}_2\langle p_1^{2i+1}p_2^j \mid j > 0, i+j=n\rangle$
- $ko^{8n-1} = \mathbb{F}_2\langle \eta r_0^i p_2^n\rangle$
- $ko^{8n-2} = \mathbb{F}_2\langle \eta^2 r_0^i p_2^n\rangle \oplus \mathbb{F}_2\langle p_1^i p_2^j s \mid i+2j = 2(n-1)\rangle$
- $ko^{8n+2} = \mathbb{F}_2\langle p_1^i p_2^j s \mid i+2j = 2n-1\rangle$

This is sufficient to determine $KO^*_{O(2)}$.

Next, we prove that the relations in Figure 5.6.4 must hold. By Proposition 5.6.16, relations in degrees not congruent to 1 or 2 mod 8 can be checked simply by verifying that they become true in $ku^*_{O(2)}$. This leaves $\beta s = \eta^2 r_0$, $p_0 s = 0$, and $r_0 s = \eta^2 p_2$. By Theorem 2.9.5, $ko^6_{O(2)} = JU_2(O(2))/JSp(O(2))$. Since $vc^2 = 2c$,

$$JU_2 = \mathbb{Z}\langle 2c, c_2, c_2^2, \ldots\rangle$$

and

$$JSp = \mathbb{Z}\langle 2c, 2c_2, 2c_2^2, \ldots\rangle.$$

Hence, $ko^6_{O(2)} = \mathbb{F}_2\langle c_2, c_2^2, \ldots\rangle$. The elements $\eta^2 r_0^i p_2$ give the c_2^i with $i > 1$, and map to 0 in $ku^*_{O(2)}$, so s can be chosen to correspond to c_2. Then, since the map

FIGURE 5.6.4. Relations among the generators of $ko^*_{O(2)}$

$\eta p_1 = 0$	$\alpha p_1 = 4p_0$	$\beta p_1 = \alpha p_0$
$\eta r_1 = 0$	$\alpha r_1 = 4r_0$	$\beta r_1 = \alpha r_0$
$\eta s = 0$	$\alpha s = 0$	$\beta s = \eta^2 r_0$
$p_0 p_2 = 0$	$p_1 p_2 = s^2$	

	p_0	p_1	r_0	r_1	s
p_0	$2p_0$	$2p_1$	0	0	0
p_1	$2p_1$	p_1^2	0	0	$p_1 s$
r_0	0	0	βp_2	αp_2	$\eta^2 p_2$
r_1	0	0	αp_2	$4p_2$	0

$\beta : ko^6_{O(2)} \longrightarrow ko^{-2}_{O(2)}$ is the natural inclusion $JU_2/JSp \longrightarrow RU/RSp$, $\beta s = \eta^2 r_0$, since this is detected by $v^2 c_2 \widetilde{\eta}^2$. Note that the complexification only determines s modulo the $\eta^2 r_0^i p_2$, but the image in $ko^{-2}_{O(2)}$ (and hence in $KO^6_{O(2)}$) determines a unique class with complexification cc_2 which satisfies $\beta s = \eta^2 r_0$.

Now $p_0 s$ maps to $vc^2 c_2 = 0$ in $ku^6_{O(2)}$, so $p_0 s$ must be detected on the 2-line of the Bockstein spectral sequence. Hence, $p_0 s$ must be a sum of terms $\eta^2 r_0^i p_2$. Multiplication by β is a monomorphism on the span of the $\eta^2 r_0^i p_2$, so it is sufficient to compute $\beta p_0 s = \eta^2 p_0 r_0 = 0$.

Similarly, $r_0 s$ is not detected on the 0-line of the Bockstein spectral sequence, so can also be written as a sum of the $\eta^2 r_0^i p_2$. Now, multiplying by β gives $\beta r_0 s = \eta^2 r_0^2$, which is detected by $v^4 c_2^2 \widetilde{\eta}^2$, as is $\beta \eta^2 p_2$. Hence $r_0 s = \eta^2 p_2$.

Finally, the relations listed are sufficient to write all products in terms of the spanning set listed for $ko^*_{O(2)}$, so our list of relations is complete. □

REMARK 5.6.19. p_0 and r_0 are $1 - \det$ and the Euler class, respectively, and p_1 and r_1 are their images in $JSp = ko^4$. This explains the similarity of the action of ko^* on them. The class p_2 refines the square of the Euler class in the sense that $r_1^2 = 4p_2$, $r_0 r_1 = \alpha p_2$ and $r_0^2 = \beta p_2$. The class s is a square root of the product $p_1 p_2 = s^2$.

Since the relation $\beta s = \eta^2 r_0$ is hidden in the Bockstein spectral sequence, some other means of detecting it is required. The representation theoretic argument just given is the simplest means, but this relation is also accesible through the Adams spectral sequence. Since the Adams spectral sequence argument can be applied in degrees where the representation theoretic description $ku^6 = JU_2/JSp$ is not available, we will give it as well. To do this, we start by decomposing $H^*BO(2)$ as an $\mathcal{A}(1)$-module,

LEMMA 5.6.20. *As an $\mathcal{A}(1)$-module, $H^*BO(2)$ is the direct sum of*

(i) $H^*BO(1) = (w_1)$
(ii) *a trivial $\mathcal{A}(1)$-module* $\mathbb{F}_2[w_2^4]$
(iii) *jokers* $\mathbb{F}_2[w_2^4] \cdot \{w_2, w_1 w_2, w_2^2, w_1 w_2^2 + w_1^3 w_2, w_1^2 w_2^2\}$
(iv) *a free $\mathcal{A}(1)$-module* $\mathbb{F}_2[w_1^4, w_2^4]\langle w_1^2 w_2, w_2^3\rangle$

Proof: It is a simple matter to verify that the sum of these modules gives all of $H^*BO(2)$ and a check of Poincaré series shows that the sum must be direct. □

PROPOSITION 5.6.21. *The Adams spectral sequence*

$$\mathrm{Ext}^{*,*}_{\mathcal{A}(1)}(\mathbb{F}_2, H^*BO(2)) \Longrightarrow ko^*_{O(2)}$$

collapses at E_2. *The natural map*

$$ko^*_{O(2)} \longrightarrow KO^*_{O(2)} \oplus H\mathbb{F}_2^* BO(2)$$

is a monomorphism.

Proof: The Bott class $b \in \mathrm{Ext}^{4,12}_{\mathcal{A}(1)}(\mathbb{F}_2, \mathbb{F}_2)$ acts monomorphically on Ext of each summand except the $\mathcal{A}(1)$ free summands. (See Appendix A for the Adams spectral sequence charts.) Hence, there can be no nonzero differentials on the classes created by the free summands. Since columns corresponding to total degrees congruent to -1 modulo 4 are all zero, there can be no differentials on classes in degrees congruent to 0 modulo 4. The remaining indecomposable classes are those in the 0-line coming from the $w_1^2 w_2^{4j+2}$ at the tops of the jokers. The Bott map sends these monomorphically into non-negative degrees, where they equal h_1^2 times odd powers of $c_2^R(\sigma_1)$. By the Bockstein spectral sequence calculation above, these latter are nonzero elements of $ko^2_{O(2)}$. The only possible targets for differentials on these classes are h_1-multiples of powers of $c_2^R(\sigma_1)$, all of which are non-zero, again by the Bockstein spectral sequence, ruling out these possible differentials as well.

The monomorphism claim is immediate from the fact that the 0-line is detected in $H\mathbb{F}_2^* BO(2)$ and multiplication by the Bott map is a monomorphism above the 0-line. □

The Adams spectral sequence shows that $\beta : ko^6_{O(2)} \longrightarrow ko^{-2}_{O(2)}$ is a monomorphism, since there are no Bott annihilated classes, i.e., classes coming from free $\mathcal{A}(1)$ summands, in codegree 6. Since the class $s \in ko^6_{O(2)}$ is detected by $cc_2 \in ku^6_{O(2)}$ in the Bockstein spectral sequence, the Adams spectral sequence edge homomorphism (induced by $ko \longrightarrow ku \longrightarrow H\mathbb{F}_2$) implies that s is detected by $w_1^2 w_2^2 \in \mathrm{Ext}^{0,-6}_{\mathcal{A}(1)}(\mathbb{F}_2, H^*BO(2)) \subset H^6 BO(2)$. This is in the bottom joker in $H^*BO(2)$, and $\beta s = \eta^2 r_0$ in that summand (see Appendix A.8.12). That r_0 there coincides with what we have called r_0 here is immediate from complexification. This implies that the image of s in $KO^6_{O(2)}$ must be $\eta^2\beta^{-1}r_0$. Now we can use the monomorphism of the preceding Proposition to complete the calculation of $ko^*BO(2)$.

Either approach makes the hidden extension $\beta s = \eta^2 r_0$ clear. One advantage of the approach we took is that we get $ko^*_{O(2)}$ directly, rather than first computing the completion $ko^*BO(2)$, then computing the pullback square.

We record the images of the natural maps from $ko^*_{O(2)}$ to $ku^*_{O(2)}$, $KO^*_{O(2)}$ and $H\mathbb{F}_2^* BO(2)$ for reference.

$ko^*_{O(2)}$	$ku^*_{O(2)}$	$KO^*_{O(2)}$	$H^*BO(2)$
p_0	vc	p_0	0
p_1	c^2	$\alpha\beta^{-1}p_0$	w_1^4
p_2	c_2^2	$\beta^{-1}r_0^2$	w_2^4
r_0	$v^2 c_2$	r_0	0
r_1	$2c_2$	$\alpha\beta^{-1}r_0$	0
s	cc_2	$\eta^2\beta^{-1}r_0$	$w_1^2 w_2^2$

5.6.C. The group $O(3)$. Specializing to $n = 3$, Proposition 5.6.1 says

COROLLARY 5.6.22. $KU^*_{O(3)} = KU^*[c, c_1]/(vc^2 - 2c)$

It will be helpful to record the following relations.

PROPOSITION 5.6.23. *In $ku^*_{O(3)}$ and $KU^*_{O(3)}$ we have*

$$\lambda_1 = 3 - vc_1 \qquad vc_2 = (1 + vc)(c_1 - c)$$
$$\lambda_3 = 1 - vc \qquad vc_3 = c(c_1 - c)$$
$$\lambda_2 = \lambda_1\lambda_3 = (3 - vc_1)(1 - vc) \qquad c_1 - c = vc_2 - v^2c_3$$

Proof: The first two relations follow from the definitions of c and c_1 as the first Chern classes of the determinant λ_3 and the defining representation λ_1. Then $\lambda_2 = \lambda_1\lambda_3$ gives the third.

The generic relation $v^2c_2 = 3 - 2\lambda_1 + \lambda_2$ gives $v^2c_2 = v(c_1 - c - vc^2 + vcc_1) = v(c_1 - c)(1 + vc)$. We can now divide by v since v acts monomorphically above degree -4. Similarly, $v^3c_3 = 1 - \lambda_1 + \lambda_2 - \lambda_3 = (1 - \lambda_1)(1 - \lambda_3)$ gives $v^3c_3 = v^2cc_1 - v^2c^2 = v^2c(c_1 - c)$, which we can divide by v^2. Finally, $c = c_1 - v^2c_2 + v^3c_3$, as usual. □

Similarly, the fact that all representations of $O(3)$ are real gives

LEMMA 5.6.24. $KO^*_{O(3)} = KO^*[p_0, q_0]/(p_0^2 - 2p_0)$ *where p_0 and q_0 have complexifications vc and vc_1 respectively. The restriction $KO^*_{O(3)} \longrightarrow KO^*_{O(2)}$ sends p_0 to p_0 and q_0 to $p_0 + r_0$.*

Proof: With $R = RO(O(3)) = RU(O(3))$, the fact that all simple representations are real gives that $KO^*_{O(3)}$ is R, $R/2$, $R/2$, 0, $2R$, 0, 0, 0 and repeat with period 8 under multiplication by β, with the evident action of KO^*. We have $R = \mathbb{Z}[vc, vc_1]/((vc)^2 - 2(vc))$, which gives the first statement. The second follows from the fact that the determinant restricts to the determinant, and that c_1 restricts to $c_1 = c + vc_2$. □

Theorem 5.6.2 gives the submodule $(w_1) \otimes \mathbb{F}_2[w_2^2] \oplus \mathbb{F}_2[w_2^2] \subset H^*BO(3)$. Calculating the complementary free submodule gives

LEMMA 5.6.25. *As an $E(1)$-module, $H^*BO(3)$ is the direct sum of*

(i) *$(w_1) \otimes \mathbb{F}_2[w_2^2]$, a sum of suspensions of $H^*BO(1) = (w_1)$*
(ii) *a trivial $E(1)$-module $F_2[w_2^2]$*
(iii) *a free $E(1)$-module $\mathbb{F}_2[w_1^2, w_2^2, w_3^2]\langle w_2, w_3\rangle$*

Proof: It is a simple matter to verify that the sum of these modules gives all of $H^*BO(3)$ and a check of Poincare series shows that the sum must be direct. Note that

$$\langle w_2\rangle = \{w_2, w_1w_2 + w_3, w_1^3w_2 + w_1^2w_3 + w_1w_2^2 + w_2w_3, w_1^2w_2^2 + w_3^2\}$$

and

$$\langle w_3\rangle = \{w_3, w_1w_3, w_1^3w_3 + w_1w_2w_3 + w_3^2, w_1^2w_2w_3 + w_1w_3^2\}.$$

□

The Chern classes aren't sufficient to generate $ku^*BO(3)$. There are torsion classes which come from the free summands generated by w_2 and w_3 under the lifts $\overline{Q}_0 : H \longrightarrow \Sigma H\mathbb{Z}$ and $\overline{Q}_1 : H\mathbb{Z} \longrightarrow \Sigma^3 ku$ of the Milnor primitives Q_0 qnd Q_1.

DEFINITION 5.6.26. Let $q_2 = \overline{Q}_1\overline{Q}_0(w_2) \in ku^6BO(3)$ and $q_3 = \overline{Q}_1\overline{Q}_0(w_2) \in ku^7BO(3)$.

PROPOSITION 5.6.27. *The classes q_2 and q_3 are nonzero classes annihilated by $(2, v)$. The class q_3 is independent of c, c_2, and c_3, while $q_2 = cc_2 - 3c_3$. These are the only nonzero 2 or v-torsion classes in $ku^6BO(3)$ and $ku^7BO(3)$.*

Proof: They are nonzero by reduction to $H^*BO(3)$, where we get $Q_1Q_0(w_2) = w_1^2w_2^2 + w_3^2$ and $Q_1Q_0(w_3) = w_1^2w_2w_3 + w_1w_3^2$. They are v-torsion because of the exact sequence

$$0 \longrightarrow H^3(BG;\mathbb{Z}) \xrightarrow{\overline{Q}_1} ku^6BG \xrightarrow{v} ku^4BG \longrightarrow H^4(BG;\mathbb{Z}) \xrightarrow{\overline{Q}_1} ku^7BG \xrightarrow{v} 0$$

from Corollary 2.9.4. They are 2-torsion because they are in the image of $\overline{Q}_0$. The element q_2 is unique because $H^3(BO(3);\mathbb{Z}) = \mathbb{Z}/2\mathbb{Z}$ generated by $\overline{Q}_0(w_2)$. For q_3, note that $H^4(BO(3);\mathbb{Z})$ is generated by classes with mod 2 reductions w_1^4, w_1w_3 and w_2^2. But w_1^4 and w_2^2 are the images of c^2 and c_2 in $ku^4BO(3)$, respectively, so $\overline{Q}_1(w_1^4) = \overline{Q}_1(w_2^2) = 0$. Thus, $q_3 = \overline{Q}_1(w_1w_3)$ is the unique non-zero class in the image of $\overline{Q}_1$.

Finally, observe that Proposition 5.6.23 gives $v(cc_2 - 3c_3) = c(1 + vc)(c_1 - c) - 3c(c_1 - c) = (c_1 - c)(c + vc^2 - 3c) = 0$, so that $cc_2 - 3c_3$ is a v-torsion class in $ku^6BO(3)$. It is nonzero because its image in $H^*BO(3)$ is $w_1^2w_2^2 + w_3^2$. Hence it equals q_2. □

REMARK 5.6.28. The class q_3 restricts to the class $q_{\{1,2,3\}}$ in $ku^*BV(3)$, where $V(3)$ is the diagonal elementary abelian subgroup of $O(3)$. Recall that $q_S = \overline{Q}_1\overline{Q}_0(x_S)$, where $x_S = \prod_{i\in S} x_i$ in $H\mathbb{F}_2^*BV_n = \mathbb{F}_2[x_1, \ldots, x_n]$.

In general, $ku^*BO(n)$ will contain $(2, v)$-annihilated classes $q_i = \overline{Q}_1\overline{Q}_0(w_i)$ for $i = 2, \ldots, n$. Their restriction to $ku^*BV(n)$ will be $q_i \mapsto \sum_{|S|=i} q_S$ since w_i restricts to $\sum_{|S|=i} x_S$. See also Remark 5.6.32.

THEOREM 5.6.29. *$ku^*BO(3) = ku^*[c, c_2, c_3, q_3]/R$, where R is an ideal containing $(vc^2 - 2c, 2(cc_2 - 3c_3), v(cc_2 - 3c_3), 2q_3, vq_3, vcc_3 - 2c_3)$.*

Proof: Since the Adams spectral sequence collapses, the decomposition of $H^*BO(3)$ as an $E(1)$-module tells us that $ku^*BO(3)$ has an associated graded

$$E^0(ku^*BO(3)) = ku^*[c, c_2]/(vc^2 - 2c) \oplus \mathbb{F}_2[c, c_2, c_3]\langle q_2, q_3\rangle.$$

Note that since $c = c_1 - vc_2 + v^2c_3$, we may use either c or c_1 in the second summand. Since $q_2 = cc_2 - 3c_3$, we conclude that $ku^*BO(3)$ is generated as a ku^*-algebra by c, c_2, c_3 and q_3. The relations listed have already been proved with the exception of the last. The most systematic way to verify it is to use the monomorphism

$ko^*BO(3) \longrightarrow KU^*BO(3) \oplus H^*BO(3)$, which we record in the following table.

$ku^*BO(3)$	$KU^*BO(3)$	$H^*BO(3)$
c	c	w_1^2
c_1	c_1	w_1^2
c_2	$v^{-1}(1+vc)(c_1-c)$	w_2^2
c_3	$v^{-1}c(c_1-c)$	w_3^2
q_2	0	$w_3^2+w_1^2w_2^2$
q_3	0	$Q_1Q_0(w_3)$

Though q_2 is redundant, it can be useful to write $cc_2 = 3c_3 + q_2$ and exploit the fact that $2q_2 = 0 = vq_2$. □

COROLLARY 5.6.30. *In the short exact sequence*

$$0 \longrightarrow TU \longrightarrow ku^*_{O(3)} \longrightarrow QU \longrightarrow 0$$

we have

$$TU = PC\langle q_2, q_3\rangle, \text{ where } PC = \mathbb{F}_2[c_1, c_2, c_3]$$

and

$$QU = RU(G)[v][c_1, c_2, c_3] \subseteq RU(G)[v, v^{-1}] = KU^*_G.$$

□

5.6.D. The group $O(n)$. Lest the reader get the wrong impression from the free summands in $H^*BO(n)$ for $n < 4$, we point out that they begin to get more complicated at $n = 4$. Let us write w_S for the product $\prod_{i\in S} w_i$.

PROPOSITION 5.6.31. *$E(1)$-free summands of $H^*BO(n)$ complementary to the summands in Theorem 5.6.2 are:*

$n = 4$:

$\mathbb{F}_2[w_1^2, w_2^2, w_3^2, w_4^2]\langle w_2, w_3, w_4, w_{234}\rangle$
$\oplus\mathbb{F}_2[w_1^2, w_2^2, w_4^2]\langle w_{24}\rangle$

$n = 5$:

$\mathbb{F}_2[w_1^2, w_2^2, w_3^2, w_4^2, w_5^2]\langle w_2, w_3, w_4, w_5, w_{234}, w_{235}, w_{245}, w_{345}\rangle$
$\oplus\mathbb{F}_2[w_1^2, w_2^2, w_4^2, w_5^2]\langle w_{24}, w_{34}\rangle$

$n = 6$:

$\mathbb{F}_2[w_1^2, w_2^2, w_3^2, w_4^2, w_5^2, w_6^2]\langle w_2, w_3, w_4, w_5, w_6, w_{26}, w_{234}, w_{235}, w_{245}, w_{345}\rangle$
$\oplus\mathbb{F}_2[w_1^2, w_2^2, w_3^2, w_4^2, w_5^2, w_6^2]\langle w_{236}, w_{246}, w_{256}, w_{346}, w_{456}, w_{23456}\rangle$
$\oplus\mathbb{F}_2[w_1^2, w_2^2, w_4^2, w_5^2, w_6^2]\langle w_{24}, w_{34}, w_{2456}\rangle$
$\oplus\mathbb{F}_2[w_1^2, w_2^2, w_4^2, w_6^2]\langle w_{46}\rangle$

REMARK 5.6.32. Each w_S generating a free $E(1)$ will give rise to a $(2, v)$-annihilated class $\overline{Q}_1\overline{Q}_0(w_S) \in ku^*BO(n)$. This follows from the same argument as in Proposition 5.6.27.

5.7. Special orthogonal groups

The representation rings are

$$RU(SO(2n+1)) = \mathbb{Z}[\lambda_1, \dots, \lambda_n]$$

with $\lambda_{n+i} = \lambda_{n+1-i}$ and

$$RU(SO(2n)) = \mathbb{Z}[\lambda_1, \dots, \lambda_{n-1}, \lambda_n^+, \lambda_n^-]/R$$

with $\lambda_{n+i} = \lambda_{n-i}$ and $\lambda_n = \lambda_n^+ + \lambda_n^-$. The ideal R is generated by one relation

$$(\lambda_n^+ + \sum_i \lambda_{n-2i})(\lambda_n^- + \sum_i \lambda_{n-2i}) = (\sum \lambda_{n-1-2i})^2$$

All the λ_i are real. $RU(SO(2n))$ is free over $RU(SO(2n+1))$ on $\{1, \lambda_n^+\}$.

The cohomology ring is $H\mathbb{F}_2^* BSO(n) = \mathbb{F}_2[w_2, \dots, w_n]$ where $w_i = w_i(\lambda_1)$.

We have already examined $SO(2) = T(1)$ in Section 5.2, where we found (writing c_1 rather than y_1 here)

$$ku^*_{SO(2)} = ku^*[c_1, \bar{c}_1]/(vc_1\bar{c}_1 = c_1 + \bar{c}_1)$$

and

$$ku^* BSO(2) = ku^*[[c_1]].$$

The maps induced in ku^* by the fibre sequence $SO(2) \stackrel{i}{\longrightarrow} O(2) \stackrel{\det}{\longrightarrow} O(1)$ are

PROPOSITION 5.7.1. *$\det^*(c) = c$, while $i^*(c) = 0$, $i^*(c_2) = c_1\bar{c}_1$ and $i^*(c_1) = i^*(c + vc_2) = c_1 + \bar{c}_1$.*

Proof: The defining representation of $O(1)$ pulls back to $\lambda_2 = \det(\lambda_1)$. The relation $\det^*(c) = c$ follows. This forces $i^*(c) = 0$. Since $i^*(\lambda_1) = \lambda_1^+ + \lambda_1^-$, $i^*(c_2) = c_2(\lambda_1^+ + \lambda_1^-) = c_1\bar{c}_1$. then $i^*(c_1) = i^*(c + vc_2) = vc_1\bar{c}_1 = c_1 + \bar{c}_1$. □

To compute $ku^*_{SO(3)}$ we start with the periodic case. Here $RU(SO(3)) = \mathbb{Z}[\lambda_1]$ with $\lambda_2 = \lambda_1$ and $\lambda_3 = 1$.

PROPOSITION 5.7.2. *In periodic K-theory, the Chern classes of the defining representation of $SO(3)$ satisfy $c_1 = vc_2$, $c_3 = 0$ and $v^2c_2 = vc_1 = 3 - \lambda_1$. Therefore, $KU^*_{SO(3)} = KU^*[c_2]$*

Proof: It is enough to observe that in representation theory, $c_1 = c_2$ and $c_3 = 0$. □

Next we compute $ku^* BSO(3)$ with the Adams spectral sequence, starting by decomposing the cohomology as usual.

LEMMA 5.7.3. *As an $E(1)$-module, $H^*BSO(3)$ is the direct sum of:*

(i) *a trivial $E(1)$-module $F_2[w_2^2]$, and*
(ii) *a free $E(1)$-module $\mathbb{F}_2[w_2^2, w_3^2]\langle w_2\rangle$*

Proof: We leave this to the reader, as usual, noting for use below that $\langle w_2\rangle = \{w_2, w_3, w_2w_3, w_3^2\}$. □

THEOREM 5.7.4.

$$ku^*_{SO(3)} = ku^*[c_2, c_3]/(2c_3, vc_3).$$

The first Chern class, $c_1 = vc_2$*. The restriction* $ku^*_{O(3)} \longrightarrow ku^*_{SO(3)}$ *sends* c *and* q_3 *to* 0*, and sends each* c_i *to* c_i*.*

Proof: By the decomposition of the cohomology, the Adams spectral sequence must collapse at E_2 and the map into mod 2 cohomology plus periodic K-theory must be a monomorphism, just as for the orthogonal groups. Thus we have an associated graded

$$E^0(ku^*_{SO(3)}) = ku^*[c_2] \oplus c_3\mathbb{F}_2[c_2, c_3].$$

The monomorphism $ku^*_{SO(3)} \longrightarrow K^*_{SO(3)} \oplus H^*BSO(3)$ sends c_2 to (c_2, w_2^2) and c_3 to $(0, w_3^2)$. It follows that $ku^*BSO(3) = ku^*[[c_2, c_3]]/(2c_3, vc_3)$. Then the pullback square gives $ku^*_{SO(3)} = ku^*[c_2, c_3]/(2c_3, vc_3)$.

In general, we expect $c_1 = vc_2 - v^2c_3$, since this is true in $SU(n)$, but here $v^2c_3 = 0$.

The restriction from $O(3)$ is computed by using the monomorphism to periodic K-theory plus mod 2 cohomology. Note that $q_2 = cc_2 - 3c_3$ restricts to c_3. □

REMARK 5.7.5. In the Bockstein spectral sequence for $ko^*_{SO(3)}$ we see exactly the same filtration shift in $\beta : ku^6_{SO(3)} \longrightarrow ku^{-2}_{SO(3)}$ as we do for $O(2)$ and the dihedral groups. This can be shown either by representation theory or by the Adams spectral sequence, just as in those cases.

CHAPTER 6

Examples for Cohomology

In this chapter we calculate ko_*^G for cyclic groups, quaternion 2-groups, the Klein 4-group V and the alternating group A_4. These are the easiest groups because the rank is small, so the deviation from a purely representation theoretic description is slight.

6.1. Description of the method

For all the groups in this chapter, we can deduce the answer from the complex case by purely algebraic means. As usual, the starting point is the short exact sequence

(1) $$0 \longrightarrow TU \longrightarrow ku_*^G \longrightarrow QU \longrightarrow 0$$

where

$$QU = \mathrm{im}(ku_*^G \longrightarrow KU_*^G)$$

and TU is the v-power torsion. For the periodic groups $TU = 0$, and for V and A_4 the torsion TU is detected in mod 2 cohomology. For all groups considered in this chapter except A_4, QU is generated by $1, v$ and the Chern classes (so that it is the modified Rees ring).

To obtain ko_G^*, we apply the η-Bockstein spectral sequence, so the reader will want to refer back to Chapter 4 for more detail. In any case, the first step is to calculate the E^2-term of the Bockstein spectral sequence. Since the above short exact sequence is intrinsic, d^1 preserves TU and we may view (1) as a short exact sequence of chain complexes. The action on TU is via Sq^2, and the action on QU is $1-\tau$ from degrees 0 mod 4, and $1+\tau$ from degrees 2 mod 4. It will transpire that, in the cases we consider here, other differentials then follow from the requirement that the spectral sequence collapses at the E^4-page and is zero above the 2-line.

6.2. The conclusion

In the real case our results can be phrased in similar terms. Because the real case is more complicated, there are two counterparts (QO and $\overline{QO}$) of QU and four counterparts (ST, T, TO and ZTU) of TU. We also need to consider two modules (τ and $Q\tau$) of η-multiples.

To start with it is natural to consider the exact sequence

(2) $$0 \longrightarrow ST \longrightarrow ko_*^G \longrightarrow QO \longrightarrow 0$$

where

$$QO = \mathrm{im}(ko_*^G \longrightarrow KO_*^G)$$

and ST is the β-power torsion. Remarkably often, QO is generated by elements in degrees ≥ 0 together with characteristic classes lifting to ko. Remarkably often,

ST is detected in ordinary cohomology. It is also sometimes convenient to consider the part detected by complex characters. For this, we consider the exact sequence

$$0 \longrightarrow T \longrightarrow ko^G_* \longrightarrow \overline{QO} \longrightarrow 0 \tag{3}$$

where

$$\overline{QO} = \mathrm{im}(ko^G_* \longrightarrow KU^G_* = RU(G)[v, v^{-1}]).$$

Since the map $ko^G_* \longrightarrow KU^G_*$ factors through KO^G_*, the ring $\overline{QO}$ is a quotient of QO, and we have a diagram

$$\begin{array}{ccccccccc} 0 & \longrightarrow & ST & \longrightarrow & ko^G_* & \longrightarrow & QO & \longrightarrow & 0 \\ & & \downarrow & & =\downarrow & & \downarrow & & \\ 0 & \longrightarrow & T & \longrightarrow & ko^G_* & \longrightarrow & \overline{QO} & \longrightarrow & 0. \end{array}$$

We define $Q\tau$ by the exact sequence

$$0 \longrightarrow Q\tau \longrightarrow QO \longrightarrow \overline{QO} \longrightarrow 0 \tag{4}$$

so that by our analysis of the Bockstein spectral sequence, $Q\tau$ consists precisely of the η-multiples in QO. From the Snake Lemma, we also have an exact sequence

$$0 \longrightarrow ST \longrightarrow T \longrightarrow Q\tau \longrightarrow 0. \tag{5}$$

Finally, we let τ denote the submodule of η-multiples, and since η maps to zero under complexification, $\tau \subseteq T$.

By definition ZTU consists of Sq^2-cycles in TU, and TO consists of Bockstein ∞-cycles in ZTU.

LEMMA 6.2.1. *There is a short exact sequence*

$$0 \longrightarrow \tau \longrightarrow T \longrightarrow TO \longrightarrow 0,$$

and if we define $S\tau := ST \cap \tau$ *then there are short exact sequences*

$$0 \longrightarrow S\tau \longrightarrow ST \longrightarrow TO \longrightarrow 0,$$

and

$$0 \longrightarrow S\tau \longrightarrow \tau \longrightarrow Q\tau \longrightarrow 0.$$

Proof: The first exact sequence follows from the Bockstein spectral sequence. The submodule τ consists of the part of the E^∞-term on the 1- and 2-lines, so the quotient consists of the ∞-cycles on the 0-line that map to zero in KU^*_G.

The other two short exact sequences follow from the Snake Lemma. □

REMARK 6.2.2. (i) Many of these groups are easy to read off from the Bockstein spectral sequence.

First, ZQU consists of the part of QU on the 0-line at $E^{1\frac{1}{2}}$, and $\overline{QO}$ consists of ∞-cycles in QU.

Similarly, ZTU is the part of TU on the 0-line at E^2 and TO consists of ∞-cycles in TU. It is also clear that τ consists of all ∞-cycles above the 0-line. The module T consists of TO together with all ∞-cycles above the 0-line.

It can be more subtle to identify which η-multiples from $H(TU)$ are detected in KO^*_G, so care is needed when finding $S\tau$, $Q\tau$, ST and QO.

(ii) For the dihedral groups $\overline{QO} \neq ZQU$ (see Lemma 8.2.1) and $TO \neq ZTU$ (for example d^3 is non-zero on ZTU_{-12}, since d^3 is non-zero on the (-12)-diagonal above the 0-line). To show $S\tau \neq 0$ we need to find an η-multiple not detected in KO, but these do not exist for any of the groups we consider

Note that β can increase η-filtration (e.g. for $V(r)$ with $r \geq 3$). This occurs when an element of $E^2(ku)$ maps to zero in $E^2(KU)$ (for example this happens for $V(3)$ and $E^2_{1,-6}$). On the other hand it can be shown in our examples that these elements are nonetheless detected. Thus for all groups we consider $Q\tau = \tau \neq 0$.

6.3. Cyclic groups

The complex connective K-theory of the cyclic groups has no v-torsion, so the Bockstein spectral sequence for their real connective K-theory is entirely determined by representation theory. Let us write $RU(C_n) = \mathbb{Z}[\alpha]/(\alpha^n - 1)$ and let $y = c_1(\alpha) \in ku^2_{C_n}$, so that $vy = 1 - \alpha$. Then we have $ku^*_{C_n} = ku^*[y]/([n](y))$ where

$$[i](y) = \frac{1 - (1 - vy)^i}{v}.$$

in other words, $v \cdot [i](y) = 1 - \alpha^i$. Recall also that multiplication by v^i is an isomorphism from $ku^{2i}_{C_n}$ to $JU^i \subseteq RU(C_n) = ku^0_{C_n}$ where $JU^i = ((1-\alpha)^i)$, is the augmentation ideal.

Here, RU_{sc} is spanned by 1, the $\alpha^i + \alpha^{n-i}$ and, if n is even, $\alpha^{n/2}$. Similarly, RU_{asc} is spanned by the $\alpha^i - \alpha^{n-i}$ and equals $\mathrm{im}(1-\tau)$ as always.

The negative codegrees are given by the periodic theory, so it suffices to describe the positive codegrees.

THEOREM 6.3.1. *Let $p_1 = p_1(\alpha)$. In positive codegrees, the real connective K-theory $ko^*_{C_n}$ is mapped monomorphically into the complex connective K-theory: $c(ko^{4i+2}_{C_n}) = (p_1)^i RU_{asc}$ and $c(ko^{4i}_{C_n}) = (p_1)^i RO$.*

We emphasize that $ko^{4i}_{C_n}$ is the ideal in $RO(C_n)$ generated by p_1^i (the ideal generated by its image $(1-\alpha)(1-\alpha^{-1})$ in $RU(C_n)$ is larger).

The proofs differ in the even and odd order cases because the even order cyclic groups have a nontrivial simple real representation. However, both proceed from the following lemmas.

LEMMA 6.3.2. *The sequence*

$$RU \xrightarrow{1-\tau} JU \xrightarrow{1+\tau} JU^2 \xrightarrow{1-\tau} JU^3 \xrightarrow{1+\tau} \cdots$$

is exact.

To prove this lemma, we use another to reduce to the first two cases. Let us take the notational liberty of writing p_1 for $c(p_1) = (1-\alpha)(1-\alpha^{-1})$ when the context makes it clear that we are talking about elements of the complex representation ring.

LEMMA 6.3.3. *Multiplication by p_1 induces an isomorphism $JU^i \longrightarrow JU^{i+2}$.*

Proof: It suffices to note that p_1 is a unit multiple of $(1-\alpha)^2$. In fact $p_1 = (-\alpha^{-1})(1-\alpha)^2$. □

Proof of Lemma 6.3.2: By Lemma 6.3.3, it will suffice to prove exactness at JU and at JU^2. At JU this is simply the generic fact that $RU_{asc} = \ker(1+\tau) = \mathrm{im}(1-\tau)$ coupled with the fact that $\mathrm{im}(1-\tau)$ is contained in JU.

At JU^2 we calculate. Since $JU^2 = (p_1)$, $\tau(p_1) = p_1$, and p_1 is regular on JU, we see that $0 = (1-\tau)(p_1 x) = p_1(1-\tau)(x)$ if and only if $(1-\tau)(x) = 0$. That is, $\ker(JU^2 \xrightarrow{1-\tau} JU^3) = p_1 RU_{sc}$. Thus it suffices to show that $p_1 RU_{sc} \subset (1+\tau)(JU) = (1+\tau)(\ker(\epsilon))$, where ϵ is the augmentation $\epsilon(\alpha^i) = 1$. This is immediate:

$$p_1 = (1-\alpha)(1-\alpha^{-1}) = 2 - \alpha - \alpha^{-1} = (1+\tau)(1-\alpha),$$

$$p_1(\alpha^i + \alpha^{n-i}) = (1+\tau)(2\alpha^i - \alpha^{i+1} - \alpha^{i-1}),$$

and when $n = 2m$,

$$p_1 \alpha^m = (1+\tau)(\alpha^m - \alpha^{m-1}).$$

□

Proof of Theorem 6.3.1: The η-Bockstein spectral sequence for $ko^*_{C_n}$ collapses to the 0-line at E^2. This follows from Lemma 6.3.2 because, along the $2i$-diagonal, the E^1 term is the truncation of the sequence in Lemma 6.3.2 which starts with JU^i. Thus, at E^2 it has collapsed to the 0-line with $E^2_{4i,0} = (p_1)^i RU_{sc} = (p_1)^i RO(C_n)$ and $E^2_{4i+2,0} = (p_1)^i RU_{asc}$. □

6.4. Quaternion groups

We let Q denote the quaternion group of order 2^{n+2} for $n \geq 1$. It was shown in [**28**, 2.4] that there is no v-torsion and that $ku^*_Q = QU$ is the modified Rees ring (i.e., it is the subring of $RU(Q)[v, v^{-1}]$ generated by $1, v$ and the Chern classes).

To ease comparison between different quaternion groups, we choose notation by reference to $Pin(2)$, viewed as a subgroup of the group $Sp(1)$ of unit quaternions. The group Q consists of the 2^{n+1} complex numbers $\rho_\theta = e^{i\theta} = \cos\theta + i\sin\theta$ and the 2^{n+1} properly quaternionic elements $\sigma_\theta = \rho_\theta j = j\cos\theta + k\sin\theta$ where $\theta = 2l\pi/2^{n+1}$ for $l = 0, 1, \ldots, 2^{n+1}-1$. We write $\rho_{0,n}, \rho_{1,n}, \ldots, \rho_{2^{n+1}-1,n}$ and $\sigma_{0,n}, \sigma_{1,n}, \ldots, \sigma_{2^{n+1}-1,n}$ for these, even omitting the n when confusion is unlikely. Note the relations $\rho_\pi = j^2 = -1$ and $j\rho_\theta j^{-1} = \rho_{-\theta}$.

The elements ρ_θ form a cyclic subgroup C of index 2. The two elements $\rho_{0,n}$ (the identity) and $\rho_{2^n,n}$ (of order 2) are central, and the other elements ρ_θ are in conjugacy classes of two $\{\rho_{i,n}, \rho_{-i,n}\}$. The elements σ_θ are in two conjugacy classes, with 2^n elements in each, one with θ consisting of the even multiples of $2\pi/2^{n+2}$ and one with odd multiples. The even conjugacy class generates a quaternion subgroup Q', and the odd conjugacy class generates a quaternion subgroup Q'', both of order 2^{n+1}. We write $s' = s_0$ and $s'' = s_{2\pi/2^{n+1}}$ for representatives of these conjugacy classes.

There are four one dimensional representations: the trivial representation ϵ, together with $\hat{Q}'$, $\hat{Q}''$, and $\hat{C} = \hat{Q}'\hat{Q}''$, where $\hat{H}$ means the representation with kernel H for a subgroup H of index 2. Since $Q' \cap Q''$ is a cyclic group C' of index 2 in C we see, for example, that $\hat{Q}'$ restricts to $\hat{C}'$ on Q''. The remaining simple representations $\psi_1, \ldots, \psi_{2^n-1}$ are of dimension 2, and restricted from $Pin(2)$. Those with an even subscript pull back from $O(2)$ and are therefore real. Those with an odd subscript are obtained from the natural representation in $Sp(1)$ by pullback along an automorphism of Q and are therefore quaternionic. It is easy to see

$$\psi_k \hat{Q}' = \psi_k \hat{Q}'' = \psi_{k+2^n}.$$

Notice that as representations of Q

$$\psi_k = \psi_{k+2^{n+1}}, \text{ and } \psi_{k+2^n} = \psi_{2^n-k}.$$

In particular

$$\psi_{2^n} = \hat{Q}' + \hat{Q}''.$$

In any case, all representations of Q are self-conjugate. It follows that RU_{sc} is of rank 2^n+3 and $RU_{asc}=0$. Furthermore $1+\tau=2$, so that $H \cong 2^{2^n+3}$.

Referring to the discussion of the Bockstein spectral sequence in Chapter 4, we must identify the ranks of $\delta_0, \delta_4 : H \longrightarrow H$ we note that $\ker(\delta_0) = RO/RU \cong 2^{2^{n-1}+3}$ (so δ_0 is of rank 2^{n-1} in positive degrees) and $\ker(\delta_4) = RSp/RU \cong 2^{2^{n-1}}$ (so δ_4 is of rank $2^{n-1}+3$ in positive degrees).

To complete the determination of the E^2-page, we must calculate the ideals JU_k, together with the conjugation action on them. From this we may calculate the homology of the exact sequence

$$RU \xrightarrow{1-\tau} JU \xrightarrow{1+\tau} JU_2 \xrightarrow{1-\tau} JU_3 \xrightarrow{1+\tau} \cdots$$

Noting that $d^2=0$ since ku_Q^* is in even degrees, we may complete the calculation. Indeed, the behaviour of d^3 in positive degrees is known, and the behaviour in negative degrees follows since E^4 is zero above the 2-line. This is enough to determine the behaviour of the Bockstein spectral sequence. We illustrate this for the first two cases Q_8 and Q_{16}.

EXAMPLE 6.4.1. *(The quaternion group of order 8)*
In this case all representations are self-conjugate, so $RU_{sc} = RU, RU_{asc} = 0$ and $H \cong 2^5$ as before.

The full character table is

	1	-1	$\pm i$	$\pm j$	$\pm k$
ϵ	1	1	1	1	1
$\hat{i}$	1	1	1	-1	-1
$\hat{j}$	1	1	-1	1	-1
$\hat{k}$	1	1	-1	-1	1
ψ	2	-2	0	0	0

We quickly calculate

	1	-1	$\pm i$	$\pm j$	$\pm k$
$c_1(\hat{i})$	0	0	0	2	2
$c_1(\hat{j})$	0	0	2	0	2
$c_1(\hat{k})$	0	0	2	2	0
$c_1(\psi)$	0	4	2	2	2
$c_2(\psi)$	0	4	2	2	2

A little more calculation shows that for $k \geq 1$ we have

JU_{2k-1}	1	-1	$\pm i$	$\pm j$	$\pm k$
	0	0	0	2^k	2^k
	0	0	2^k	0	2^k
	0	0	2^k	2^k	0
	0	4^k	2^k	2^k	2^k

and

JU_{2k}	1	-1	$\pm i$	$\pm j$	$\pm k$
	0	0	0	0	2^{k+1}
	0	0	0	2^{k+1}	0
	0	0	2^{k+1}	0	0
	0	4^k	2^k	2^k	2^k

In particular, multiplication by $c_2(\psi) = (04222)$ is an isomorphism $JU_n \longrightarrow JU_{n+2}$ for $n \geq 1$.

As observed above, all representations are self-conjugate. Thus $1 + \tau = 2$ and $JU^{asc}_{2k-1} = 0$, whilst $1 - \tau = 0$ and $JU^{sc}_{2k} = JU_{2k}$. For the E^2-term of the Bockstein spectral sequence we use the fact that $JU_{2k}/2JU_{2k-1} = 2^2$, as can be read off from the displays above.

The Bockstein spectral sequence is displayed in Figure 6.4.1 below. Note that the rank of the δ_0 differential is 1 in both positive and negative degrees. The rank of δ_4 is 4 in positive degrees and 1 in negative degrees.

We conclude that the connective *ko*-cohomology is as follows.

THEOREM 6.4.2. *The ko-cohomology of the quaternion group Q_8 is as follows,*

n	$ko^{Q_8}_n$
$8k+7 \geq 0$	0
$8k+6 \geq 0$	$RU/R0 \cong 2$
$8k+5 \geq 0$	$RSp/RU \cong 2$
$8k+4 \geq 0$	RSp
$8k+3 \geq 0$	0
$8k+2 \geq 0$	$RU/RSp \cong 2^4$
$8k+1 \geq 0$	$RO/RU \cong 2^4$
$8k+0 \geq 0$	RO
$-8k+7 \leq -1$	0
$-8k+6 \leq -1$	2
$-8k+5 \leq -1$	2
$-8k+4 \leq -1$	JSp^{2k-1}
$-8k+7 \leq -1$	0
$-8k+2 \leq -1$	2
$-8k+1 \leq -1$	2
$-8k+0 \leq -1$	JSp^{2k}

Proof: The output of the Bockstein spectral sequence has $\overline{QO}$ in negative degrees given by JU'_{4k-2} and JU'_{4k}, which are defined by exact sequences

$$0 \longrightarrow JU'_{4k} \longrightarrow JU_{4k} \xrightarrow{\delta_0} 2 \longrightarrow 0$$

and

$$0 \longrightarrow JU'_{4k-2} \longrightarrow JU_{4k-2} \xrightarrow{\delta_4} 2 \longrightarrow 0.$$

We need only observe that these are powers of JSp. Indeed, JU'_2 must contain all first Pontrjagin classes, and hence all of JSp; since this is of index 2, we find

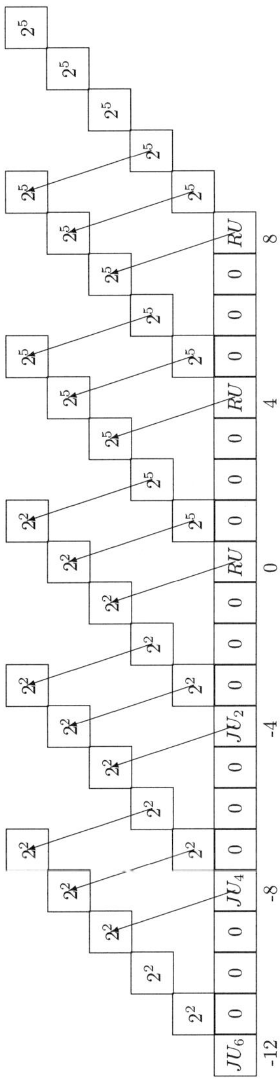

FIGURE 6.4.1. The E^2-page for Q_8

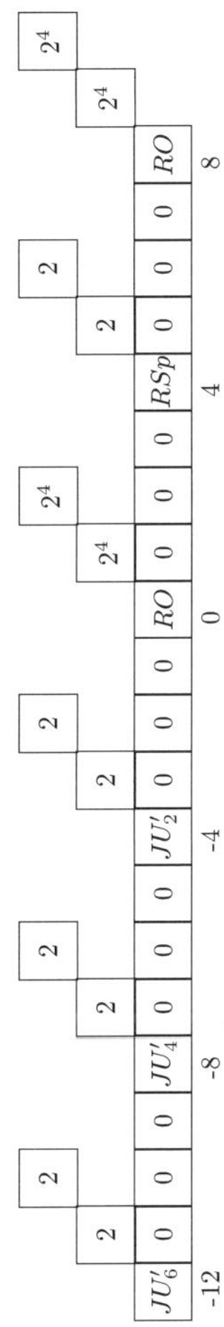

FIGURE 6.4.2. The E^∞-page for Q_8

$JU'_2 = JSp$. Then JU'_{2n} must contain JSp^n, which is of index 2 in JU_{2n}. □

EXAMPLE 6.4.3. *(The quaternion group of order 16)*

The group Q_{16} has three 2-dimensional simple representations, two symplectic and one real. Thus $RU_{sc} = RU \cong \mathbb{Z}^7$, $RU_{asc} = 0$ and $H \cong 2^7$.

The full character table is

	1	-1	j	$j\rho$	ρ	ρ^2	ρ^3
ϵ	1	1	1	1	1	1	1
$\hat{\rho}$	1	1	-1	-1	1	1	1
$\hat{j}$	1	1	1	-1	-1	1	-1
$\widehat{j\rho}$	1	1	-1	1	-1	1	-1
ψ_1	2	-2	0	0	$\sqrt{2}$	0	$-\sqrt{2}$
ψ_2	2	2	0	0	0	-2	0
ψ_3	2	-2	0	0	$-\sqrt{2}$	0	$\sqrt{2}$

One verifies that $\det(\psi_1) = \det(\psi_3) = \epsilon$ and $\det(\psi_2) = \hat{\rho}$. From this we may calculate the Chern classes.

	1	-1	j	$j\rho$	ρ	ρ^2	ρ^3
$c_1(\hat{\rho})$	0	0	2	2	0	0	0
$c_1(\hat{j})$	0	0	0	2	2	0	2
$c_1(\widehat{j\rho})$	0	0	2	0	2	0	2
$c_1(\psi_1)$	0	4	2	2	$2-\sqrt{2}$	2	$2+\sqrt{2}$
$c_1(\psi_2)$	0	0	2	2	2	4	2
$c_1(\psi_3)$	0	4	2	2	$2+\sqrt{2}$	2	$2-\sqrt{2}$
$c_2(\psi_1)$	0	4	2	2	$2-\sqrt{2}$	2	$2+\sqrt{2}$
$c_2(\psi_2)$	0	0	0	0	2	4	2
$c_2(\psi_3)$	0	4	2	2	$2+\sqrt{2}$	2	$2-\sqrt{2}$

One may then proceed to calculate the ideals JU_k. After finding JU_1, JU_2, JU_3 and JU_4, we observe that multiplication by

$$p_* := c_2(\psi_3) - c_2(\psi_2) = (0, 4, 2, 2, \sqrt{2}, -2, -\sqrt{2})$$

gives an isomorphism $JU_i \longrightarrow JU_{i+2}$ for $i = 1$ and $i = 2$; it follows that the same holds for all i, and hence $JU_{2n+1} = (p_*)^n JU_1$ and $JU_{2n+2} = (p_*)^n JU_2$ for $n \geq 0$. For later purposes, note that p_* is not an element of JSp; one may also verify that multiplication by $c_2(\psi_3)$ does not give a similar periodicity at the level of ku_Q^*.

From this it is easy to verify that

$$JU_{2k}/2JU_{2k-1} = 2^4 \text{ for } k \geq 1.$$

The Bockstein spectral sequence is displayed in Figure 6.4.3 below. Note that the rank of the δ_0 differential is 2 in both positive and negative degrees. The rank of δ_4 is 5 in positive degrees, and 2 in negative degrees.

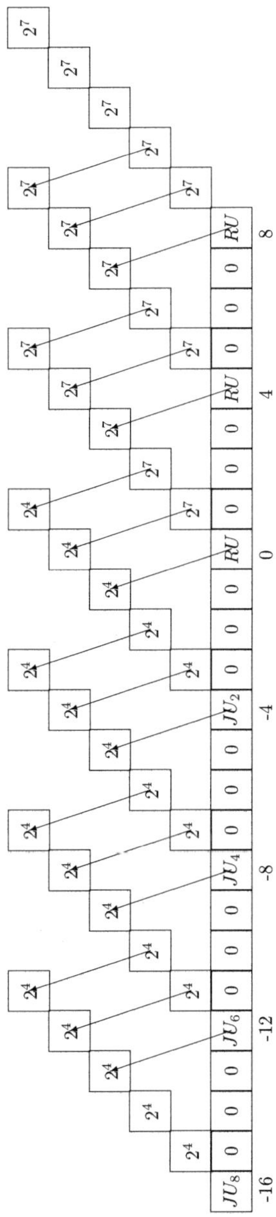

FIGURE 6.4.3. The E^2-page for Q_{16}

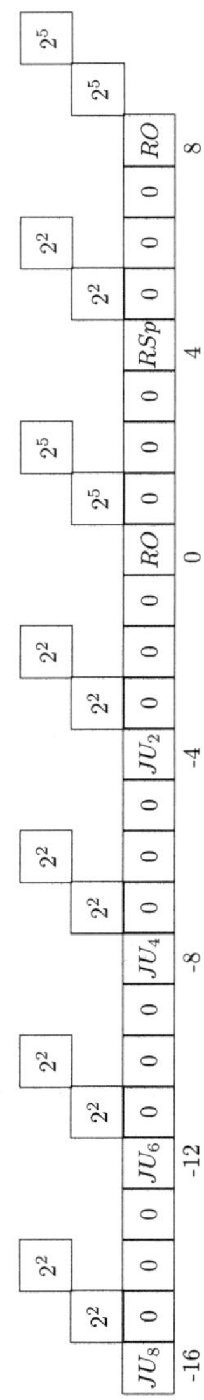

FIGURE 6.4.4. The E^∞-page for Q_{16}

We conclude that the connective ko-cohomology is as follows.

THEOREM 6.4.4. *The ko-cohomology of the generalized quaternion group Q_{16} is as follows*

n	$ko_n^{Q_{16}}$
$8k+7 \geq 0$	0
$8k+6 \geq 0$	$RU/R0 \cong 2^2$
$8k+5 \geq 0$	$RSp/RU \cong 2^2$
$8k+4 \geq 0$	RSp
$8k+3 \geq 0$	0
$8k+2 \geq 0$	$RU/RSp \cong 2^5$
$8k+1 \geq 0$	$RO/RU \cong 2^5$
$8k+0 \geq 0$	RO
$-8k+7 \leq -1$	0
$-8k+6 \leq -1$	2^2
$-8k+5 \leq -1$	2^2
$-8k+4 \leq -1$	JSp^{2k-1}
$-8k+7 \leq -1$	0
$-8k+2 \leq -14$	2^2
$-8k+1 \leq -15$	2^2
$-8k+0 \leq -1$	JSp^{2k}

Proof: The Bockstein spectral sequence gives the above results immediately, except that in degrees $-8k+4$ and $-8k$ it gives the ideals JU'_{4k-2} and JU'_{4k}, which are defined by exact sequences

$$0 \longrightarrow JU'_{4k} \longrightarrow JU_{4k} \xrightarrow{\delta_0} 2^2 \longrightarrow 0$$

and

$$0 \longrightarrow JU'_{4k+2} \longrightarrow JU_{4k+2} \xrightarrow{\delta_4} 2^2 \longrightarrow 0.$$

Evidently, $p_1(\psi_1)RO \subseteq ko^Q_{-4} = JU'_2$. One may verify that $p_1(\psi_1)RO = JSp$ is of index 4 in JU_2. Hence $JSp^2 \subseteq ko^Q_{-8} = JU'_4$. One verifies JSp^2 is of index 4 in JU_4 and hence we have equality. One checks that $p_*^2 = (16,4,4,2,4,2) \in JSp^2$, and so the equality in higher degrees follows. □

It is natural to conjecture the connective ko-cohomology in general.

CONJECTURE 6.4.5. The ko-cohomology of the generalized quaternion group Q of order 2^{n+2} is as follows

n	ko^Q_n
$8k+7 \geq 0$	0
$8k+6 \geq 0$	$RU/R0 \cong 2^{2^{n-1}}$
$8k+5 \geq 0$	$RSp/RU \cong 2^{2^{n-1}}$
$8k+4 \geq 0$	RSp
$8k+3 \geq 0$	0
$8k+2 \geq 0$	$RU/RSp \cong 2^{2^{n-1}+3}$
$8k+1 \geq 0$	$RO/RU \cong 2^{2^{n-1}+3}$
$8k+0 \geq 0$	RO
$-8k+7 \leq -1$	0
$-8k+6 \leq -1$	2^{n-1}
$-8k+5 \leq -1$	2^{n-1}
$-8k+4 \leq -1$	JSp^{2k-1}
$-8k+7 \leq -1$	0
$-8k+2 \leq -14$	2^{n-1}
$-8k+1 \leq -15$	2^{n-1}
$-8k+0 \leq -1$	JSp^{2k}

To verify this, we need to check a number of things. First, we need to check that

$$JU_{2k}/2JU_{2k-1} = 2^n \text{ for } k \geq 1.$$

It then follows that the stable values above the 0-line are

$$\cdots \overset{2^{n-1}}{\longleftarrow} 2^n \overset{2^{n-1}}{\longleftarrow} 2^n \overset{2^{n-1}}{\longleftarrow} 2^n+3 \overset{2^{n-1}+3}{\longleftarrow} 2^n+3 \overset{2^{n-1}}{\longleftarrow} 2^n+3 \overset{2^{n-1}+3}{\longleftarrow} \cdots$$

This gives the indicated values in positive degrees and in degrees 1 and 2 mod 4. To verify the values in degrees $-4k \leq -4$, we need to observe that the index of JSp^r in JU_{2r} is precisely 2^{n-1}.

These results should be proved by the methods of Chapter 8. Unfortunately, the difference in second Chern classes seems to make the calculations a little harder in the quaternion case.

6.5. Klein 4-group

The full details are given in Chapter 9, but we give a brief summary here to make sense of our calculations for the alternating group. This may provide a helpful preparation for the more intricate discussion of the general case.

To start with we have the exact sequence

$$0 \longrightarrow TU \longrightarrow ku^*_V \longrightarrow QU \longrightarrow 0$$

where QU is the Rees ring (generated by $1, v$ and the two first Chern classes), and TU is a free module on a generator of degree -6 over the polynomial ring $PC = \mathbb{F}_2[y_1, y_2]$ on the Chern classes y_1, y_2 of degree -2 [**28**, Chapter 4].

6.5.A. The Rees ring. The Rees ring is quite easy to deal with since all representations are self-conjugate, so that $1+\tau = 2$, $1-\tau = 0$ and $RU_{asc} = 0$. Furthermore, all the representations are real so that $\delta_0 = 0$.

In more detail, if $V = \{e, a, b, c\}$ and we write characters as 4-tuples, JU has $\mathbb{Z}$-basis given by the three first Chern classes, $(0,2,0,2)$, $(0,0,2,2)$, $(0,2,2,0)$. However the higher powers are diagonal in the sense that for $k \geq 2$ JU^k has a $\mathbb{Z}$-basis $(0,2^k,0,0)$, $(0,0,2^k,0)$, $(0,0,0,2^k)$. Accordingly, $JU^2/2JU \cong 2$, but for $k \geq 2$, $JU^{k+1}/2JU^k = 0$.

Thus, the contribution of QU to the E^2-page is entirely on the 0 line below degree -4.

6.5.B. The torsion. First, TU is free over $PC = \mathbb{F}_2[y_1, y_2]$ on the generator $q_{12} = y_1y_2(y_1+y_2)$. Since $Sq^2 q_{12} = 0$, the Sq^2 action on TU is simply a shift of that on PC. To understand the Sq^2 action on PC, note that it makes PC into a chain complex; as such, it is isomorphic to a suitably graded version of the Koszul complex for $PP = \mathbb{F}_2[z_1, z_2]$, where $z_i = y_i^2$. Indeed 1 gives the generator in homological degree 0, y_1, y_2 are the two generators in homological degree 1 and y_1y_2 is the generator in homological degree 2.

Since z_1, z_2 is a regular sequence in PP, we see that the Sq^2-homology of PC consists of $\mathbb{F}_2$ generated by 1, and the PP-module ZPC of Sq^2 cycles is the free module on 1 and $Sq^2(y_1y_2) = q_{12}$. The corresponding statements for TU are obtained by multiplication with q_{12} (which simply shifts degrees down by 6).

6.5.C. The Bockstein spectral sequence. This is enough to determine the $E^{1\frac{1}{2}}$-term of the Bockstein spectral sequence (i.e., with the connecting homomorphism of the the short exact sequence $0 \longrightarrow TU \longrightarrow ku^*_V \longrightarrow QU \longrightarrow 0$ of chain complexes as a late d^1).

The Bockstein spectral sequence is displayed in full in Figure 9.12.2. The remaining differentials are forced by the requirement that the E^∞-term is zero above the 2-line.

6.5.D. Conclusion. We state our conclusion in terms of the exact sequence

$$0 \longrightarrow ST \longrightarrow ko^*_V \longrightarrow QO \longrightarrow 0$$

with QO the image in KO^*_V.

As always, QO agrees with KO^*_V in degrees ≥ 0. In negative degrees it is generated by JSp in degree -4. As for TO, we have

$$ST = TO = ZTU \cong PP(-6) \oplus PP(-12).$$

It is notable that there are no η-multiples in negative degrees,

6.6. The alternating group A_4

As in the complex case, this group has a number of interesting features.

To start with we have the exact sequence

$$0 \longrightarrow TU \longrightarrow ku^*_{A_4} \longrightarrow QU \longrightarrow 0.$$

The first thing to note is that QU is not generated (over $RU(A_4)[v]$) by Chern classes (the subalgebra generate by Chern classes is of index 2 in degree -6). The module TU is free over the double of $H^*(BA_4; \mathbb{F}_2)$ on a generator of degree -6.

6.6.A. The enlarged modified Rees ring. The group A_4 has three 1-dimensional simple modules, $\epsilon, \alpha, \alpha^2$, and one 3-dimensional simple module τ. Evidently ϵ and τ are self-conjugate, and α, α^2 is a conjugate pair.

The full character table is

	e	$(12)(34)$	(123)	(132)
ϵ	1	1	1	1
α	1	1	ω	ω^2
α	1	1	ω^2	ω
τ	3	-1	0	0

One may calculate $c_2(\tau) = c_1(\tau) = 3 - \tau$, and $c_3(\tau) = 0$. It is convenient to record the first few terms of the Chern filtration. For purposes of calculation, we need only the second and third columns (the first is zero, and the fourth is conjugate to the third).

In the following tables giving $\mathbb{Z}$-bases for homogeneous parts of QU, we take

$$\Gamma = c_1(\tau) - c_1(\alpha) - c_1(\alpha^2) = c_2(\tau) - c_1(\alpha)c_2(\alpha)$$

and $\omega = e^{2\pi i/3}$.

JU_1	e	$(12)(34)$	(123)	(132)
$c_1(\alpha)$	0	0	$1-\omega$	$1-\omega^2$
$c_1(\alpha)$	0	0	$1-\omega^2$	$1-\omega$
Γ	0	4	0	0

It is probably useful to comment that the first row can be replaced by $(0, 0, 3, 3)$. The remaining pieces fall into a single pattern.

JU_{2k}	e	$(12)(34)$	(123)	(132)
	0	0	3^k	3^k
	0	0	$3^k\omega$	$3^k\omega^2$
	0	2^{2k}	0	0

For $k \geq 2$ we write $JU^+_{2k-1} := QU_{4k-2}$, which has JU_{2k-1} as the subgroup of index 2 generated by the first two rows and twice the third.

JU^+_{2k-1}	e	$(12)(34)$	(123)	(132)
	0	0	3^k	3^k
	0	0	$3^{k-1}\omega$	$3^{k-1}\omega^2$
	0	2^{2k-1}	0	0

From this one may easily verify the following statement.

LEMMA 6.6.1. *The homology of the sequence*

$$\xleftarrow{1-\tau} JU_6 \xleftarrow{1+\tau} JU_5^+ \xleftarrow{1-\tau} JU_4 \xleftarrow{1+\tau} JU_3^+ \xleftarrow{1-\tau} JU_2 \xleftarrow{1+\tau} JU_1 \xleftarrow{1-\tau} RU \xleftarrow{1+\tau} RU \xleftarrow{1-\tau}$$

is 2^2 at RU in degree $4k \geq 0$, and 2 at JU_2. It is otherwise exact.

Thus, the contribution of QU to the E^2-page is entirely on the 0 line below degree -4.

6.6.B. The torsion. The double of $H^*(BA_4;\mathbb{F}_2)$ is

$$DH := \mathbb{F}_2[\mu_4,\nu_6,\pi_6]/(\mu^3=\nu^2+\nu\pi+\pi^2),$$

and $Sq^2\nu = 0, Sq^2\pi = \mu^2, Sq^2\mu = \nu$. The torsion submodule TU is detected in $H^*(BA_4;\mathbb{F}_2)$, where it is the free DH-module on ν.

LEMMA 6.6.2. *The Sq^2-homology of TU is $\mathbb{F}_2$ in degree -6. The Sq^2-cycles are given by the short exact squence*

$$0 \longrightarrow \nu\cdot\mathbb{F}_2[\mu^2,\nu] \longrightarrow ZTU \longrightarrow \pi\nu^2\cdot\mathbb{F}_2[\mu^2,\nu] \longrightarrow 0$$

(as an $\mathbb{F}_2[\mu^2,\nu]$-module, it is free on one generator of degree -6 and one of degree -18).

Proof: Any monomial can be uniquely expressed in the form $a(\mu,\nu)+b(\mu,\nu)\cdot\pi$, and Sq^2 preserves the polynomial subalgebra $\mathbb{F}_2[\mu,\nu]$. From this we see that there is a short exact sequence

$$0 \longrightarrow \mathbb{F}_2[\mu,\nu] \longrightarrow DH \longrightarrow \mathbb{F}_2[\mu,\nu]\cdot\pi \longrightarrow 0$$

of Sq^2-complexes. Since $Sq^2\nu = Sq^2\mu^2 = 0$, the action of Sq^2 on $\mathbb{F}_2[\mu,\nu]$ is linear over $\mathbb{F}_2[\mu^2,\nu]$, and is thus multiplication by ν

$$\mu\cdot\mathbb{F}_2[\mu^2,\nu] \xrightarrow{\nu} \mathbb{F}_2[\mu^2,\nu].$$

The cycles are thus $\mathbb{F}_2[\mu^2,\nu]$ and the homology is $\mathbb{F}_2[\mu^2,\nu]/(\nu)\cong\mathbb{F}_2[\mu^2]$. The long exact sequence in Sq^2-homology has connecting homomorphisms

$$Z(\pi\cdot\mathbb{F}_2[\mu,\nu]) = \pi\cdot\mathbb{F}_2[\mu^2,\nu] \xrightarrow{\mu^2} \mathbb{F}_2[\mu^2] = H(\mathbb{F}_2[\mu,\nu])$$

and

$$H(\pi\cdot\mathbb{F}_2[\mu,\nu]) = \pi\cdot\mathbb{F}_2[\mu^2] \xrightarrow{\mu^2} \mathbb{F}_2[\mu^2] = H(\mathbb{F}_2[\mu,\nu]).$$

The second is a monomorphism with cokernel $\mathbb{F}_2$. The former has kernel generated by $\pi\nu$. □

6.6.C. The Bockstein spectral sequence. This is enough to determine the $E^{1\frac{1}{2}}$-term of the η-Bockstein spectral sequence. The connecting homomorphism of the the short exact sequence $0\longrightarrow TU \longrightarrow ku^*_{A_4} \longrightarrow QU \longrightarrow 0$ of chain complexes must cancel the copies of 2 on the 4-diagonal and the 6-diagonal. This gives the E^2-term, and the only remaining differential is the d^3 from the periodic case in positive degrees.

The Bockstein spectral sequence is displayed in Figure 6.6.5.

6.6.D. Conclusion. We state our conclusion in terms of the exact sequence

$$0 \longrightarrow ST \longrightarrow ko^*_V \longrightarrow QO \longrightarrow 0$$

with QO the image in KO^*_V.

As always, QO agrees with KO^*_V in degrees ≥ 0. In negative degrees it is generated by RU_{asc} in degree -2 and JU_2^{sc} in degree -4. It is striking that unlike the complex case, we do not require an additional generator in degree -6.

As for the torsion part, we have

$$ST = TO = ZTU \cong \mathbb{F}_2[\mu^2,\nu](-6)\oplus\mathbb{F}_2[\mu^2,\nu](-18).$$

It is notable that there are no η-multiples in negative degrees,

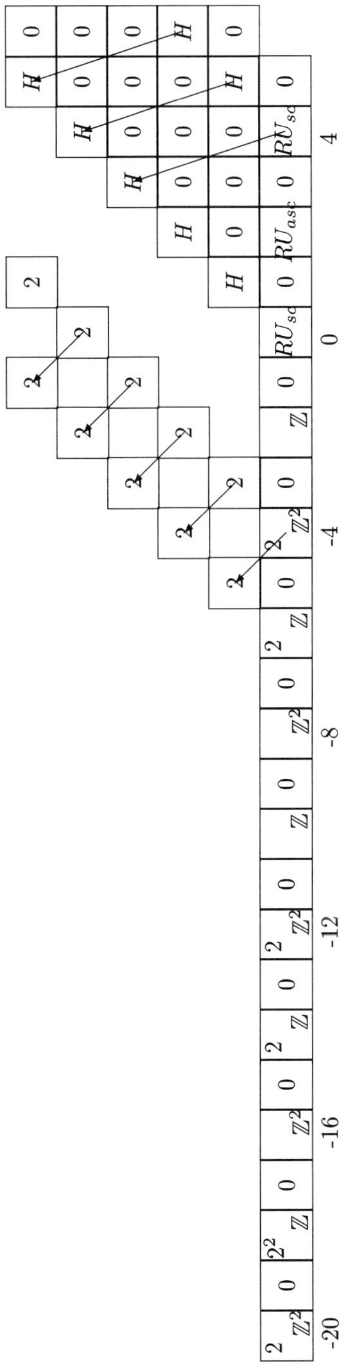

FIGURE 6.6.5. The $E^{1\frac{1}{2}}$-page for A_4

CHAPTER 7

Examples for Homology

In this chapter we calculate $ko_*(BG)$ for cyclic groups, generalized quaternion 2-groups, the Klein 4-group and the alternating group A_4. There is quite sufficient complexity in these cases to illustrate our methods, and we also observe a number of interesting patterns.

7.1. Comparison of methods

There are several methods for calculating $ko_*(BG)$. The traditional route is to use a direct calculation by an Adams spectral sequence; this is a natural method if we have not already calculated $ku^*(BG), ku_*(BG)$ or $ko^*(BG)$, but if one of these is known, it involves an enormous duplication of effort. It also gives limited information about the structure of $ko_*(BG)$ as a module over $ko^*(BG)$, tends to involve many unnatural choices, and requires us to cope with infinite cohomological dimension. We generally avoid this method, although we have used it to check some results, and there is one additive extension for dihedral groups that we were not able to resolve by other means.

Instead, there are two methods, one working from $ko^*(BG)$ and one working from $ku_*(BG)$. Our favoured method is to work from $ko^*(BG)$: we can then use the local cohomology spectral sequence

$$E_2^{*,*} = H_I^*(ko^*(BG)) \Rightarrow ko_*(BG)$$

described in Section 2.5. To start with, this is a finite spectral sequence: local cohomology vanishes above the Krull dimension r, so it only has r columns and collapses after the E_r page. By its very construction, it gives $ko_*(BG)$ as a $ko^*(BG)$-module up to a finite filtration, and displays a Gorenstein-type duality. These are theoretical advantages. However, our experience shows practical advantages too. Firstly, the differentials are often forced just from the fact that $ko_*(BG)$ is connective. Secondly, it seems that at E_∞ it is often the case that the non-zero entries are only on the H_I^0-column, the H_I^1-column and the H_I^r-column. The H_I^0- and H_I^1-columns come from cyclic subgroups (in the same sort of way that representation theory is determined by cyclic groups), and the H_I^r-column is detected in ordinary cohomology. This is a very useful filtration.

In low ranks our favoured method is very effective. However in high ranks the local cohomology calculations seem to be rather complicated. The alternative method is to assume $ku_*(BG)$ is known (this will also have involved local cohomology calculations, but the modules involved for the complex case are less intricate). In that case we can apply the η-Bockstein spectral sequence. As usual there are only three differentials, and the first is completely algebraic. On the other hand the forest of non-zero entries make it hard to infer differentials from sparsity and connectivity arguments: it is necessary to understand $ku_*(BG)$ rather well. This method also has the disadvantage that the module structure over $ko^*(BG)$ is less accessible.

For the groups we treat in this chapter, we are able to use our preferred method in all cases. Indeed, it is only for elementary abelian groups of high rank that we are forced to use the Bockstein spectral sequence method.

7.2. Our preferred method

To apply the local cohomology spectral sequence, we need to calculate the local cohomology of the ring ko^*_G. We assume that the output of the earlier steps is a short exact sequence

$$0 \longrightarrow ST \longrightarrow ko^*_G \longrightarrow QO \longrightarrow 0$$

where QO is the image of ko^*_G in KO^*_G. This gives a long exact sequence in local cohomology. We may treat QO as a module over itself, and invariance under base change shows it has the same local cohomology as it does as a ko^*_G-module. Because QO is a 1-dimensional graded ring, its local cohomology is in codegrees 0 and 1. On the other hand, in the examples we have seen, ST always has depth ≥ 2 so that its local cohomology is zero in degrees 0 and 1 (this is analogous to the theorem that the group cohomology ring has depth ≥ 1, but we have not been able to give a general proof).

Assembling this information, the long exact sequence degenerates to show

$$H^0_I(ko^*_G) = H^0_I(QO),$$
$$H^i_I(ko^*_G) = H^i_I(ST) \text{ for } i \geq 3$$

and there is an exact squence

$$0 \longrightarrow H^1_I(ko^*_G) \xrightarrow{\delta} H^1_I(QO) \longrightarrow H^2_I(ST) \longrightarrow H^2_I(ko^*_G) \longrightarrow 0,$$

where δ is the connecting homomorphism. We find it convenient to treat δ as a late d_1, so that we display the $E_{1\frac{1}{2}}$ term of the spectral sequence with $H^1_I(QO)$ as its (-1)-column and $H^2_I(QO)$ as its (-2)-column.

When it comes to calculating the local cohomology of TO, we use the same type of method as for group cohomology rings. On the other hand for QO we use the exact sequence

$$0 \longrightarrow Q\tau \longrightarrow QO \longrightarrow \overline{QO} \longrightarrow 0$$

where $\overline{QO}$ is the image of QO in KU^*_G. Here $Q\tau$ consists of the η-multiples in representation theory, and is generally not hard to deal with. For $\overline{QO}$ we use character theory and linear algebra: although the methods are elementary, we have found the actual calculations very time consuming, perhaps because they are laborious to automate.

7.3. Cyclic groups

For cyclic groups we have seen that $TO = 0$, so we need only calculate $H^*_I(QO)$, and the local cohomology spectral sequence collapses to give a short exact sequence

$$0 \longrightarrow \Sigma H^1_I(QO) \longrightarrow ko_*(BC) \longrightarrow H^0_I(QO) \longrightarrow 0,$$

which need not be additively split.

Furthermore, in the exact sequence

$$0 \longrightarrow Q\tau \longrightarrow QO \longrightarrow \overline{QO} \longrightarrow 0,$$

we showed in Section 6.3 that $Q\tau = \tau$ consists of all the η-multiples, which is precisely the 2-torsion of KO^C_* in positive degrees. This is 2^a in degrees 1 and 2 mod 8, where a is the number of simple real representations (namely 1 if the group is of odd order and 2 if it is of even order). In any case, τ is bounded below, so that

$$H^*_I(\tau) = H^0_I(\tau) = \tau.$$

Our calculation of $\overline{QO}$ shows that we may calculate local cohomology using the principal ideal (p_1), so that we have an exact sequence

$$0 \longrightarrow H^0_I(\overline{QO}) \longrightarrow \overline{QO} \longrightarrow \overline{QO}[\frac{1}{p_1}] \longrightarrow H^1_I(\overline{QO}) \longrightarrow 0.$$

Since the character of p_1 is vanishes only at the identity, $H^0_I(\overline{QO})$ consists of all relevant multiples of the regular representation (that is all multiples in degrees $8k \geq 0$ and even multiples in degrees $8k+4 \geq 0$), which forms a copy of $\mathbb{Z}$ in each degree $4k \geq 0$.

Turning to first local cohomology, there are three cases: degrees 0 mod 8, degrees 4 mod 8 and degrees 2 mod 4. For emphasis we denote the first two by $H^1_I(\overline{QO}_{sc})$ and the final one by $H^1_I(\overline{QO}_{asc})$. Since the local cohomology spectral sequence collapses, we can summarize the situation as follows, where $a = 1$ if C is of odd order and $a = 2$ if C is of even order.

$$ko^C_i = \begin{cases} H^1_I(\overline{QO}_{sc})_{8k+8} & \text{if } i = 8k+7 \geq 0 \\ 0 & \text{if } i = 8k+6 \geq 0 \\ H^1_I(\overline{QO}_{asc})_{8k+6} & \text{if } i = 8k+5 \geq 0 \\ \mathbb{Z}' & \text{if } i = 8k+4 \geq 0 \\ H^1_I(\overline{QO}_{sc})_{8k+4} & \text{if } i = 8k+3 \geq 0 \\ 2^a & \text{if } i = 8k+2 \geq 0 \\ 2^a \cdot H^1_I(\overline{QO}_{asc})_{8k+2} & \text{if } i = 8k+1 \geq 0 \\ \mathbb{Z} & \text{if } i = 8k+0 \geq 0 \end{cases}$$

In fact the extension in degrees $8k+1$ is non-split whenever possible (i.e., when n is divisible by 4), so that ko^{8k+1}_C is always the sum of a cyclic group of order 2 and another cyclic group.

We may illustrate this for cyclic groups of order 2,3 and 4.

EXAMPLE 7.3.1. *(Cyclic of order 2)* The case of the cyclic group of order 2 is a bit too simple to show typical behaviour, but it is a good group to start with. First, we note that both simple representations are real, so all representations are self-conjugate. Thus $RO = RU_{sc} \cong \mathbb{Z}^2$, $RU_{asc} = 0$, and $\overline{QO} = \overline{QO}_{sc}$ so that the contribution of $H^1_I(\overline{QO})$ is zero in degrees 1 and 5 mod 8.

The calculations are given in detail in Section 12.2 of Chapter 12. The notation is consistent with that used here, so the reader can refer to Chapter 12 for further details. The conclusion is that $H^1_I(\overline{QO}_{sc})_{8k+4} = [2^{4k+3}]$ and $H^1_I(\overline{QO}_{sc})_{8k+8} = [2^{4k+4}]$, where $[a]$ denotes a cyclic group of order a.

EXAMPLE 7.3.2. *(Cyclic of order 3)* This case is fairly typical of cyclic groups of odd order. Of the simple representations, there is one real one, and one conjugate pair of complex representations. Thus $RU_{sc} \cong \mathbb{Z}^2$, $RU_{asc} = \mathbb{Z}$.

In calculating the ideals JU_k we use characters, and write $\omega = e^{2\pi i/3}$. One verifies that JU_1 is spanned by the characters (033) and $(0, 1-\omega, 1-\omega^2)$, and that JU_2 is spanned by (033) and $(0, 3\omega, 3\omega^2)$. The other ideals can be recovered since multiplication by $p_1 = (033)$ gives an isomorphism $JU_k \longrightarrow JU_{k+2}$ for $k \geq 1$. Since τ is given by complex conjugation on characters and p_1 is real, it is easy to pick out the self-conjugate and anti self-conjugate parts.

Starting with degrees congruent to 0 mod 4 and identifying the self-conjugate part as a subgroup of $JU_{sc} = \mathbb{Z}$, the modules under multiplication by p_1 consist of

$$\cdots \mathbb{Z} \xrightarrow{3} \mathbb{Z} \xrightarrow{3} \mathbb{Z} \xrightarrow{\cong} 3\mathbb{Z} \xrightarrow{\cong} 9\mathbb{Z} \xrightarrow{\cong} \cdots,$$

where the last unmodified $\mathbb{Z}$ is in degree 0. It follows that $H^1_I(\overline{QO}_{sc})_{4k} = [3^k]$ for $k \geq 0$. Similarly, we find $H^1_I(\overline{QO}_{asc})_{4k-2} = [3^k]$ for $k \geq 0$. We then read off the answer from the collapsed local cohomology spectral sequence.

n	$ko_n(BC(3))$
$8k+7 \geq 7$	$[3^{2k+2}]$
$8k+6 \geq 6$	0
$8k+5 \geq 5$	$[3^{2k+2}]$
$8k+4 \geq 4$	$\mathbb{Z}'$
$8k+3 \geq 3$	$[3^{2k+1}]$
$8k+2 \geq 2$	2
$8k+1 \geq 1$	$[3^{2k+1}] \oplus 2$
$8k+0 \geq 0$	$\mathbb{Z}$
15	81
14	0
13	81
12	$\mathbb{Z}'$
11	27
10	2
9	$27 \oplus 2$
8	$\mathbb{Z}$
7	9
6	0
5	9
4	$\mathbb{Z}'$
3	3
2	2
1	$3 \oplus 2$
0	$\mathbb{Z}$

EXAMPLE 7.3.3. *(Cyclic of order 4)* This case is fairly typical of cyclic groups of even order. Of the simple representations, there are two real ones, and one conjugate pair of complex representations. Thus $RU_{sc} \cong \mathbb{Z}^3$, $RU_{asc} = \mathbb{Z}$.

To identify the ideals JU_k we work with characters and write $i = e^{2\pi i/4}$. One verifies that JU_1 is spanned by $(0202), (0, 1-i, 2, 1+i)$ and (0040), and that JU_2 is spanned by $(0242), (0, 2i, 4, -2i)$ and (0080). Multiplication by $p_1 = (0242)$ gives an isomorphism $JU_k \longrightarrow JU_{k+2}$ for $k \geq 1$. Since τ is given by complex conjugation on characters and p_1 is real, it is easy to pick out the self-conjugate and anti self-conjugate parts.

Starting with degrees congruent to 2 mod 4 and identifying the anti self-conjugate part as a subgroup of $JU_{asc} = \mathbb{Z}$, the modules under multiplication by p_1 consist of

$$\cdots \mathbb{Z} \xrightarrow{2} \mathbb{Z} \xrightarrow{2} \mathbb{Z} \xrightarrow{\cong} 2\mathbb{Z} \xrightarrow{\cong} 4\mathbb{Z} \xrightarrow{\cong} \cdots,$$

where the last unmodified $\mathbb{Z}$ is in degree -2. Thus $H^1_I(\overline{QO}_{asc})_{4k-2} = [2^k]$ for $k \geq 0$.

The self-conjugate part (in degrees 0 mod 4) involves a little more calculation. Since multiplication by p_1 gives an isomorphism $JU^{sc}_{2k} \longrightarrow JU^{sc}_{2k+2}$ for $k \geq 1$, we identify $\overline{QO}_{sc}[1/p_1]_{2n}$ with JU^{sc}_2 for all n, and local cohomology groups vanish below degree 0. For the remaining groups, we have exact sequences

$$RO \xrightarrow{p_1^{2k+1}} JU^{sc}_2 \longrightarrow H^1_I(\overline{QO}_{sc})_{8k} \longrightarrow 0$$

and

$$RSp \xrightarrow{p_1^{2k+2}} JU^{sc}_2 \longrightarrow H^1_I(\overline{QO}_{sc})_{8k+4} \longrightarrow 0.$$

It is easy to see how the orders grow, since the determinant of p_1 as a selfmap of $JU^{sc}_2 \cong \mathbb{Z}^2$ is 2^6. A little more linear algebra gives the exact structures of the local cohomology groups. First $H^1_I(\overline{QO}_{sc})_0 = 0$, and

$$H^1_I(\overline{QO}_{sc})_{8k} = [2^{4k+1}] \oplus [2^{2k-1}] \text{ for } k \geq 1.$$

Similarly

$$H^1_I(\overline{QO}_{sc})_{8k+4} = [2^{4k+3}] \oplus [2^{2k+1}] \text{ for } k \geq 0.$$

We then read off the answer from the collapsed local cohomology spectral sequence, except for the additive extension in degrees $8k+1$, which we discuss after tabulating the conclusion.

n	$ko_n(BC(4))$
$8k+7 \geq 7$	$[2^{4k+5}] \oplus [2^{2k+1}]$
$8k+6 \geq 6$	0
$8k+5 \geq 5$	$[2^{2k+2}]$
$8k+4 \geq 4$	$\mathbb{Z}'$
$8k+3 \geq 3$	$[2^{4k+3}] \oplus [2^{2k+1}]$
$8k+2 \geq 2$	2^2
$8k+1 \geq 1$	$[2^{2k+1}] \oplus 2^2$
$8k+0 \geq 0$	$\mathbb{Z}$
15	$512 \oplus 8$
14	0
13	16
12	$\mathbb{Z}'$
11	$128 \oplus 8$
10	2
9	$8 \oplus 2^2$
8	$\mathbb{Z}$
7	$32 \oplus 2$
6	0
5	4
4	$\mathbb{Z}'$
3	$8 \oplus 2$
2	2^2
1	$4 \oplus 2$
0	$\mathbb{Z}$

Turning to the extensions in degree $8k+1$, we see from the Atiyah-Hirzebruch spectral sequence that $ko_1(BC) = 4 \oplus 2$, so that the extension is non-split. Now consider multiplication by the Bott element, which gives a diagram

$$\begin{array}{ccccccccc} 0 & \longrightarrow & [2^{2k+3}] & \longrightarrow & ko_{8k+9}(BC) & \longrightarrow & 2^2 & \longrightarrow & 0 \\ & & \uparrow & & \uparrow{\scriptstyle\beta} & & \uparrow & & \\ 0 & \longrightarrow & [2^{2k+1}] & \longrightarrow & ko_{8k+1}(BC) & \longrightarrow & 2^2 & \longrightarrow & 0 \end{array}$$

The calculation shows that the map of kernels is a monomorphism and the map of quotients is an isomorphism. It follows that if the extension for $ko_{8k+1}(BC)$ is split, the same is true for $ko_{8k+9}(BC)$. It therefore remains to observe that the extension for $ko_9 BC$ is split. Somewhat surprisingly, this follows from the displayed diagram with $k = 0$. This corresponds to the fact that $[2^{2k+1}] \longrightarrow [2^{2k+3}]$ always induces the zero map on $\mathrm{Ext}_{\mathbb{Z}}(2, \cdot)$.

7.4. The quaternion group of order 8

7.4.A. Generic preamble. For quaternion groups we have seen that $TO = 0$, so we need only calculate $H_I^*(QO)$, and the local cohomology spectral sequence collapses to give a short exact sequence

$$0 \longrightarrow \Sigma H_I^1(QO) \longrightarrow ko_*(BQ) \longrightarrow H_I^0(QO) \longrightarrow 0,$$

which we shall see splits additively.

We will continue the discussion on the assumption that Conjecture 6.4.5 is true. Since we have verified them for the groups of order 8 and 16, the results are unconditional in those cases. In the exact sequence

$$0 \longrightarrow Q\tau \longrightarrow QO \longrightarrow \overline{QO} \longrightarrow 0,$$

we showed in Section 6.4 that $Q\tau = \tau$ consists of all the η-multiples. This is concentrated in degrees congruent to 1 or 2 mod 4. In positive degrees, it is the 2-torsion of KO_*^Q (namely 2^a in degrees 1 and 2 mod 8 and 2^b in degrees 5 and 6 mod 8, where $a = 2^{n-1} + 3$ is the number of simple real representations and $b = 2^{n-1}$ is the number of quaternionic representations. In negative degrees it is 2^b in all degrees 1 and 2 mod 4.

7.4.B. The local cohomology of τ. As before, local cohomology may be calculated using a single element. In fact, we may use the element p_*^2 of degree -8, so that we have an exact sequence

$$0 \longrightarrow H_I^0(\tau) \longrightarrow \tau \longrightarrow \tau[\frac{1}{p_*^2}] \longrightarrow H_I^1(\tau) \longrightarrow 0.$$

For the quaternion group of order 8, p_* lies in RSp, and so in effect one may use p_* rather than p_*^2. Since the local cohomology spectral sequence collapses and $ko_*(BQ)$ is connective, we deduce that the negative local cohomology of τ is zero. Thus $\tau[1/p_*]$ is 2^b in degrees 1 and 2 mod 4, where $b = 2^{n-1}$ is the number of quaternionic simple representations, and the limit is achieved in τ_{-2} and τ_{-3}.

We now turn to the specific case of Q_8. Since we may use p_* itself, for $k \geq 0$ we have an exact sequence

$$0 \longrightarrow H_I^0(\tau)_{4k+\epsilon} \longrightarrow \tau_{4k+\epsilon} \xrightarrow{p_*^{k+1}} 2^b \longrightarrow H_I^1(\tau)_{4k+\epsilon} \longrightarrow 0,$$

where $\epsilon = 1$ or 2.

Extending the above observations, p_* must surjective from τ_1 (either because $H^1_I(\tau)_1$ would contribute torsion to $ko_0(BQ)$, or by direct calculation). The corresponding result follows for τ_2. On the other hand, since p_* is divisible by 2, the map is zero for $k \geq 1$.

This completes the calculation, so that

$$H^0_I(\tau)_n = \begin{cases} 2^3 & \text{for } n = 1, 2 \\ 2^4 & \text{for } n \equiv 1, 2 \bmod 8,\ n \geq 9 \\ 2 & \text{for } n \equiv 5, 6 \bmod 8\ n \geq 5 \\ 0 & \text{otherwise} \end{cases}$$

and

$$H^1_I(\tau)_n = \begin{cases} 2 & \text{for } n \equiv 1, 2 \bmod 4,\ n \geq 5 \\ 0 & \text{otherwise} \end{cases}$$

7.4.C. The local cohomology of $\overline{QO}$. Our calculation of $\overline{QO}$ shows that for Q_8 we may calculate local cohomology using the principal ideal (p_*), so that we have an exact sequence

$$0 \longrightarrow H^0_I(\overline{QO}) \longrightarrow \overline{QO} \longrightarrow \overline{QO}[\frac{1}{p_*}] \longrightarrow H^1_I(\overline{QO}) \longrightarrow 0.$$

Since the character of p_* vanishes only at the identity, $H^0_I(\overline{QO})$ consists of all relevant multiples of the regular representation (that is all multiples in degrees $8k \geq 0$ and even multiples in degrees $8k + 4 \geq 0$), which forms a copy of $\mathbb{Z}$ in each degree $4k \geq 0$.

Since the local cohomology spectral sequence collapses, and since the only additive extensions involve η-multiples in the quotient, we can summarize the situation as follows, where $a_k = a = 2^{n-1} + 3$ and $b_k = b = 2^{n-1}$ for k sufficiently large.

$$ko^Q_i = \begin{cases} H^1_I(\overline{QO})_{8k+8} & \text{if } i = 8k + 7 \geq 0 \\ 2^{b_k} \oplus 2^{b_k} & \text{if } i = 8k + 6 \geq 0 \\ 2^{b_k} & \text{if } i = 8k + 5 \geq 0 \\ \mathbb{Z}' \oplus 2^{b_k} & \text{if } i = 8k + 4 \geq 0 \\ H^1_I(\overline{QO})_{8k+4} & \text{if } i = 8k + 3 \geq 0 \\ 2^{a_k} & \text{if } i = 8k + 2 \geq 0 \\ 2^{a_k} \cdot 2^{b_k} & \text{if } i = 8k + 1 \geq 0 \\ \mathbb{Z} \oplus 2^{b_k} & \text{if } i = 8k + 0 \geq 8 \end{cases}$$

Turning to the actual calculation of first local cohomology, there are two cases: degrees 0 and 4 mod 8. We have exact sequences

$$RO \xrightarrow{p_*^{2k+1}} JSp \longrightarrow H^1_I(\overline{QO})_{8k} \longrightarrow 0$$

and

$$RSp \xrightarrow{p_1^{2k+2}} JSp \longrightarrow H^1_I(\overline{QO})_{8k+4} \longrightarrow 0.$$

It is easy to see how the orders grow, since the determinant of p_* as a selfmap of $JSp \cong \mathbb{Z}^4$ is 2^5. A little more linear algebra gives the exact structures of the local cohomology groups. First $H^1_I(\overline{QO})_0 = 0$, and

$$H^1_I(\overline{QO})_{8k} = [2^{4k+2}] \oplus [2^{2k}]^2 \oplus [2^{2k-2}] \text{ for } k \geq 1.$$

Similarly

$$H^1_I(\overline{QO})_{8k+4} = [2^{4k+3}] \oplus [2^{2k+2}]^2 \oplus [2^{2k}] \text{ for } k \geq 0.$$

We then read off the answer from the collapsed local cohomology spectral sequence. There are additive extensions to resolve in degrees congruent to 1 mod 4. Using the Adams Spectral Sequence one may see that they split in degrees 5 and 9 [**18**, Theorems 4 and 5], and the fact that those in higher degrees split follows since multiplication by the Bott element is an isomorphism from that point on.

n	$ko_n(BQ_8)$
$8k+7 \geq 7$	$[2^{4k+6}] \oplus [2^{2k+2}]^2 \oplus [2^{2k}]$
$8k+6 \geq 6$	2
$8k+5 \geq 5$	$2 \oplus 2$
$8k+4 \geq 4$	$\mathbb{Z}' \oplus 2$
$8k+3 \geq 3$	$[2^{4k+3}] \oplus [2^{2k+2}]^2 \oplus [2^{2k}]$
$8k+2 \geq 10$	2^4
$8k+1 \geq 9$	$2 \oplus 2^4$
$8k+0 \geq 8$	$\mathbb{Z} \oplus 2$
15	$[2^{10}] \oplus 16^2 \oplus 4$
14	2
13	$2 \oplus 2$
12	$\mathbb{Z}' \oplus 2$
11	$128 \oplus 16^2 \oplus 4$
10	2^4
9	$2 \oplus 2^4$
8	$\mathbb{Z} \oplus 2$
7	$64 \oplus 4^2 \oplus 1$
6	2
5	$2 \oplus 2$
4	$\mathbb{Z}' \oplus 2$
3	$8 \oplus 4^2 \oplus 1$
2	2^3
1	2^3
0	$\mathbb{Z}$

7.5. The quaternion group of order 16

Next, we describe the calculation for the quaternion group of order 16. To make the account readable, there is a little repetition from the discussion for the group of order 8, but the calculation soon introduces some differences.

7.5.A. The local cohomology of τ. Once again, local cohomology may be calculated using a single element, but this time p_* does not lie in RSp, so we must use p_*^2. In any case, we have an exact sequence

$$0 \longrightarrow H^0_I(\tau) \longrightarrow \tau \longrightarrow \tau[\frac{1}{p_*^2}] \longrightarrow H^1_I(\tau) \longrightarrow 0.$$

Since the local cohomology spectral sequence collapses and $ko_*(BQ)$ is connective, we deduce that the negative local cohomology of τ is zero. Thus $\tau[1/p_*^2]$ is 2^b

in degrees 1 and 2 mod 4, where $b = 2^{n-1}$ is the number of quaternionic simple representations, and the limit is achieved in τ_{-2}, τ_{-3}, τ_{-6} and τ_{-7}.

Thus for $k \geq 0$ we have exact sequences

$$0 \longrightarrow H^0_I(\tau)_{8k+\epsilon} \longrightarrow \tau_{8k+\epsilon} \xrightarrow{p_*^{2k+2}} 2^b \longrightarrow H^1_I(\tau)_{8k+\epsilon} \longrightarrow 0,$$

where $\epsilon = 1, 2, 5$ or 6. Extending the above observations, p_*^2 must be surjective from τ_1 (either because $H^1_I(\tau)_1$ would contribute torsion to $ko_0(BQ)$, or by direct calculation). In higher degrees, we must make specific calculations with characters, which show that the map from τ_2 is also a surjection, the maps from τ_5, τ_6, τ_9 and τ_{10} have rank 1, and after that the maps are zero.

This completes the calculation, so that

$$H^0_I(\tau)_n = \begin{cases} 2^3 & \text{for } n = 1, 2 \\ 2^4 & \text{for } n = 9, 10 \\ 2^5 & \text{for } n \equiv 1, 2 \bmod 8,\ n \geq 17 \\ 2 & \text{for } n = 5, 6 \\ 2^2 & \text{for } n \equiv 5, 6 \bmod 8\ n \geq 13 \\ 0 & \text{otherwise} \end{cases}$$

and

$$H^1_I(\tau)_n = \begin{cases} 2 & \text{for } n = 5, 6, 9, 10 \\ 2^2 & \text{for } n \equiv 1, 2 \bmod 4,\ n \geq 13 \\ 0 & \text{otherwise} \end{cases}$$

7.5.B. The local cohomology of $\overline{QO}$. Our work on $\overline{QO}$ shows that we may calculate local cohomology using the principal ideal (p_*^2), so that we have an exact sequence

$$0 \longrightarrow H^0_I(\overline{QO}) \longrightarrow \overline{QO} \longrightarrow \overline{QO}[\frac{1}{p_*^2}] \longrightarrow H^1_I(\overline{QO}) \longrightarrow 0.$$

Since the character of p_* vanishes only at the identity, $H^0_I(\overline{QO})$ consists of all relevant multiples of the regular representation (that is all multiples in degrees $8k \geq 0$ and even multiples in degrees $8k + 4 \geq 0$), which forms a copy of $\mathbb{Z}$ in each degree $4k \geq 0$.

Since the local cohomology spectral sequence collapses, we can summarize the situation as follows, where $a_k = a = 2^{n-1} + 3$ and $b_k = b = 2^{n-1}$ for k sufficiently large.

$$ko^Q_i = \begin{cases} H^1_I(\overline{QO})_{8k+8} & \text{if } i = 8k + 7 \geq 0 \\ 2^{b_k} & \text{if } i = 8k + 6 \geq 0 \\ 2^{b_k} \cdot 2^{b_k} & \text{if } i = 8k + 5 \geq 0 \\ \mathbb{Z}' \oplus 2^{b_k} & \text{if } i = 8k + 4 \geq 0 \\ H^1_I(\overline{QO})_{8k+4} & \text{if } i = 8k + 3 \geq 0 \\ 2^{a_k} & \text{if } i = 8k + 2 \geq 0 \\ 2^{a_k} \cdot 2^{b_k} & \text{if } i = 8k + 1 \geq 0 \\ \mathbb{Z} \oplus 2^{b_k} & \text{if } i = 8k + 0 \geq 0 \end{cases}$$

Turning to the actual calculation of first local cohomology, there are two cases: degrees 0 and 4 mod 8. We have exact sequences

$$RO \xrightarrow{p_1^{2k+2}} JSp^2 \longrightarrow H^1_I(\overline{QO})_{8k} \longrightarrow 0$$

and

$$RSp \overset{p_1^{2k+2}}{\longrightarrow} JSp \longrightarrow H_I^1(\overline{QO})_{8k+4} \longrightarrow 0.$$

It is easy to see how the orders grow, since the determinant of p_1^2 as a selfmap of $JSp \cong \mathbb{Z}^6$ or $JSp^2 \cong \mathbb{Z}^6$ is 2^{12}. A little more linear algebra gives the exact structures of the local cohomology groups. We then read off the answer from the collapsed local cohomology spectral sequence. There are several unresolved additive extensions.

n	$ko_n(BQ_{16})$
$8k+7 \geq 7$	$\langle 2^{12k+12} \rangle$
$8k+6 \geq 14$	2^2
$8k+5 \geq 13$	$2^2 \cdot 2^2$
$8k+4 \geq 12$	$\mathbb{Z}' \oplus 2^2$
$8k+3 \geq 3$	$\langle 2^{12k+8} \rangle$
$8k+2 \geq 18$	2^5
$8k+1 \geq 17$	$2^2 \cdot 2^5$
$8k+0 \geq 16$	$\mathbb{Z} \oplus 2^2$
15	$[2^{11}] \oplus 16^2 \oplus 8 \oplus 4$
14	2^2
13	$2^2 \cdot 2^2$
12	$\mathbb{Z}' \oplus 2^2$
11	$[2^8] + 16^2 + 8 + 2$
10	2^4
9	$2 \cdot 2^4$
8	$\mathbb{Z} \oplus 2$
7	$128 \oplus 4^2 \oplus 2$
6	2
5	$2 \cdot 2$
4	$\mathbb{Z}' \oplus 2$
3	$16 \oplus 4^2$
2	2^3
1	2^3
0	$\mathbb{Z}$

REMARK 7.5.1. Adams spectral sequence calculations show that the extensions in degrees of the form $4k+1$ all split. In degrees of the form $4k-1$ the Adams spectral sequence leaves some unresolved additive extensions. It is a matter of linear algebra to complete the exact calculation of these groups by the local cohomology spectral sequence, but we have not taken time to do this beyond degree 15.

7.6. Klein 4-group

The case of the Klein 4-group is given in detail in Section 12.3 of Chapter 12. The notation is consistent with that used here, so the reader might like to refer to Section 12.3 now for more details and the conclusion of the calculation. We will summarize the process here.

As usual we start from the short exact sequences

$$0 \longrightarrow T \longrightarrow ko_V^* \longrightarrow \overline{QO} \longrightarrow 0$$

where $\overline{QO}$ is the image in periodic complex K-theory, and

$$0 \longrightarrow \tau \longrightarrow T \longrightarrow TO \longrightarrow 0$$

where τ consists of the η-multiples. In this case $TO = ZQU$ and $\tau = Q\tau$.

The module τ is bounded below, so torsion, and

$$H_I^*(\tau) = H_I^0(\tau) = \tau.$$

Indeed, τ is 2^4 in positive degrees congruent to 1 or 2 mod 8, and otherwise it is zero.

Next, character calculations allow the calculation of the local cohomology of $\overline{QO}$.

$$H^1_{JO}(\overline{QO})_i = \begin{cases} [2] & i = -4 \\ [2^{4k+1}]^2 \oplus [2^{4k}] & i = 8k \geq 0 \\ [2^{4k+4}]^2 \oplus [2^{4k+3}] & i = 8k+4 \geq 0 \\ 0 & \text{otherwise} \end{cases}$$

Finally, $TU = PC(-6)$, with generator q_{12}. The Sq^2 action makes TU into a Koszul complex for the maximal ideal of PP. Thus $TO = ZPC(-6) = PP(-6) \oplus PP(-12)$, with generators q_{12} and q_{12}^2, where $q_{12} = y_1y_2(y_1+y_2)$. Accordingly,

$$H_I^*(TO) = H_I^2(TO) = PP^\vee(2) \oplus PP^\vee(-4).$$

This gives the $E_{1\frac{1}{2}}$-term of the local cohomology spectral sequence. In addition to the d_1 coming from the connecting homomorphism, there is a single non-zero differential: $d_2 : (H_I^0)_1 \longrightarrow (H_I^2)_2$, and $ko_*(BV)$ can be read off. It is notable that the behaviour of the spectral sequence is forced from the fact that $ko_*(BV)$ is connective.

7.7. The alternating group A_4

In the usual way we start with the exact squence

$$0 \longrightarrow T \longrightarrow ko^*_{A_4} \longrightarrow \overline{QO} \longrightarrow 0$$

from Section 6.6. In this case $STO = ZTU$ and $Q\tau = \tau$.

7.7.A. Local cohomology of TO. The local cohomology of TO is easily dealt with. Indeed, the local cohomology can be calculated as a module over the polynomial ring $\mathbb{F}_2[\mu^2, \nu]$, where μ^2 has degree -8 and ν has degree -6. As a module over this, TO is free on one generator of degree -6 and one of degree -18. Accordingly the local cohomology is

$$H_I^*(TO) = H_I^2(TO) = \mathbb{F}_2[\mu^2, \nu]^\vee \otimes (\Sigma^8\mathbb{F}_2 \oplus \Sigma^{-4}\mathbb{F}_2)$$

7.7.B. Local cohomology of QO. First, note that the torsion τ is bounded below (indeed it is a copy of 2^2 in degrees ≥ 0 which are 1 or 2 mod 8). Thus

$$H_I^*(\tau) = H_I^0(\tau) = \tau.$$

It remains to calculate $H_I^*(\overline{QO})$. First recall that all representations are real, so all representations are self-conjugate. We observe that local cohomology may be calculated from the principal ideal (p_1), where $p_1 = p_1(\tau) = (0433)$, so that there is an exact sequence

$$0 \longrightarrow H_I^0(\overline{QO}) \longrightarrow \overline{QO} \longrightarrow \overline{QO}[\frac{1}{p_1}] \longrightarrow H_I^1(\overline{QO}) \longrightarrow 0.$$

As usual, $H^0_I(\overline{QO})$ consists of characters concentrated at the identity, which are the relevant multiples of the regular representation (namely all multiples in degrees $8k \geq 0$ and even multiples in degrees $8k+4 \geq 4$), which forms a copy of $\mathbb{Z}$ in each degree $4k \geq 0$. We now turn to $H^1_I(\overline{QO})$.

7.7.C. First local cohomology of $\overline{QO}$. In degrees congruent to 2 mod 4 we work in $RU_{asc} \cong \mathbb{Z}$, which consists of multiples of $\alpha - \alpha^2$, and evidently p_1 acts as multiplication by 3. Identifying the anti-self-conjugate part as a subgroup of $RU_{asc} = \mathbb{Z}$, the modules under multiplication by p_1 consists of

$$\cdots \mathbb{Z} \xrightarrow{3} \mathbb{Z} \xrightarrow{3} \mathbb{Z} \xrightarrow{\cong} 3\mathbb{Z} \xrightarrow{\cong} 9\mathbb{Z} \xrightarrow{\cong} \cdots,$$

where the last unmodified $\mathbb{Z}$ is in degree -2. Hence we see $H^1_I(\overline{QO}_{asc})_{4k-2} = [3^k]$ for $k \geq 0$.

Once again degrees divisible by 4 are a little more complicated. We recorded the ideals JU_k in Section 6.6, and it is visible that multiplication by $p_1 : JU_k \longrightarrow JU_{k+2}$ is an isomorphism for $k \geq 4$. Accordingly, the local cohomology is zero below degree -4, and we may make identifications

$$\overline{QO}[1/p_1]_{8k} \cong JU_4$$

and

$$\overline{QO}[1/p_1]_{8k+4} \cong JU_6.$$

In these terms, we may calculate the local cohomology. In degree -4 we have the exact sequence

$$JU_2 \xrightarrow{p_1^2} JU_6 \longrightarrow H^1_I(\overline{QO})_{-4} \longrightarrow 0.$$

For the remaining groups we have exact sequences

$$RO \xrightarrow{p_1^{2k+2}} JU_4 \longrightarrow H^1_I(\overline{QO}_{sc})_{8k} \longrightarrow 0$$

and

$$RSp \xrightarrow{p_1^{2k+4}} JU_6 \longrightarrow H^1_I(\overline{QO}_{sc})_{8k+4} \longrightarrow 0.$$

With some linear algebra we find

$$H^1_I(\overline{QO})_i = \begin{cases} [2^{4k}] \oplus [3^{2k}] & \text{if } i = 8k \geq 0 \\ [2^{4k-1}] \oplus [3^{2k-1}] & \text{if } i = 8k-4 \geq 0 \\ 2 & \text{if } i = -4 \end{cases}$$

7.7.D. Local cohomology spectral sequence. We have now identified the $E_{1\frac{1}{2}}$-term of the local cohomology spectral sequence displayed in Figure 7.7.D. Turning to the differentials, we note that all potential differentials are detected 2-locally. For these, we note that BA_4 is a 2-local retract of $BV(2)$. We will show that this means we may deduce all differentials from the known behaviour of the local cohomology spectral sequence for $V(2)$. The reader should refer forward to Section 12.3 of Chapter 12 for a description of the behaviour of the LCSS for $V(2)$; we deduce the behaviour for A_4 from it.

LEMMA 7.7.1. *The d_2 differentials are all zero except for*

$$d_2 : H^0_I(ku^*_{A_4})_1 \longrightarrow H^2_I(TO_{A_4})_2,$$

which is of rank 1.

FIGURE 7.7.1. The $E^{1\frac{1}{2}}$-page of the local cohomology spectral sequence for $ko_*(BA_4)$

H^2_{JO}	H^1_{JO}	H^0_{JO}	
2^3 ←	$[2^{11}] \oplus [3^5]$	$\mathbb{Z}$	20
0	0	0	
2	$[3^5]$	2^2	
0	0	2^2	
2^2 ←	$[2^8] \oplus [3^4]$	$\mathbb{Z}$	16
0	0	0	
2^2	$[3^4]$	0	
0	0	0	
2 ←	$[2^7] \oplus [3^3]$	$\mathbb{Z}$	12
0	0	0	
2	$[3^3]$	2^2	
0	0	2^2	
2^2 ←	$[2^4] \oplus [3^2]$	$\mathbb{Z}$	8
0	0	0	
0	$[3^2]$	0	
0	0	0	
2 ←	$[2^3] \oplus [3]$	$\mathbb{Z}$	4
0	0	0	
2	$[3]$	2^2	
0	0	2^2	
0	0	$\mathbb{Z}$	0
0	0		
0	0		
0	0		
2 ←	2		−4

Proof: The named differential from degree 1 must be an epimorphism because $ko_0(BA_4) \cong \mathbb{Z}$. We now argue that all other d_2 differentials are zero.

First, since ko_* is a summand, the only potential differentials on the H^0_I column are the d_2s from the copies of 2^2 in degrees 1 and 2 mod 4.

By naturality of the spectral sequence, we have a commutative square

$$\begin{array}{ccc} H^0_I(ku_0^{A_4}) & \xrightarrow{d_2} & H^2_I(TO_{A_4}) \\ \downarrow & & \downarrow \\ H^0_I(ku_0^{V(2)}) & \xrightarrow{d_2} & H^2_I(TO_{V(2)}). \end{array}$$

We claim that the right-hand vertical is a monomorphism. In more detail, the TO-QO short exact sequence is natural, so by transfer the map $TO_{A_4} \longrightarrow TO_{V(2)}$ is a split monomorphism of modules over $ZH^*(BA_4;\mathbb{F}_2)$-modules. Since the local cohomology of $TO_{V(2)}$ may be calculated using $H^*(BA_4;\mathbb{F}_2)$-module structure, the right-hand vertical is a monomorphism as claimed. Since all the d_2-differentials for $V(2)$ are zero except from degree 1, the same follows for A_4. □

LEMMA 7.7.2. *The d_1 differentials*

$$d_1 : H^1_I(QO)_{4k} \longrightarrow H^2_I(TO_{A_4})_{4k},$$

are all of maximal rank (namely of rank 1 for $k = -1$ or $k \geq 1$, and zero otherwise).

Proof: First we observe that by sparsity, the only possible nonzero d_1 differentials are those named. The codomains of these d_1s are all elementary abelian, so d_1 vanishes on multiples of 2 and induces

$$d_1 : H^1_I(QO_{A_4})_{4k}/2 \longrightarrow H^2_I(TO_{A_4})_{4k}.$$

Since the domain is of lower rank (0 or 1) than the codomain, it suffices to show this map is a monomorphism. For this we consider the commutative square

$$\begin{array}{ccc} H^1_I(QO_{A_4})_{4k}/2 & \xrightarrow{d_1} & H^2_I(TO_{A_4})_{4k} \\ \downarrow & & \downarrow \\ H^1_I(QO_{V(2)})_{4k}/2 & \xrightarrow{d_1} & H^2_I(TO_{V(2)})_{4k}. \end{array}$$

We show in Lemma 12.3.2 that the bottom horizontal is a monomorphism for $k \geq -1$, so it suffices to show that the left-hand vertical is a monomorphism. This follows from the fact that by transfer, $ko^*_{A_4} \longrightarrow ko^*_{V(2)}$ is a 2-local split monomorphism as in Lemma 7.7.1. □

7.7.E. Conclusion. We may now read off $ko_*(BA_4)$ from the spectral sequence. It is convenient to write

$$\sum_n a_n t^n = \frac{t^{-4} + t^8}{(1-t^8)(1-t^6)}$$

for the Hilbert series of the dual of $H^2_I(TO)$ as calculated in Subsection 7.7.A above.

n	$ko_n(BA_4)$
$8k+7 \geq 7$	$[2^{4k+3}] \oplus [3^{2k+2}]$
$8k+6 \geq 6$	$2^{a_{8k+8}-1}$
$8k+5 \geq 5$	$[3^{2k+2}]$
$8k+4 \geq 4$	$\mathbb{Z}' \oplus 2^{a_{8k+6}}$
$8k+3 \geq 3$	$[2^{4k+2}] \oplus [3^{2k+1}]$
$8k+2 \geq 2$	$2^2 \oplus 2^{a_{8k+4}-1}$
$8k+1 \geq 9$	$2^2 \oplus [3^{2k+1}]$
$8k+0 \geq 8$	$\mathbb{Z} \oplus 2^{a_{8k+2}}$
15	$[2^7] \oplus [3^4]$
14	2
13	$[3^4]$
12	$\mathbb{Z}' \oplus 2^2$
11	$[2^6] \oplus [3^3]$
10	2^2
9	$2^2 \oplus [3^3]$
8	$\mathbb{Z} \oplus 2$
7	$[2^3] \oplus [3^2]$
6	2
5	$[3^2]$
4	$\mathbb{Z}'$
3	$[2^2] \oplus [3]$
2	2^2
1	$2 \oplus [3]$
0	$\mathbb{Z}$

CHAPTER 8

Dihedral Groups

In this chapter we apply the methods described in Chapters 6 and 7 in a more substantial class of examples by calculating ko^*_D, $ko^*(BD)$ and $ko_*(BD)$ where D is a dihedral group D of order 2^{n+2} for $n \geq 1$. Even for cohomology, the precise calculation (see Subsection 8.2.F) is not easy to summarize, but we find that the image $\overline{QO}$ in periodic complex K-theory is generated in negative degrees by JSp in degree -4 and JU_4 in degree -8, which sits well with the fact that all representations are real, and the simple representations are of dimension 1 or 2. The calculation of the homology is given in Corollary 8.5.9.

As before, the starting point is the known value of ku^*_D from [**28**]. Applying the Bockstein spectral sequence, we obtain ko^*_D, using the Adams Spectral Sequence to resolve one additive extension in ko^6_D. The completion involved in obtaining $ko^*(BD)$ is barely noticable. Finally, we apply the local cohomology spectral sequence to deduce $ko_*(BD)$.

8.1. Recollections

To ease comparison between different dihedral groups, we choose notation by reference to $O(2)$. We will use the notation of Subsection 5.6.B.

8.1.A. The group D. The group D consists of the 2^{n+1} rotations r_θ and the 2^{n+1} reflections $s_{\theta/2}$, where $\theta = 2k\pi/2^{n+1}$ for $k = 0, 1, \ldots, 2^{n+1} - 1$. We write $r_{0,n}, r_{1,n}, \ldots, r_{2^{n+1}-1,n}$ and $s_{0,n}, s_{1,n}, \ldots, s_{2^{n+1}-1,n}$ for these, even omitting the n when confusion is unlikely. The rotations form a cyclic subgroup C. The two elements $r_{0,n}$ (the identity) and $r_{2^n,n}$ (of order 2) are central, and the other rotations are in conjugacy classes of two $\{r_{i,n}, r_{-i,n}\}$. The reflections s_θ are in two conjugacy classes, with 2^n elements in each, one with θ consisting of the even multiples of $2\pi/2^{n+2}$ and one with odd multiples. The even conjugacy class generates a dihedral subgroup D', and the odd conjugacy class generates a dihedral subgroup D'', both of order 2^{n+1}. We write $s' = s_0$ and $s'' = s_{2\pi/2^{n+2}}$ for representatives of these conjugacy classes.

There are four one dimensional representations: the trivial representation ϵ, together with $\hat{D}'$, $\hat{D}''$, and $\delta = \hat{C} = \hat{D}'\hat{D}''$, where $\hat{H}$ means the representation with kernel H for a subgroup H of index 2. Since $D' \cap D''$ is a cyclic subgroup of index 2 in C we see, for example, that $\hat{D}'$ restricts to δ on D''. The remaining simple representations $\sigma_1, \ldots, \sigma_{2^n-1}$ are of dimension 2, and restricted from $O(2)$. Accordingly all representations are real. It is easy to see

$$\sigma_k \hat{D}' = \sigma_k \hat{D}'' = \sigma_{k+2^n}.$$

Notice that as representations of D

$$\sigma_k = \sigma_{k+2^{n+1}}, \text{ and } \sigma_{k+2^n} = \sigma_{2^n-k}.$$

In particular

$$\sigma_{2^n} = \hat{D}' + \hat{D}''.$$

The cohomology ring of D is

$$H^*(BD;\mathbb{F}_2) = \mathbb{F}_2[w_1, w_2, x]/(x(w_1 + x))$$

where x and $w_1 + x$ are the first Stiefel-Whitney classes of $\hat{D}'$ and $\hat{D}''$. Notice that although w_1 and w_2 restrict from $O(2)$, the class $x = x_n$ depends on n, and x_n restricts to 0 on D' and to w_1 on D''. The cohomology ring has Hilbert series $1 + 2t + 3t^2 + 4t^3 + \cdots = 1/(1-t)^2$ as if it were polynomial on two generators. It is useful to introduce notation for the double $DH^*(BD;\mathbb{F}_2) = \mathbb{F}_2[c_1, c_2, X]/(X(c_1 + X))$ where $c_1 = c_1(\sigma_1) = w_1^2, c_2 = c_2(\sigma_1) = w_2^2$ and $X = c_1(\hat{D}') = x^2$.

8.1.B. The complex equivariant connective K-theory, ku_D^*. We recall from [**28**, 2.5] that there is an exact sequence

$$0 \longrightarrow TU \longrightarrow ku_D^* \longrightarrow QU \longrightarrow 0$$

where QU is the image in $KU_D^* = RU(D)[v, v^{-1}]$, and TU is the v-power torsion. Indeed QU is the modified Rees ring, generated by $RU(D)[v]$ and the Chern classes of representations. Furthermore, TU is detected in mod 2 cohomology and in fact TU is the free $DH^*(BD;\mathbb{F}_2)$-submodule of $H^*(BD;\mathbb{F}_2)$ generated by the degree -6 class c_1c_2.

8.2. The real equivariant connective K-theory, ko_D^*

We now wish to apply the Bockstein spectral sequence to ku_D^* to deduce ko_D^*. We have a fairly complete description of $d^1 : ku_D^n \longrightarrow ku_D^{n+2}$. Indeed, it preserves torsion so the short exact sequence

$$0 \longrightarrow TU \longrightarrow ku_D^* \longrightarrow QU \longrightarrow 0$$

may be viewed as a short exact sequence of chain complexes. The differential on TU is Sq^2 and that on QU is $1 - \tau$ from degrees divisible by 4 and $1 + \tau$ from other even degrees. The short exact sequence of chain complexes gives a long exact sequence of cohomology. We will calculate cycles and boundaries for TU and QU, and then the connecting homomorphism. In fact we display $H(TU; Sq^2)$ and $H(QU)$ in a Bockstein spectral sequence chart we refer to as the $E^{1\frac{1}{2}}$-page, treating the connecting homomorphism as a late d^1. There can be no d^2 because both TU and QU are in even degrees.

8.2.A. The action of Sq^2 on TU. From the Cartan formula we find $Sq^1 w_1 = w_1^2$, $Sq^1 x = x^2 = w_1 x$, $Sq^2 w_1 = Sq^2 x = 0$ and $Sq^2 w_2 = w_2^2$, and the Wu formula shows $Sq^1 w_2 = w_1 w_2$. This determines the Sq^2 action.

Indeed, Sq^2 is linear over the double $DDH^*(BD;\mathbb{F}_2)$ of $DH^*(BD;\mathbb{F}_2)$, namely the subalgebra generated by $p_1 := c_1^2, p_2 := c_2^2$ and $X^2 = Xc_1$. To give the action on $DH^*(BD;\mathbb{F}_2)$ we therefore only need to calculate

$$Sq^2(c_1) = c_1^2, Sq^2(X) = Xc_1, Sq^2(c_2) = c_1c_2, Sq^2(Xc_2) = 0,$$

giving the pattern

$$
\begin{array}{lllllllll}
1 \\
\\
c_1 & X \\
\downarrow & \downarrow \\
c_1^2 & Xc_1 & c_2 \\
 & & \downarrow \\
c_1^3 & Xc_1^2 & c_1c_2 & Xc_2 \\
\downarrow & \downarrow \\
c_1^4 & Xc_1^3 & c_1^2c_2 & Xc_1c_2 & c_2^2 \\
 & & \downarrow & \downarrow \\
c_1^5 & Xc_1^4 & c_1^3c_2 & Xc_1^2c_2 & c_1c_2^2 & Xc_2^2 \\
\downarrow & \downarrow & & & \downarrow & \downarrow \\
c_1^6 & Xc_1^5 & c_1^4c_2 & Xc_1^3c_2 & c_1^2c_2^2 & Xc_1c_2^2 & c_2^3 \\
 & & \downarrow & \downarrow & & & \downarrow \\
c_1^7 & Xc_1^6 & c_1^5c_2 & Xc_1^4c_2 & c_1^3c_2^2 & Xc_1^2c_2^2 & c_1c_2^3 & Xc_2^3 \\
\downarrow & \downarrow & & & \downarrow & \downarrow \\
c_1^8 & Xc_1^7 & c_1^6c_2 & Xc_1^5c_2 & c_1^4c_2^2 & Xc_1^3c_2^2 & c_1^2c_2^3 & Xc_1c_2^3 & c_2^4
\end{array}
$$

The Sq^2-homology is concentrated in degrees $0; -6, -8; -14, -16; -22, -24; \ldots$, being represented by the free $\mathbb{F}_2[p_2]$-submodule on 1 and Xc_2, where $p_2 = c_2^2$. The cycles are as illustrated

$$
\begin{array}{lllllllll}
1 \\
\\
\cdot & \cdot \\
\\
c_1^2 & Xc_1 & \cdot \\
\\
\cdot & \cdot & c_1c_2 & Xc_2 \\
\\
c_1^4 & Xc_1^3 & \cdot & \cdot & c_2^2 \\
\\
\cdot & \cdot & c_1^3c_2 & Xc_1^2c_2 & \cdot & \cdot \\
\\
c_1^6 & Xc_1^5 & \cdot & \cdot & c_1^2c_2^2 & Xc_1c_2^2 & \cdot \\
\\
\cdot & \cdot & c_1^5c_2 & Xc_1^4c_2 & \cdot & \cdot & c_1c_2^3 & Xc_2^3 \\
\\
c_1^8 & Xc_1^7 & \cdot & \cdot & c_1^4c_2^2 & Xc_1^3c_2^2 & \cdot & \cdot & c_2^4
\end{array}
$$

This is the direct sum of a free module over $DDH^*(BD; \mathbb{F}_2)$ on 1 together with the ideal of $DDH^*(BD; \mathbb{F}_2)$ generated by c_1c_2 and Xc_2. It will be sufficient to consider the structure as a module over $PP = \mathbb{F}_2[p_1, p_2]$, where $p_1 = c_1^2$ and $p_2 = c_2^2$; as a PP-module the Sq^2-cycles are free on $1, Xc_1, c_1c_2$ and Xc_2.

We are interested in the $DH^*(BD; \mathbb{F}_2)$-submodule TU, which is free on c_1c_2. Since c_1c_2 is a Sq^2-cycle, we may simply translate the whole discussion down by 6 degrees to obtain the following conclusion.

LEMMA 8.2.1. *(i) The Sq^2-homology of TU is concentrated in degrees*

$$-6; -12, -14; -20, -22; -28, -30; \ldots,$$

being represented by the free $\mathbb{F}_2[p_2]$*-submodule on* c_1c_2 *and* $Xc_1c_2^2$*, where* $p_2 = c_2^2$.

(ii) The module of Sq^2*-cycles of* TU *a free module over* $PP = \mathbb{F}_2[p_1, p_2]$ *on* c_1c_2 *(in degree* -6*),* $Xc_1^2c_2$ *(in degree* -10*), and the two classes* $c_1^2c_2^2$ *and* $Xc_1c_2^2$ *(in degree* -12*).* □

8.2.B. The Bockstein spectral sequence and representation theory. We must consider the complex representation ring $RU(D)$. To motivate the questions we study, we describe the Bockstein spectral sequence. The main simplifying factor in this example is that all representations are self-conjugate, so that $1+\tau = 2$ and $1 - \tau = 0$.

In discussing the η-Bockstein spectral sequence, we refer to the line of slope 1 meeting the 0-line at n as the *n-diagonal*; at E^1 all the groups on the n-diagonal are the same. Considering the Bockstein spectral sequence in positive degrees, we see that d^1 is multiplication by 2 from the n-diagonal when $n \equiv 2 \bmod 4$, and it is zero if $n \equiv 0 \bmod 4$. Thus E^2 lies on n-diagonals with n divisible by 4, $d^2 = 0$ and d^3 is an isomorphism from the n-diagonal when $n \equiv 4 \bmod 8$ (except for the differential from the 0-line). Thus, writing R for $RU(D)$, we obtain

$s=3$				$\cdot$		$\cdot$		$\cdot$		$\cdot$		$\cdot$		$\cdot$
$s=2$			$R/2$		$\cdot$		$\cdot$		$\cdot$		$R/2$		$\cdot$	
$s=1$		$R/2$		$\cdot$		$\cdot$		$\cdot$		$R/2$		$\cdot$		$\cdot$
$s=0$	R		$\cdot$		$2R$		$\cdot$		R		$\cdot$		$2R$	
E^∞	0		2		4		6		8		10		12	

Since QU is torsion free, we again see that after d^1, QU contributes 0 except in degrees divisible by 4, and the contribution to the $E^{1\frac{1}{2}}$-page is

$s=3$	$J_6/2J_5$		$\cdot$		$J_4/2J_3$		$\cdot$		$J_2/2J_1$	
$s=2$		$\cdot$		$J_4/2J_3$		$\cdot$		$J_2/2J_1$		$\cdot$
$s=1$	$\cdot$		$J_4/2J_3$		$\cdot$		$J_2/2J_1$		$\cdot$	
$s=0$		J_4		$\cdot$		J_2		$\cdot$		R
$E^{1\frac{1}{2}}$		-8		-6		-4		-2		0

where J_n denotes the ideal JU_n.

8.2.C. The connecting homomorphism. By the $E^{1\frac{1}{2}}$-page there are no further d^1s on $H(TU)$, and the classes of $H(TU)$ are widely enough separated that they cannot be related by d^2 or d^3. Since QU is torsion free, it follows that $H(QU)$ is entirely on n-diagonals with n divisible by 4. Thus the classes on the n-diagonal with $n = -6, -14, -22, \ldots$ cannot support a differential, and cannot receive a d^3 from $H(QU)$. It follows that the connecting homomorphism must be surjective

$$d^1 : H(QU)_{-(8k+4)} \twoheadrightarrow H(TU)_{-(8k+6)} \cong \mathbb{F}_2.$$

8.2.D. The behaviour of d^3. The above discussion determines the E^3-page, and the fact that $\eta^3 = 0$ means that d^3 must be exact off the 0-line. For brevity, we write $h_{-2n} = H(QU)_{-2n}$ and $\overline{h}_{-(8k+4)} = \ker(d^1)$, so that we have a long exact sequence

$$\cdots \longleftarrow h_{-24} \longleftarrow \overline{h}_{-20} \oplus 2 \longleftarrow h_{-16} \longleftarrow \overline{h}_{-12} \oplus 2 \longleftarrow h_{-8} \longleftarrow \overline{h}_{-4} \longleftarrow 0.$$

We write $Z\overline{h}_n$ for the cycles in this complex in degree n, so that $Z\overline{h}_{-8k}$ is a submodule of h_{-8k}, $Z\overline{h}_{-4} = 0$ and $Z\overline{h}_{-8k-4}$ is a submodule of $\overline{h}_{-8k-4} \oplus 2$.

Furthermore, since the spectral sequence commutes with the action of the Bott element, the component of $d^3 : h_{-8k} \longrightarrow \overline{h}_{8k-4}$ is zero.

In terms of the filtration we have

ko^D_n	n	
0	$8k+7$	≥ 7
0	$8k+6$	≥ 6
0	$8k+5$	≥ 5
$2R$	$8k+4$	≥ 4
0	$8k+3$	≥ 3
$R/2R$	$8k+2$	≥ 2
$R/2R$	$8k+1$	≥ 1
R	$8k+0$	≥ 0
0	-1	
0	-2	
0	-3	
$2J_1$	-4	
0	-5	
$Z\overline{h}_{-8k} \cdot 2^{2k-1}$	$-8k+2$	≤ -6
$Z\overline{h}_{-8k}$	$-8k+1$	≤ -7
$2^{2k-2} \oplus J^*_{4k}$	$-8k+0$	≤ -8
0	$-8k-1$	≤ -9
$Z\overline{h}_{-8k-4} \cdot 2^{2k}$	$-8k-2$	≤ -10
$Z\overline{h}_{-8k-4}$	$-8k-3$	≤ -11
$(2^{2k} \oplus \overline{J}_{4k+2})^*$	$-8k-4$	≤ -12
0	$-8k-5$	≤ -13

Here J^*_{4k} is the inverse image of $Z\overline{h}_{-8k}$ under the quotient map $J_{4k} \longrightarrow J_{4k}/2J_{4k-1}$. Similarly $(2^{2k} \oplus \overline{J}_{4k+2})^*$ is the inverse image of $Z\overline{h}_{-(8k+4)}$ under the quotient map

$$2^{2k} \oplus \overline{J}_{4k+2} \longrightarrow 2 \oplus \overline{J_{4k+2}/2J_{4k-1}} = 2 \oplus \overline{h}_{-(8k+4)}.$$

The dot in dimensions $-8k \pm 2$ refers to extensions we will determine in the next section.

8.2.E. The Bockstein spectral sequence for the dihedral group of order 8. This seems a convenient point to give details of the concrete example of D_8. We display the Bockstein spectral sequence in Figure 8.2.1.

Amongst the features to observe are the torsion groups in degrees -6, -10, -14, $\ldots$, the survival of JU_{4k} along with the fact that there is torsion in degree $-8k \leq 16$, and the fact that there are η-multiples in arbitrarily negative degrees (in degrees $-8k+1$, $-8k+2$ for $k \geq 1$).

8.2.F. The connective real K-cohomology, ko^*_D. This is a convenient point to summarize the outcome of the calculation, although it depends on a number of representation theoretic calculations deferred to Section 8.3.

The resulting values are tabulated below.

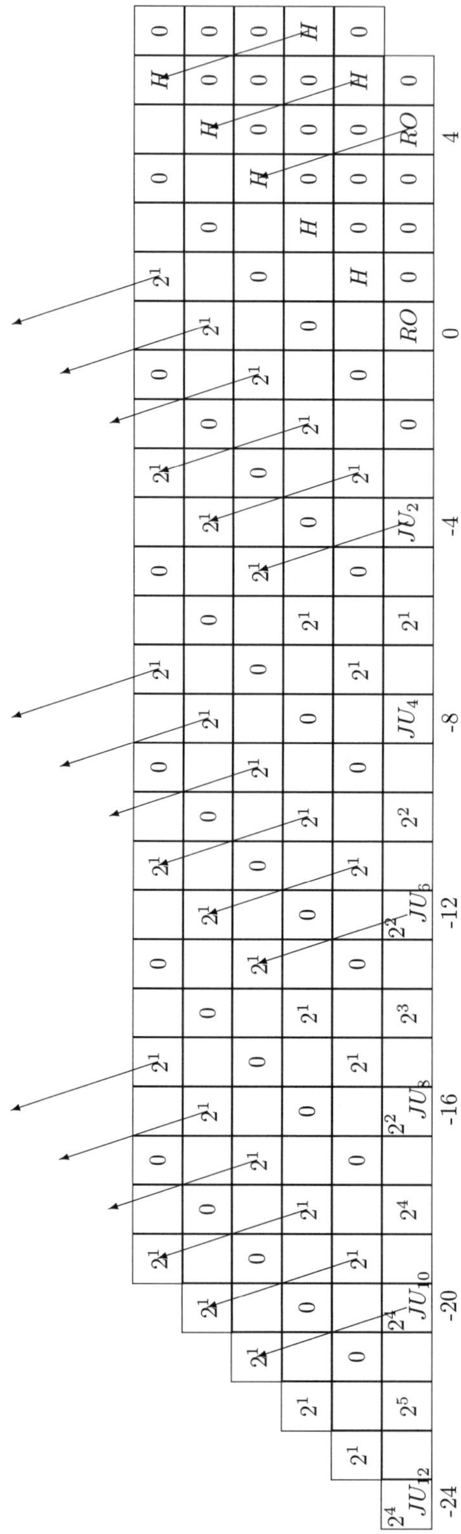

FIGURE 8.2.1. The E^2-page for the dihedral group D_8

ko_n^D	n	
0	$8k+7$	≥ 7
0	$8k+6$	≥ 6
0	$8k+5$	≥ 5
$2R$	$8k+4$	≥ 4
0	$8k+3$	≥ 3
$R/2R$	$8k+2$	≥ 2
$R/2R$	$8k+1$	≥ 1
R	$8k+0$	≥ 0
0	-1	
0	-2	
0	-3	
$2J_1$	-4	
0	-5	
$2^{2^n-1} \oplus 2^{2k-1}$	$-8k+2$	≤ -6
2^{2^n-1}	$-8k+1$	≤ -7
$2^{2k-2} \oplus J_{4k}$	$-8k+0$	≤ -8
0	$-8k-1$	≤ -9
2^{2k}	$-8k-2$	≤ -10
0	$-8k-3$	≤ -11
$2^{2k} \oplus 2J_{4k+1}$	$-8k-4$	≤ -12
0	$-8k-5$	≤ -13

The first ingredient is that we show in Proposition 8.3.10 that the pattern of d^3s consists of isomorphisms and zero maps.

LEMMA 8.2.2. *The differential d^3 is zero from degrees divisible by 8. Hence $Z\overline{h}_{-8k} = h_{-8k}$ and $Z\overline{h}_{-8k-4} = 0$.*

We then insert the values of the groups $h_* = H_*(QU)$, and it remains to show that the unresolved additive extensions all split. This follows from the result in degree -6, since the potential elements of order 4 are all multiples of the element in degree -6 by monomials in c_1^2 and $Xc_1 = c_1(\hat{D}')c_1$.

PROPOSITION 8.2.3. *Multiplication by the Bott map $\beta : ko_D^6 \longrightarrow ko_D^{-2}$ is a monomorphism. Hence, ko_D^6 is an elementary 2-group and all remaining additive extensions split.*

Proof: We decompose H^*BD into irreducible $\mathcal{A}(1)$-modules and determine that the analogous statement is true at E_2 of the Adams spectral sequence, and that the differentials also map monomorphically, so that the result holds at E_∞. Details follow the statement of Lemma 8.2.4 below.

Note that this exactly mimics the case of $O(2)$ (Subsection 5.6.B). The additional summands in the following decomposition of H^*BD contribute

(i) an additional copy of $ko^*BO(1)$ from the additional 1-dimensional representations of D not pulled back from O(2),
(ii) some addtional $(2, \eta, \alpha, \beta)$ torsion connected to Xc_1, and
(iii) classes which support differentials enforcing the relation among powers of σ_1.

LEMMA 8.2.4. *As an $\mathcal{A}(1)$-module, H^*BD is the direct sum of*

(i) $\{w_1^i \mid i > 0\} \cong \widetilde{H}^*BO(1)$,
(ii) $\{x^i \mid i > 0\} \cong \widetilde{H}^*BO(1)$,
(iii) *a trivial* $\mathcal{A}(1)$*-module* $\mathbb{F}_2[w_2^4]$,
(iv) *jokers* $\mathbb{F}_2[w_2^4] \cdot \{w_2, w_1w_2, w_2^2, w_1w_2^2 + w_1^3w_2, w_1^2w_2^2\}$,
(v) *inverted question marks* $\mathbb{F}_2[w_2^4] \cdot \{xw_2, xw_2^2 + x^3w_2, x^2w_2^2\}$,
(vi) *augmentation ideals*
$\mathbb{F}_2[w_2^4] \cdot \{xw_2^3, x^2w_2^3, x^3w_2^3, xw_2^4, x^2w_2^4, x^3w_2^4 + x^5w_2^3, x^4w_2^4\}$,
(vii) *free* $\mathcal{A}(1)$*-modules* $\mathbb{F}_2[w_1^4, w_2^4]\langle w_1^2w_2, w_2^3\rangle$
(viii) *free* $\mathcal{A}(1)$*-modules* $\mathbb{F}_2[x^4, w_2^4]\langle x^2w_2, x^4w_2^3\rangle$

Proof: It is a tedious but simple matter to verify that the sum of these modules gives all of H^*BD. A check of Poincaré series shows that the sum must be direct. □

Proof of Proposition 8.2.3, continued: See Appendix A for the Ext modules of all the $\mathcal{A}(1)$-modules in Lemma 8.2.4. The Bott map β acts monomorphically in $\mathrm{Ext}_{\mathcal{A}(1)}$ for the first five summands. For the augmentation ideals, the Bott map acts monomorphically except on the classes in Ext^0, but these are all in codegrees 12 and higher. Similarly, the free modules contribute only to codegrees 12 and higher. Thus the Bott map acts monomorphically in total codegrees less than 12 at the E_2 term of the Adams spectral sequence for ku_D^*.

It then suffices to observe that there will be no more boundaries in codegree -2 than in codegree 6, since all elements in codegree -3 are in the image of the Bott map. □

8.3. Representation theory

A substantial amount of detailed information about the representation ring is required for the calculation, and it is convenient to collect it in a separate purely algebraic section.

8.3.A. The Chern classes. We fix n, and abbreviate $r_{s,n}$ by omitting n, so that r_s denotes rotation by $2s\pi/2^{n+1}$. We may display the character table

	1	1	$2(\times 2^n - 1)$	2^n	2^n
	r_0	r_{2^n}	$r_j (1 \le j \le 2^n - 1)$	s'	s''
1	1	1	1	1	1
δ	1	1	1	-1	-1
$\hat{D}'$	1	1	$(-1)^j$	1	-1
$\hat{D}''$	1	1	$(-1)^j$	-1	1
$\sigma_k (1 \le k \le 2^n - 1)$	2	$2(-1)^k$	$2\cos(\frac{2jk\pi}{2^{n+1}})$	0	0

Writing

$$a(j) = 2 - 2\cos(\frac{2j\pi}{2^{n+1}})$$

we may tabulate the first and second Chern classes.

	1	1	$2(\times 2^n - 1)$	2^n	2^n
	r_0	r_{2^n}	$r_j (1 \le j \le 2^n - 1)$	s'	s''
$c_1(\delta)$	0	0	0	2	2
$c_1(\hat{D}')$	0	0	$1-(-1)^j$	0	2
$c_1(\hat{D}'')$	0	0	$1-(-1)^j$	2	0
$c_1(\sigma_k)(1 \le k \le 2^n - 1)$	0	$2(1-(-1)^k)$	$a(jk)$	2	2
$c_2(\sigma_k)(1 \le k \le 2^n - 1)$	0	$2(1-(-1)^k)$	$a(jk)$	0	0

We are now in a position to consider the Chern filtration

$$RU \supseteq JU_1 \supseteq JU_2 \supseteq JU_3 \supseteq \cdots,$$

where JU_1 is the augmentation ideal (generated by first Chern classes), JU_2 is generated by JU_1^2 and second Chern classes, and

$$JU_k = JU_1 JU_{k-1} + JU_2 JU_{k-2}$$

for $k \geq 3$.

It turns out we can largely separate the last two columns. We introduce the following class functions.

	1	1	$2(\times 2^n - 1)$	2^n	2^n
	r_0	r_{2^n}	$r_j (1 \le j \le 2^n - 1)$	s'	s''
$\overline{\delta}$	0	0	0	2	2
$\overline{D}$	0	0	$1-(-1)^j$	0	2
i'	0	0	0	0	2
i''	0	0	0	2	0
t	0	0	$1-(-1)^j$	0	0
$c(k)(1 \le k \le 2^n - 1)$	0	$2(1-(-1)^k)$	$a(jk)$	0	0

Note that

$$\overline{\delta} - c_1(\delta),\ \overline{D} - c_1(\hat{D}'),\ c(k) = c_2(\sigma_k) = c_1(\sigma_k) - c_1(\delta),$$

so that $\overline{\delta}$, $\overline{D}$ and $c(k)$ are characters in JU_1. The class functions t, i' and i'' are not characters in general, but all non-trivial products of them are characters since, for example $(i')^2 = c_1(\delta)c_1(\hat{D}')$.

For subsequent calculations it is often convenient to have $\mathbb{Z}$-bases for ideals. A set X of characters generates a $\mathbb{Z}$-submodule $\mathbb{Z}X$, and an R-submodule RX, where $R = RU(D)$; it is useful to know if we have $\mathbb{Z}X = RX$. If Y is another set of characters, we write XY for the set of products xy with $x \in X$ and $y \in Y$. We then have both $\mathbb{Z}[\mathbb{Z}X \cdot \mathbb{Z}Y] = \mathbb{Z}XY$ and $\mathbb{Z}[RX + RY] = R[XY]$, and henceforth we write simply AB (rather than $\mathbb{Z}[AB]$) for the abelian group generated by products ab if A and B are $\mathbb{Z}$-modules or R-modules.

8.3.B. The Chern filtration. The most familiar basis of the representation ring is in terms of the simple modules, but it is convenient to use the second Chern classes

$$c(i) = c_2(\sigma_i) = 1 - \sigma_i + \delta$$

both because of their good relation to the Chern filtration, and because of the fact that they vanish on the reflections. Although

$$\sigma_{2^n} = \hat{D}' + \hat{D}''$$

is not simple, it is convenient to treat it with the other Chern classes as the following easily verified formulae make clear.

LEMMA 8.3.1. (i) *We have the formula*

$$c(k)c(l) = 2c(k) + 2c(l) - c(k+l) - c(k-l)$$

(ii) $c(0) = 0$
(iii) $c(2^n) = t^2$ □

Thus we have

$$\begin{aligned} R = R_n = RU(D_{2^{n+2}}) &= \mathbb{Z}\{1, \delta, \sigma_1, \dots, \sigma_{2^n-1}, \hat{D}', \hat{D}''\} \\ &= \mathbb{Z}\{1, \overline{\delta}, c(1), c(2), \dots, c(2^n), \overline{D}\} \end{aligned}$$

It is convenient to separate the $c(i)$ from the other part, so we take

$$\mathcal{C} = \mathbb{Z}\{c(1), c(2), \dots, c(2^n)\}.$$

The formulae for elements give the following statements about subgroups.

COROLLARY 8.3.2. (i) $\mathcal{C} \cdot \mathcal{C} \subseteq \mathcal{C}$
(ii) $\overline{\delta} \cdot \mathcal{C} = 0$
(iii) $\overline{D} \cdot \mathcal{C} = t \cdot \mathcal{C}$ □

The other key to simplification is the behaviour of the one dimensional representations.

LEMMA 8.3.3. (i) *The square of* $\mathbb{Z}\{\overline{\delta}, c_1(\hat{D}'), c_1(\hat{D}'')\}$ *is 'diagonal', in the sense that*

$$(\mathbb{Z}\{\overline{\delta}, c_1(\hat{D}'), c_1(\hat{D}''\})^2 = \mathbb{Z}\{c(2^n), 2i', 2i\}.$$

(ii) $\overline{D}c(s) = c(s) - c(2^n - s) + c(2^n)$. □

Now separate the last two columns, taking

$$J_k' = JU_k \cap \mathcal{C} \text{ and } J_k'' = JU_k \cap \mathbb{Z}\{i', i''\}$$

The ideal JU_1 is a bit more complicated, but the higher terms in the filtration sit very simply with respect to the character columns.

LEMMA 8.3.4. *For* $k \geq 2$,

$$JU_k = J_k' \oplus J_k''$$

and

$$J_k'' = \mathbb{Z}\{2^{k-1}i', 2^{k-1}i''\}.$$

Proof: We begin by noting that the values of all characters in JU on the reflections are even integers, showing that $\mathbb{Z}\{2^{k-1}i', 2^{k-1}i''\}$ is an upper bound for JU_k''. It then follows from Lemma 8.3.3 (i) that J_k'' is as stated.

Again from the fact that all characters in JU are even on the reflections, it follows that from any element of JU_k we can subtract a suitable sum of elements of J_k'' to obtain an element of $\mathcal{C}$. It follows that $JU_k = J_k' + J_k''$; it is clear that $J_k' \cap J_k'' = 0$ so the sum is direct. □

We may now summarize the behaviour of the filtration relative to this decomposition.

COROLLARY 8.3.5.

(i) *$JU_1 = \mathcal{C} + \mathbb{Z}\{\overline{\delta}, c_1(\hat{D}'), c_1(\hat{D}'')\}$ (the sum is not direct).*
(ii) *$JU_2 = \mathcal{C} \oplus \mathbb{Z}\{2i', 2i''\}$.*
(iii) *For $k \geq 3$, we have $JU_k = J_k' \oplus \mathbb{Z}\{2^{k-1}i', 2^{k-1}i''\}$.*
(iv) *We have $J_1' = J_2' = \mathcal{C}$. For $k \geq 1$, we have*

$$J_{k+2}' = \mathcal{C} \cdot J_k' + \mathcal{C} \cdot J_{k+1}' + \overline{D} \cdot J_{k+1}'$$

Proof: The statement for JU_1 is clear and that for JU_k for $k \geq 2$ was part of the lemma. The fact that $\overline{\delta} \cdot \mathcal{C} = 0$ will give the recurrence. The definition of JU_2 gives

$$JU_2 = JU_1^2 + \mathcal{C} = \mathcal{C}^2 + \overline{D} \cdot \mathcal{C} + (\mathbb{Z}\{\overline{\delta}, c_1(\hat{D}'), c_1(\hat{D}'')\})^2 + \mathcal{C}.$$

Lemma 8.3.3 gives the result. □

For the calculation of the rest it is useful to introduce

$$\mathcal{C}_0 = \mathbb{Z}\{c(s) - c(2^n - s) \mid 1 \leq s \leq 2^{n-1} - 1\},$$

$$\mathcal{C}_* = \{x \in \mathcal{C} \mid x \text{ is supported on odd rotations }\}.$$

LEMMA 8.3.6. (i) *$\mathcal{C}_* = \mathcal{C}_0 + c(2^n)$.*
(ii) *$\overline{D} \cdot \mathcal{C} = \mathcal{C}_*$*
(iii) *$\overline{D} \cdot \mathcal{C}_* = 2 \cdot \mathcal{C}_*$*

Proof: The formula $\overline{D}c(s) = c(s) - c(2^n - s) + c(2^n)$ shows $\overline{D}\mathcal{C}$ is contained in $\mathcal{C}_0 + c(2^n)$; the special case $s = 2^{n-1}$ shows $c(2^n) \in \overline{D}\mathcal{C}$.

The fact that $\overline{D}$ is multiplication by 2 on $\mathcal{C}_*$ shows that $\mathcal{C}_0 + c(2^n)$ is rationally equal to $\mathcal{C}_*$. The fact that the simple representations give a $\mathbb{Z}$-basis, together with the explicit generators of $\mathcal{C}_0$ establishes the integral statement. □

PROPOSITION 8.3.7. *For $k \geq 2$*

$$J_{2k-1}' = \mathcal{C}^k + \mathcal{C}_* \cdot \mathcal{C}^{k-2}$$

and

$$J_{2k}' = \mathcal{C}^k + 2 \cdot \mathcal{C}_* \cdot \mathcal{C}^{k-2}.$$

Proof: First we claim

$$J_{2k-1}' = \mathcal{C}^k + \mathcal{C}_* \cdot \mathcal{C}^{k-2} + 4 \cdot \mathcal{C}_* \cdot \mathcal{C}^{k-3} + 4^2 \cdot \mathcal{C}_* \cdot \mathcal{C}^{k-4} + \cdots + 4^{k-2} \cdot \mathcal{C}_*$$

and

$$J'_{2k} = \mathcal{C}^k + 2\cdot\mathcal{C}_*\cdot\mathcal{C}^{k-2} + 2\cdot 4\cdot\mathcal{C}_*\cdot\mathcal{C}^{k-3} + 2\cdot 4^2\cdot\mathcal{C}_*\cdot\mathcal{C}^{k-4} + \cdots + 2\cdot 4^{k-2}\cdot\mathcal{C}_*$$

For this we need only observe the formula holds for J'_3 and J'_4, and that it is stable under the recursion.

To see the trailing terms are redundant, note that

$$4\cdot\mathcal{C}_*\cdot\mathcal{C}^i = \overline{D}^2\cdot\mathcal{C}_*\cdot\mathcal{C}^i = c(2^n)\cdot\mathcal{C}_*\cdot\mathcal{C}^i \subseteq \mathcal{C}_*\cdot\mathcal{C}^{i+1}.$$

□

8.3.C. A principal reduction. In the previous section, the only formulae which involve n are those involving the representations $\hat{D}'$ and $\hat{D}''$. This is reflects the fact that we could work in

$$\begin{aligned} R = RU(O(2)) &= \mathbb{Z}\{1,\delta,\sigma_1,\sigma_2\ldots\} \\ &= \mathbb{Z}\{1,\overline{\delta},c(1),c(2),\ldots\} \end{aligned}$$

The separation of rotations and reflections is more complete for $O(2)$, so that if we take

$$\mathcal{C} = \mathbb{Z}\{c(1),c(2),\ldots\}$$

there is no interaction with the rest of the augmentation ideal since $\overline{\delta}\cdot\mathcal{C}=0$. In this case

$$JU_k = J^*_k \oplus \mathbb{Z}\{2^{k-1}\overline{\delta}\}$$

for all $k\geq 1$, and furthermore $J^*_{2s-1} = J^*_{2s} = \mathcal{C}^s$.

Notice that the naming of objects is consistent under restriction, so that in calculating powers of $\mathcal{C}$, we may work in $RU(O(2))$.

LEMMA 8.3.8. *An element $x=\sum_{k\geq 1} n_k c(k)$ lies in $\mathcal{C}^2$ if and only if*

$$\sum_k n_k k^2 = 0.$$

Accordingly,

$$\mathcal{C}^2 = c(1)\cdot\mathcal{C}$$

and hence for $k\geq 2$,

$$c(1)\cdot J^*_k = J^*_{k+2},$$

so that JU has principal reduction generated by $c^* := c(1)+\overline{\delta}^2$.

Proof: Obviously $c(1)\cdot\mathcal{C}\subseteq\mathcal{C}^2$, so it remains to establish the reverse containment.

We know $c(k)c(l) = 2c(k)+2c(l)-c(k+l)-c(k-l)$, so we can verify that $\mathcal{C}^2$ satisfies the linear relation by the observation

$$2k^2+2l^2-(k+l)^2-(k-l)^2=0.$$

Now consider an element x satisfying $\sum_k n_k k^2 = 0$. We show it lies in $c(1)\cdot\mathcal{C}$. Indeed, suppose k is the largest number with $n_k=0$. If $k\geq 2$ we may add n_k copies of

$$c(1)c(k-1) = 2c(1)+2c(k-1)-c(k-2)-c(k)$$

to reduce k. Repeating this, we reach $x=n_1c(1)+n_2c(2)$ with $n_1+4n_2=0$, and since $c(1)^2=4c(1)-c(2)$ this is in $c(1)\cdot\mathcal{C}$ as required. □

Combining this with our calculations 8.3.7 for dihedral groups.

COROLLARY 8.3.9. *In $R_n = RU(D_{2^{n+2}})$ we have*

$$\mathcal{C}^2 = c(1) \cdot \mathcal{C}$$

and hence for $k \geq 2$,

$$c(1)J'_k = J'_{k+2},$$

so that JU has principal reduction generated by $c^* = c(1) + \overline{\delta}^2$.

Remembering that the terms in the above are $\mathbb{Z}$-modules rather than ideals, we may deduce the subquotients we need.

PROPOSITION 8.3.10. *For the dihedral group $D = D_{2^{n+2}}$, the subquotients giving the homology of QU in negative degrees are as follows:*

(i) $JU_2/2JU_1 = (\mathcal{C}/c(2^n))/2 \oplus 2 \cong 2^{2^n-1} \oplus 2$

(ii) $JU_{2k}/2JU_{2k-1} = \mathcal{C}^k/(2\mathcal{C}^k + 2 \cdot \mathcal{C}_* \mathcal{C}^{k-2}) \cong 2^{2^n-1}$ *for $k \geq 2$. The dimension of this vector space is independent of k and multiplication by $c(1)$ gives an isomorphism*

$$c(1) : JU_{2k}/2JU_{2k-1} \xrightarrow{\cong} JU_{2k+2}/2JU_{2k+1}.$$

REMARK 8.3.11. Part (ii) highlights the importance of the stable quotient, so we adopt the notation

$$\overline{J'} := JU_4/2JU_3 = \mathcal{C}^2/(2\mathcal{C}^2 + 2\mathcal{C}_*).$$

Proof: For the first part, we note

$$JU_2 = [\mathcal{C}/c(2^n)] \oplus [c(2^n) \oplus 2i' \oplus 2i''].$$

There is a corresponding splitting

$$JU_1 = [\mathcal{C}/c(2^n)] \oplus [c_1(\delta) \oplus c_1(\hat{D}') \oplus c_1(\hat{D}'')]$$

since $c(2^n) = c_1(\hat{D}') + c_1(\hat{D}'') - c_1(\delta)$. The result follows since the matrix

$$\begin{pmatrix} 0 & 1 & 1 \\ 1 & 0 & 1 \\ 1 & 1 & 0 \end{pmatrix}$$

expressing the second part of the basis of $2JU_1$ in terms of the second part of the basis of JU_2 has determinant 2.

In Subsection 8.3.C we saw that $\mathcal{C}^2 = c(1) \cdot \mathcal{C}$, so that $\mathcal{C}^{k+1} = c(1) \cdot \mathcal{C}^k$ for $k \geq 1$. Similarly, since $\mathcal{C}_* = \overline{D} \cdot \mathcal{C}$ we have

$$\mathcal{C}_* \cdot \mathcal{C}^{k-1} = \overline{D} \cdot \mathcal{C}^k = \overline{D} \cdot c(1) \cdot \mathcal{C}^{k-1} = c(1) \cdot \mathcal{C}_* \cdot \mathcal{C}^{k-2}$$

for $k \geq 2$. Since $c(1)$ does not vanish on any rotation, it is a regular element, and the isomorphism statement follows.

It remains to find the dimension for $k = 2$, namely the dimension of

$$\overline{J'} := \mathcal{C}^2/(2\mathcal{C}^2 + 2 \cdot \mathcal{C}_*).$$

Evidently this is a $2^{n-\epsilon(n)}$-dimensional quotient of the 2^n-dimensional mod 2 vector space $\mathcal{C}^2/2\mathcal{C}^2$, where $\epsilon(n)$ is the dimension of

$$[\mathcal{C}^2 + \mathcal{C}_*]/\mathcal{C}^2 \cong \mathcal{C}_*/\mathcal{C}_* \cap \mathcal{C}^2.$$

First we note that $\epsilon(n) \neq 0$ since $c(1) - c(2^n - 1)$ is not an element of $\mathcal{C}^2$ by 8.3.8. In fact we show $\epsilon(n) = 1$. We give two proofs of this, one working in a number ring as would be suggested by character calculations, and one using formulae for multiplication of Chern classes. The latter gives specific bases useful for later parts of the calculation.

8.3.D. Proof that $\epsilon(n) = 1$ by characters. This subsection can be omitted except for those readers wanting to relate results to character calculations. The results we apply later are proved again in Subsection 8.3.E.

We are concerned with four sets of characters: $\mathcal{C}$, $\mathcal{C}_*$, $\mathcal{C}^2$ and $(\mathcal{C}^2)_* = \mathcal{C}^2 \cap \mathcal{C}_*$. It is convenient to focus on the values of characters on r_1, so we write $e : RU(D_{2^{n+2}}) \longrightarrow \mathbb{R}$ for evaluation at r_1.

LEMMA 8.3.12. *e is monomorphic on $\mathcal{C}_*$, and thus*

$$\mathcal{C}_*/\mathcal{C}_* \cap \mathcal{C}^2 = \mathcal{C}_*/(\mathcal{C}^2)_* \cong e(\mathcal{C}_*)/e((\mathcal{C}^2)_*).$$

Proof: The values of a character on odd powers of r_1 are given by Galois automorphisms. □

Evidently $\mathcal{C}^2 \supseteq \mathcal{C} \cdot \mathcal{C}_*$, so that $(\mathcal{C}^2)_* \supseteq \mathcal{C} \cdot \mathcal{C}_*$ and hence

$$e((\mathcal{C}^2)_*) \supseteq e(\mathcal{C}) \cdot e(\mathcal{C}_*).$$

It therefore suffices to show that

$$|e(\mathcal{C}_*)/e(\mathcal{C}) \cdot e(\mathcal{C}_*)| \leq 2.$$

Write

$$\lambda(s) = 2\cos(\pi s)$$

for brevity, and recall that

$$c(k)(r_1) = 2 - 2\cos(\pi k/2^n) = 2 - \lambda(k/2^n).$$

The lattice $e(\mathcal{C})$ is generated by the numbers $c(k)(r_1) = 2 - \lambda(k/2^n)$ for $k = 1, 2, \cdots, 2^n$. Note that $\lambda(1/2) = 0$ so that $c(2^{n-1})(r_1) = 2$, and $\lambda((2^n - k)/2^n) = -\lambda(k/2^n)$ so that

$$e(\mathcal{C}) = \mathbb{Z}\{2, \lambda(k/2^n) \mid k = 1, 2, \cdots, 2^{n-1} - 1\}.$$

Similarly $e(\mathcal{C}_*)$ is generated by the numbers $c(2^n)(r^1) = 4$ and

$$c(k)(r_1) - c(2^n - k)(r_1) = \lambda(k/2^n) - \lambda((2^n - k)/2^n) = 2\lambda(k/2^n),$$

so that

$$e(\mathcal{C}_*) = \mathbb{Z}\{4, 2\lambda(k/2^n) \mid k = 1, 2, \cdots, 2^{n-1} - 1\}.$$

We now estimate $e(\mathcal{C}) \cdot e(\mathcal{C}_*)$. Since $2 \in e(\mathcal{C})$ any even multiple of a basis element is in the product. For the rest we use the addition formula

$$\lambda(s)\lambda(t) = \lambda(s + t) + \lambda(s - t).$$

Evidently

$$\lambda(1/2) \cdot 2\lambda(1/2) = 4$$

and more generally

$$\lambda(k/2^n) \cdot 2\lambda(k/2^n) = 4 - 2\lambda(2k/2^n).$$

This shows the product contains all the even basis elements. If i and j are odd with $i < j$, then we may write $i = \sigma - \tau$ and $j = \sigma + \tau$ and see that the sum of the basis elements

$$2\lambda(i/2^n) + 2\lambda(j/2^n) = \lambda(\sigma/2^n) \cdot 2\lambda(\tau/2^n)$$

lies in the product. □

8.3.E. Multiplication by $c(1)$. We begin with the crudest information: the determinant of $c(1)$ as a self-map of $\mathcal{C}$.

LEMMA 8.3.13. *For the dihedral group D of order 2^{n+2}, map $c(1) : \mathcal{C} \longrightarrow \mathcal{C}$ has determinant 2^{n+2}.*

Proof: We may write down the matrix $c(1)$ on $\mathcal{C}$ with respect to the basis $c(1)$, $c(2)$, …, $c(2^n)$. It is given by the matrix

$$M_{2^n} = \begin{pmatrix} 4 & -1 & 0 & 0 & & & & & & 0 \\ 1 & 2 & -1 & 0 & 0 & & & & & 0 \\ 2 & -1 & 2 & -1 & 0 & 0 & & & & 0 \\ 2 & 0 & -1 & 2 & -1 & 0 & & & & 0 \\ 2 & 0 & 0 & -1 & 2 & -1 & & & & 0 \\ \vdots & & & & & & & & & \vdots \\ 2 & 0 & 0 & 0 & 0 & & -1 & 2 & -1 & 0 \\ 2 & 0 & 0 & 0 & 0 & & & -1 & 2 & -1 \\ 2 & 0 & 0 & 0 & 0 & & & 0 & -2 & 2 \end{pmatrix}$$

noting the first two rows and the last are slightly outside the general pattern. One may consider the corresponding $k \times k$ matrix M_k for any $k \geq 3$, and prove by induction that it has determinant $4k$. The matrix M_{2^n} of $c(1)$ has determinant 2^{n+2}. □

Turning to more detailed information, it is useful to be more precise about $\mathcal{C}^2$ and its quotient $\overline{J'} = \mathcal{C}^2/(2\mathcal{C}^2 + 2\mathcal{C}_*)$. We begin with something essentially proved in 8.3.8.

LEMMA 8.3.14. *The ideal $\mathcal{C}^2$ in $RU(O(2))$ is spanned by the elements $\overline{c}(k) = c(k) - k^2 c(1)$ for $k \geq 2$.* □

LEMMA 8.3.15. *The ideal $\mathcal{C}^2$ in $RU(D)$ is spanned by the elements $\overline{c}(k) = c(k) - k^2 c(1)$ for $2 \leq k \leq 2^n$, together with $2^{n+2}c(1)$.*

Proof: Since $\sigma_{2^n+k} = \sigma_{2^n-k}$, the ideal $\mathcal{C}^2$ contains

$$\overline{c}(2^n + k) - \overline{c}(2^n - k) = [(2^n + k)^2 - (2^n - k)^2]c(1) = 2^{n+2}kc(1).$$

In particular, the elements named in the statement all lie in $\mathcal{C}^2$, and we let I denote their span. Thus all the elements $\overline{c}(k)$ for $2 \leq k \leq 2^{n+1}$ are contained in I. Finally, since $\sigma_k = \sigma_{2^{n+1}+k}$,

$$\overline{c}(k + 2^{n+1}) = c(k) - (k + 2^{n+1})^2 c(1) = \overline{c}(k) - (2^{2n+2} + 2^{n+2})c(1),$$

so this also lies in I. □

Now recall that $\mathcal{C}_*$ is spanned by $c(2^n)$ together with the differences $c(k) - c(2^n - k)$ for $1 \leq k \leq 2^{n-1} - 1$. The case $n = 1$ is a little different, but it is easy to verify in this case that $c(2) \notin \mathcal{C}^2$, so that $\overline{J'} = \mathcal{C}^2/(2\mathcal{C}^2 + 2\mathcal{C}_*)$ is elementary abelian of rank 1.

LEMMA 8.3.16. *If $n \geq 2$,*

$$c(k) - c(2^n - k) \in \mathcal{C}^2 \text{ if and only if } s \text{ is even.}$$

Furthermore $2\mathcal{C}^2 + 2\mathcal{C}_*$ is spanned by $2^{n+2}c(1)$ and $2\bar{c}(k)$ for $2 \leq k \leq 2^n$, so that we have the explicit basis $\{\bar{c}(k) \mid 2 \leq k \leq 2^n\}$ for $\overline{J'} = \mathcal{C}^2/(2\mathcal{C}^2 + 2\mathcal{C}_*)$.

Proof: We make the calculation

$$\begin{aligned}\bar{c}(k) - \bar{c}(2^n - k) &= [c(k) - k^2c(1)] - [c(2^n - k) - (2^n - k)^2c(1)] \\ &= [c(k) - c(2^n - k)] + (2^{2n} - 2^{n+1}k)c(1).\end{aligned}$$

This shows exactly when $\bar{c}(k) - \bar{c}(2^n - k)$ lies in $\mathcal{C}^2 + \mathcal{C}_*$. It also shows that all the differences are equal modulo $\mathcal{C}^2 + \mathcal{C}_*$, giving the dimension of $\overline{J'}$. □

Now we calculate the exact index of nilpotency of $c(1)^2$ in its action on $\overline{J'}$.

LEMMA 8.3.17.

$$c(1)^{2^n} \notin 2\mathcal{C}^2 + 2\mathcal{C}_*$$
$$c(1)^{2^n+1} \in 2\mathcal{C}^2 + 2\mathcal{C}_*$$

Proof: Working modulo $2\mathcal{C}^2$ we see $c(1)^2 = 4c(1) - c(2)$, $c(1)^4 = 4c(2) - c(4)$, ..., $c(1)^{2^n} = 4c(2^{n-1}) - c(2^n)$. Similarly, working modulo $2\mathcal{C}^2 + 2\mathcal{C}_*$ we find

$$c(1)^{2^n+1} = [4c(2^{n-1}) - c(2^n)]c(1) = c(2^n)c(1) = c(2^n - 1) + c(2^n + 1) = 0.$$

□

Before proceeding, it is useful to know the module structure in terms of our basis.

LEMMA 8.3.18. *In the $RU(D)$-module $\overline{J'} = \mathcal{C}^2/(2\mathcal{C}^2 + 2\mathcal{C}_*)$ we have*

$$c(1)\bar{c}(k) = \bar{c}(k-1) + \bar{c}(k+1) + k\bar{c}(2)$$
$$c(1)^2\bar{c}(k) = \bar{c}(k-2) + \bar{c}(k+2) + k\bar{c}(3). \quad \square$$

We next calculate the kernels of the maps

$$c(1)^{2k} : RU(D)/2 \longrightarrow \overline{J'}$$

for all k. Note that this coincides with the kernel of the map $(p^*)^k$ where $p^* = c(1)^2 + \bar{\delta}^4$.

PROPOSITION 8.3.19. *For $k \leq 2^{n-1}$, the kernel of $c(1)^{2k}$ is of dimension $2k+2$. It has basis consisting of two characters supported on the reflections, together with $c(1)^{2i}$ for $2^{n-1} - k + 1 \leq i \leq 2^{n-1}$ and $c(1)^{2i}c(1)$ for $2^{n-1} - k \leq i \leq 2^{n-1} - 1$.*

Proof: First we need to know $(c(1)^2)^{2^{n-1}}c(1) = 0$. It will then be clear that the elements named are all annihilated by $c(1)^{2k}$. To see that they are not annihilated by a lower power and exhaust the kernel we need to know that the elements $c(1)^{2i}$ for $1 \leq i \leq 2^n$ together with $c(1)^{2i}c(1)$ for $1 \leq i \leq 2^n - 1$ give a basis of $\overline{J'}$. Since there are the right number of them, we need only show they are independent. For this we need to make some calculations.

It is clear from 8.3.18 that the even and odd classes do not interact, so that we need only show the elements $c(1)^{2i}$ for $1 \leq i \leq 2^n$ are independent and the elements $c(1)^{2i}c(1)$ for $1 \leq i \leq 2^n - 1$ are independent.

The required facts follow from explicit formulae for the elements in question. $\square$

The following lemmas give formulae for the elements, but the pattern is probably best appreciated from the following tables for D_{128} (i.e., $n = 5$). The ith row gives the vector of $c(1)^{2i}$ or $c(1)^{2i}c(1)$ with respect to the basis $\overline{c}(2), \overline{c}(3), \ldots, \overline{c}(2^n)$, although in the odd case an extra column is included for $\overline{c}(1) = 0$ to emphasize the pattern. A non-zero coefficient is indicated by the entry s where $2^s \leq i < 2^{s+1}$.

$c(1)^{2i}$	2	4	6	8	10	12	14	16	18	20	22	24	26	28	30	$32 = 2^n$
$i = 1$	0															
2		1														
3	1		1													
4				2												
5			2		2											
6		2		2		2										
7	2		2		2		2									
8								3								
9							3		3							
10						3		3		3						
11					3		3		3		3					
12				3		3		3		3		3				
13			3		3		3		3		3		3			
14		3		3		3		3		3		3		3		
15	3		3		3		3		3		3		3		3	
$2^{n-1} = 16$																4

$c(1)^{2i}c(1)$	1	3	5	7	9	11	13	15	17	19	21	23	25	27	29	$31 = 2^n - 1$
$i = 1$	0	0														
2		1	1													
3	1	1	1	1												
4				2	2											
5			2	2	2	2										
6		2	2			2	2									
7	2	2	2	2	2	2	2	2								
8								3	3							
9							3	3	3	3						
10						3	3			3	3					
11					3	3	3	3	3	3	3	3				
12				3	3							3	3			
13			3	3	3	3					3	3	3	3		
14		3	3			3	3			3	3			3	3	
15	3	3	3	3	3	3	3	3	3	3	3	3	3	3	3	3
$2^{n-1} = 16$																

Some may prefer the pattern represented algebraically.

LEMMA 8.3.20. *For* $i = 2^s + 2t$ *with* $0 \leq 2t < 2^s$

$$c(1)^{2i} = \overline{c}(2^{s+1} - 4t) + \overline{c}(2^{s+1} - 4t + 4) + \overline{c}(2^{s+1} - 4t + 8) + \cdots + \overline{c}(2^{s+1} + 4t)$$

For $i = 2^s + 2t - 1$ *with* $0 \leq 2t - 1 < 2^s$

$$c(1)^{2i} = \overline{c}(2^{s+1} - 4t + 2) + \overline{c}(2^{s+1} - 4t + 6) + \overline{c}(2^{s+1} - 4t + 10) + \cdots + \overline{c}(2^{s+1} + 4t - 2). \quad \square$$

LEMMA 8.3.21. *For* $i = 2^s + t$ *with* $0 \leq t < 2^s$

$$c(1)^{2i}c(1) = \sum_{j=0}^{t} \binom{t}{j} [\overline{c}(2^{s+1} - 2t - 1 + 2j) + \overline{c}(2^{s+1} - 2t + 1 + 2j)].$$

$$(c(1)^2)^{2^{n-1}} c(1) = 0. \quad \square$$

8.4. The ku-homology of D, $ku_*(BD)$

In this section we describe the calculation in the complex case. This is easier than the calculation of $ko_*(BD)$, so the argument is useful preparation for the reader. However we also use information on the differentials in the local cohomology spectral sequence in the complex case to determine them in the real case. The exercise is also worthwhile because [**28**] only gave the case $n = 1$ (i.e., $D = D_8$).

8.4.A. Overview. First we begin as before with the short exact sequence

$$0 \longrightarrow TU \longrightarrow ku_D^* \longrightarrow QU \longrightarrow 0,$$

where $TU = \Sigma^{-6} DH^*(BD; \mathbb{F}_2)$ is a free module over the double of $H^*(BD; \mathbb{F}_2)$ on the generator c_1c_2 in degree -6, and QU is the modified Rees ring with $RU(D)$ in each positive even degree and JU_k in degree $-2k$.

We now use the local cohomology spectral sequence to calculate $ku_*(BD)$. As usual, we begin by calculating the local cohomology of TU and QU, and then treat the connecting homomorphism as a d^1 differential.

8.4.B. Local cohomology. It is easy to calculate

$$H_{\mathfrak{m}}^*(TU) = H_{\mathfrak{m}}^2(TU) = \Sigma^{-2} DH^*(BD; \mathbb{F}_2)^\vee.$$

We now turn to the harder part of the calculation. The difficult parts of this are already part of our study of the Chern filtration in Section 8.3 above. First, by the reduction result 8.3.9, we may restrict ourselves to calculating with a principal ideal.

COROLLARY 8.4.1. *The element*

$$c^* = c(1) + \overline{\delta}^2$$

of degree -4 *has the property that*

$$H_{JU}^*(QU) = H_{(c^*)}^*(QU). \quad \square$$

This shows we have a short exact sequence

$$0 \longrightarrow H_{JU}^0(QU) \longrightarrow QU \longrightarrow QU[1/c^*] \longrightarrow H_{JU}^1(QU) \longrightarrow 0.$$

Furthermore, by 8.3.9, multiplication by c^* is an isomorphism from degrees ≤ -4. Thus we find the local cohomology is zero in degrees ≤ -4. For degrees $4k \geq 0$, we have the exact sequence

$$0 \longrightarrow H^0_{JU}(QU)_{4k} \longrightarrow RU(D) \xrightarrow{(c^*)^{k+1}} JU_2 \longrightarrow H^1_{JU}(QU)_{4k} \longrightarrow 0.$$

For degree -2, we have the exact sequence

$$0 \longrightarrow H^0_{JU}(QU)_{-2} \longrightarrow JU_1 \xrightarrow{c^*} JU_3 \longrightarrow H^1_{JU}(QU)_{-2} \longrightarrow 0,$$

and, for degrees $4k+2 \geq 4$, we have the exact sequence

$$0 \longrightarrow H^0_{JU}(QU)_{4k+2} \longrightarrow RU(D) \xrightarrow{(c^*)^{k+2}} JU_3 \longrightarrow H^1_{JU}(QU)_{4k+2} \longrightarrow 0.$$

It is easy to see $H^0_{JU}(QU)$ is zero in negative degrees, and $H^0_{JU}(QU)_{2l} \cong \mathbb{Z}$ for all $l \geq 0$, consisting of multiples of the regular representation. It is not too hard to calculate the order of $H^1_{JU}(QU)$ in each degree, from the determinant of the matrix for c^* as a map from the quotient mod H^0, but finding the precise structure is more of a challenge.

PROPOSITION 8.4.2. *The orders of the homogeneous parts of $H^1_{JU}(QU)$ are as follows*

$$|H^1_{JU}(QU)_{4k}| = \begin{cases} 2^{(n+6)k+2} & \text{if } k \geq 0 \\ 1 & \text{if } k < 0 \end{cases}$$

$$|H^1_{JU}(QU)_{4k+2}| = \begin{cases} 2^{(n+6)k+5} & \text{if } k \geq 0 \\ 2 & \text{if } k = -1 \\ 1 & \text{if } k < -1. \end{cases}$$

REMARK 8.4.3. The exact structure for the case D_8 is as follows [**28**, 3.5.7]

$$H^1_{JU}(QU(D_8))_{4k} = \begin{cases} [2^{2k+1}]^3 \oplus [2^{k-1}] & \text{if } k \geq 2 \\ [8]^3 & \text{if } k = 1 \\ [2]^2 & \text{if } k = 0 \\ 0 & \text{if } k \leq -1. \end{cases}$$

$$H^1_{JU}(QU(D_8))_{4k+2} = \begin{cases} [2^{2k+2}]^2 \oplus [2^{2k+1}] \oplus [2^k] & \text{if } k \geq 1 \\ [4]^2 \oplus [2] & \text{if } k = 0 \\ [2] & \text{if } k = -1 \\ 0 & \text{if } k < -1. \end{cases}$$

Proof: Consider the part in degrees which are multiples of 4, and assume the order in degrees ≤ 0 has been established. Since c^* is injective on JU_2, we see that the order in degree $4k+4$ is the order in degree $4k$ times $|JU_2/c^*JU_2|$. Since c^* respects the decomposition $JU_2 = J_2' \oplus J_2''$, and acts on J_2'' as $\overline{\delta}^2$, we see the order is 2^4 times $|J_2'/c^*J_2'| = |\mathcal{C}/c(1)\mathcal{C}|$. Since the determinant of $c(1)$ is 2^{n+2} by 8.3.13, the order in degree $4k+4$ is 2^{n+6} times bigger.

In degrees of the form $4k+2$ we assume the value is as given in degree 4, and deduce as above that stepping up 4 degrees multiplies the order by 2^{n+6}. Stepping up from degree -2 to degree $+2$ is different because we have the diagram

$$\begin{array}{ccccccccc} 0 & \longrightarrow & JU_1 & \xrightarrow{c^*} & JU_3 & \longrightarrow & H^1_{JU}(QU)_{-2} & \longrightarrow & 0 \\ & & \downarrow & & c^* \downarrow & & \downarrow & & \\ 0 & \longrightarrow & RU(D)/\ker(c^*) & \xrightarrow{(c^*)^2} & JU_3 & \longrightarrow & H^1_{JU}(QU)_{+2} & \longrightarrow & 0. \end{array}$$

Assuming, $H^1_{JU}(QU)_{-2}$ is finite, the stepping up is reduced by the order of the quotient $RU(D)/(JU_1+\ker(c^*)) \cong RU(D)/(JU_1+\mathrm{reg})$. This quotient is generated by the image of the unit, and the smallest multiple of it which is zero is the group order. The step up is thus two to the power $(n+6)-(n+2)=4$. It therefore remains to verify

$$|JU_3/c^*JU_1| = 2 \text{ and } |JU_2/c^*RU(D)| = 2^2.$$

This could be done directly, but instead, we use the local cohomology spectral sequence to deduce the answer from the connectivity of $ku_*(BD)$ and the values of $H^2_{\mathfrak{m}}(TU)$ in degrees -2 and 0. Indeed, since $ku_*(BD)$ is zero in negative degrees, the differential

$$d^1 : H^1_{JU}(QU)_{2k-2} \longrightarrow H^2_{\mathfrak{m}}(TU)_{2k-2}$$

must be an isomorphism for $k=0,1$. □

8.4.C. The local cohomology spectral sequence. The reader may find it helpful to refer to Figure 8.4.2, where the local cohomology spectral sequence for $ku_*(BD_8)$ is displayed.

LEMMA 8.4.4. *The differential*

$$d^1 : H^1_{JU}(QU)_{2k-2} \longrightarrow H^2_{\mathfrak{m}}(TU)_{2k-2}$$

has rank 1 if $k=0$, 2 if $k=1$ and 3 for $k \geq 2$.

Proof: It suffices to find the orders of $\tilde{ku}_{2k-4}(BD)$. This torsion is a quotient A of the dual of $DH^*(BD;\mathbb{F}_2)$, so that it is a module over ku_* via the quotient $ku_* \longrightarrow \mathbb{F}_2$. By [**28**, 3.2.1] this will contribute $\mathrm{Ext}^2_{ku_*}(A, ku_*)$ to the known value of $ku^*(BD)$. We conclude that as graded abelian groups $A = \Sigma^2 DH^*(BD;\mathbb{F}_2)$, giving the stated ranks. □

We may now read the answer off from the spectral sequence.

COROLLARY 8.4.5. *The ku-homology of dihedral 2-groups is tabulated below, where an unbracketed 2^k means an elementary abelian group of order 2^k, $[2^k]$ means a cyclic group of order 2^k and $\langle 2^k \rangle$ means an undetermined abelian group of order 2^k.*

$ku_n(BD)$	n	
$\langle 2^{(n+6)(k+1)-1} \rangle$	$4k+3$	≥ 3
$\mathbb{Z} \oplus 2^{2k+1}$	$4k+2$	≥ 2
$\langle 2^{(n+6)k+2} \rangle$	$4k+1$	≥ 1
$\mathbb{Z} \oplus 2^{2k}$	$4k+0$	≥ 0

More specifically, $ku_1(BD) \cong [2] \oplus [2]$, and the rank of the groups in odd degrees is monotonic increasing, and it is 2^n+2 in large degrees.

REMARK 8.4.6. The exact structure for $ku_*(BD_8)$ is given in [**28**, 3.5.1]. As above $ku_1(BD_8) \cong [2] \oplus [2]$ and in other degrees it is as tabulated below.

$ku_n(BD_8)$	n	
$[2^{2k+2}]^3 \oplus [2^k]$	$4k+3$	≥ 3
$\mathbb{Z} \oplus 2^{2k+1}$	$4k+2$	≥ 2
$[2^{2k+1}]^3 \oplus [2^{k-1}]$	$4k+1$	≥ 5
$\mathbb{Z} \oplus 2^{2k}$	$4k+0$	≥ 0

8.5. The ko-homology of D, $ko_*(BD)$

After Section 8.3, the calculation of ko^*_D presented in Subsection 8.2.F is fully justified, and we have provided the representation theoretic calculations we need to go forward. In the present section we calculate $ko_*(BD)$ from this information, using the local cohomology spectral sequence. The strategy is the same as for the complex case presented in Section 8.4, but more complicated in detail.

8.5.A. Overview. We let QO denote the image of ko^*_D in KO^*_D, and $\overline{QO}$ the image of ko^*_D in KU^*_D. We then define ST and T by the short exact sequences in the diagram

$$\begin{array}{ccccccccc} 0 & \longrightarrow & ST & \longrightarrow & ko^D_* & \longrightarrow & QO & \longrightarrow & 0 \\ & & \downarrow & & =\downarrow & & \downarrow & & \\ 0 & \longrightarrow & T & \longrightarrow & ko^D_* & \longrightarrow & \overline{QO} & \longrightarrow & 0. \end{array}$$

From the Bockstein spectral sequence we know the structure of ST and T. In this case $ST = TO$ and the submodule τ of η-multiples maps isomorphically to $Q\tau = T/TO$, so that we have an isomorphism

$$T = TO \oplus \tau.$$

The rank of $D_{2^{n+2}}$ is independent of n, but the number of conjugacy classes increases. Accordingly, the contribution from ordinary cohomology is easy to understand but the representation theory is harder. Thus module TO is the easiest to understand, followed by τ, with $\overline{QO}$ being the hardest. We take the easiest first.

8.5.B. The module TO detected in mod 2 cohomology. This part of the calculation is entirely independent of n. We have an exact sequence

$$0 \longrightarrow TO \longrightarrow ZTU \longrightarrow \Sigma^{-12}\mathbb{F}_2[p_2] \longrightarrow 0,$$

where

$$ZTU = \ker(Sq^2 : TU \longrightarrow TU),$$

and $p_2 = c_2^2$. As summarized in 8.2.1, we have calculated

$$ZTU = (\Sigma^{-6}\mathbb{F}_2 \oplus \Sigma^{-10}\mathbb{F}_2 \oplus \Sigma^{-12}\mathbb{F}_2^2) \otimes PP$$

where $PP = \mathbb{F}_2[p_1, p_2]$ with $p_i = c_i^2$.

COROLLARY 8.5.1. *The local cohomology of TO is all in degree 2 where there is an extension*

$$0 \longrightarrow \Sigma^{-4}\mathbb{F}_2[p_2]^\vee \longrightarrow H^2_{\mathfrak{m}}(TO) \longrightarrow (\Sigma^6\mathbb{F}_2 \oplus \Sigma^2\mathbb{F}_2 \oplus \mathbb{F}_2^2) \otimes PP^\vee \longrightarrow 0,$$

and the extension is split over $\mathbb{F}_2[p_2]$.

FIGURE 8.4.2. The $E^{1\frac{1}{2}}$-page of the local cohomology spectral sequence for $ku_*(BD_8)$.

$2^{12} \leftarrow$	$[2^{11}]^3 \oplus [2^4]$	$\mathbb{Z}$	20
0	0	0	
$2^{11} \leftarrow$	$[2^{10}]^2 \oplus [2^9] \oplus [2^4]$	$\mathbb{Z}$	
0	0	0	
$2^{10} \leftarrow$	$[2^9]^3 \oplus [2^3]$	$\mathbb{Z}$	16
0	0	0	
$2^9 \leftarrow$	$[2^8]^2 \oplus [2^7] \oplus [2^3]$	$\mathbb{Z}$	
0	0	0	
$2^8 \leftarrow$	$[2^7]^3 \oplus [2^2]$	$\mathbb{Z}$	12
0	0	0	
$2^7 \leftarrow$	$[2^6]^2 \oplus [2^5] \oplus [2^2]$	$\mathbb{Z}$	
0	0	0	
$2^6 \leftarrow$	$[2^5]^3 \oplus [2]$	$\mathbb{Z}$	8
0	0	0	
$2^5 \leftarrow$	$[2^4]^2 \oplus [2^3] \oplus [2]$	$\mathbb{Z}$	
0	0	0	
$2^4 \leftarrow$	$[2^3]^3 \oplus [1]$	$\mathbb{Z}$	4
0	0	0	
$2^3 \leftarrow$	$[2^2]^2 \oplus [2] \oplus [1]$	$\mathbb{Z}$	
0	0	0	
$2^2 \leftarrow$	$[2]^2 \oplus [1] \oplus [1]$	$\mathbb{Z}$	0
0	0		
$2 \leftarrow$	$[2]$		
			-4
H^2_{JU}	H^1_{JU}	H^0_{JU}	

Proof: Since PP is polynomial on generators of degrees -4 and -8, its local cohomology is entirely in cohomological degree 2, where it is $\Sigma^{12}PP^\vee$, and similarly, the local cohomology of $\mathbb{F}_2[p_2]$ is entirely in cohomological degree 1, where it is $\Sigma^8\mathbb{F}_2[p_2]^\vee$. The result now follows from the long exact sequence giving TO as a submodule of ZTU. $\square$

8.5.C. The module τ of η-multiples. The Bockstein spectral sequence implies that τ is in degrees 1 and 2 mod 8. In relevant degrees greater than 0 it is $\overline{R} = R/2$ (a mod 2 vector space of rank $2^n + 3$), and in relevant degrees less than 0 it is a mod 2 vector space $\overline{J'} = \mathcal{C}^2/(2\mathcal{C}^2 + 2\mathcal{C}_*)$ of rank $2^n - 1$.

$$\begin{array}{ccccccccccc} \cdot & \overline{J'} & \cdots & \cdot & \overline{J'} & \cdots & \cdot & \overline{R} & \cdots & \cdot & \overline{R} \\ \overline{J'} & \cdot & \cdots & \overline{J'} & \cdot & \cdots & \overline{R} & \cdot & \cdots & \overline{R} & \cdot \\ -15 & -14 & \cdots & -7 & -6 & \cdots & 1 & 2 & \cdots & 9 & 10 \end{array}$$

From the Bockstein spectral sequence we see that

$$p^* = (c^*)^2 = c(1)^2 + \overline{\delta}^4$$

survives to ko^*_D, and thus by 8.3.9 provides a reduction of the augmentation ideal. For the calculation of local cohomology, we need to know how τ behaves under multiplication by p^*, which we can read off from Proposition 8.3.19

LEMMA 8.5.2. *The action of p^* on τ is an isomorphism in negative degrees (i.e., when both domain and codomain are $\overline{J'}$), and the kernel of $(p^*)^k : \overline{R} \longrightarrow \overline{J'}$ is of dimension $2k+2$ for $1 \le k \le 2^{n-1}$ and of dimension 2^n+3 for $k \ge 2^{n-1}+1$.*

Proof: The statement about negative degrees follows from 8.3.9, and the rest follows from 8.3.19. $\square$

COROLLARY 8.5.3. *The local cohomology of τ is all in cohomological degrees 0 and 1 where it has Hilbert series*

$$PS(H^0_{\mathfrak{m}}(\tau)) = t(1+t)\left[4 + 6t^8 + 8t^{16} + 10t^{24} + \cdots + (2^n+2)t^{8(2^{n-1}-1)}\right] + t^{2^{n+2}} \cdot \frac{(2^n+3)t(1+t)}{1-t^8}$$

and

$$PS(H^1_{\mathfrak{m}}(\tau)) = t(1+t)\left[0 + 2t^8 + 4t^{16} + 6t^{24} + \cdots + (2^n-2)t^{8(2^{n-1}-1)}\right] + t^{2^{n+2}} \cdot \frac{(2^n-1)t(1+t)}{1-t^8}.$$

In both cases the Bott element acts monomorphically. In H^0, the R-module is a submodule of $\overline{R}$ in each degree. In H^1, the R-module is a quotient of $\overline{J'}$ in each degree.

Proof: Since (p^*) is a reduction of the augmentation ideal, the local cohomology is defined by the exact sequence

$$0 \longrightarrow H^0_{\mathfrak{m}}(\tau) \longrightarrow \tau \longrightarrow \tau[1/p_2] \longrightarrow H^0_{\mathfrak{m}}(\tau) \longrightarrow 0.$$

The module $\tau[1/p_2]$ is $\overline{J'}$ in every degree congruent to 1 or 2 mod 8. The map $\tau \longrightarrow \tau[1/p_2]$ is (a) an isomorphism in negative degrees, (b) surjective in degrees 1 and 2, (c) in degrees $8k+1$ and $8k+2$ for $0 \leq k \leq 2^{n-1}$, it has cokernel of dimension $2k-2$ and kernel of dimension $2k+2$ and (c) the map is zero in degrees $\geq 2^{n+2}$. $\square$

8.5.D. The part detected by characters $\overline{QO}$. The Bockstein spectral sequence calculation shows that the image $\overline{QO}$ of $ko^*(BD)$ in $KU^*(BD)$ is entirely in degrees which are multiples of 4. The pattern is slightly different in degrees of the form $8k$ and in degrees of the form $8k+4$. We write $\overline{QO}_{ev}$ for the former: in degrees ≥ 0 it is $R = RU(D) = RO(D)$, and in degree $-8k < 0$ it is JU_{4k}. We write $\overline{QO}_{od}$ for the latter: in postive degrees it is $2R = 2RU(D) = RSp(D)$, and in degree $-8k-4 < 0$ it is $2JU_{4k+1}$.

By the reduction result 8.3.9, we may restrict ourselves to calculating with a principal ideal.

COROLLARY 8.5.4. *The element*

$$p^* = c(1)^2 + \overline{\delta}^4$$

of degree -8 has the property that

$$H^*_{JO}(\overline{QO}) = H^*_{(p^*)}(\overline{QO}). \quad \square$$

This shows we have a short exact sequence

$$0 \longrightarrow H^0_{JO}(\overline{QO}) \longrightarrow \overline{QO} \longrightarrow \overline{QO}[1/p^*] \longrightarrow H^1_{JO}(\overline{QO}) \longrightarrow 0.$$

Furthermore, by Corollary 8.3.9, multiplication by p^* is an isomorphism from degrees ≤ -8. Thus we find the local cohomology is zero in degrees ≤ -8. For degrees $8k \geq 0$, we have the exact sequence

$$0 \longrightarrow H^0_{JO}(\overline{QO})_{8k} \longrightarrow RO(D) \stackrel{(p^*)^{k+1}}{\longrightarrow} JU_4 \longrightarrow H^1_{JO}(\overline{QO})_{8k} \longrightarrow 0.$$

For degree -4, we have the exact sequence

$$0 \longrightarrow H^0_{JO}(\overline{QO})_{8k+4} \longrightarrow 2JU_1 \stackrel{p^*}{\longrightarrow} 2JU_5 \longrightarrow H^1_{JO}(\overline{QO})_{-4} \longrightarrow 0,$$

and, for degrees $8k+4 \geq 4$, we have the exact sequence

$$0 \longrightarrow H^0_{JO}(\overline{QO})_{8k+4} \longrightarrow RSp(D) \stackrel{(p^*)^{k+2}}{\longrightarrow} 2JU_5 \longrightarrow H^1_{JO}(\overline{QO})_{8k+4} \longrightarrow 0.$$

It is easy to see $H^0_{JO}(\overline{QO})$ is zero in negative degrees, and $H^0_{JO}(\overline{QO})_{4l} \cong \mathbb{Z}$ for all $l \geq 0$, consisting of multiples of the regular representation. It is not too hard to calculate the order of $H^1_{JO}(\overline{QO})$ in each degree, from the determinant of the matrix for p^* as a map from the quotient mod H^0, but finding the precise structure is more of a challenge.

PROPOSITION 8.5.5. *The orders of the homogeneous parts of $H^1_{JO}(\overline{QO})$ are as follows*

$$|H^1_{JO}(\overline{QO})_{8k}| = \begin{cases} 2^{(2n+12)k+2} & \text{if } k \geq 0 \\ 1 & \text{if } k < 0 \end{cases}$$

$$|H^1_{JO}(\overline{QO})_{8k+4}| = \begin{cases} 2^{(2n+12)k+n+11} & \text{if } k \geq 0 \\ 2 & \text{if } k = -1 \\ 1 & \text{if } k < -1. \end{cases}$$

REMARK 8.5.6. The exact structure for the case D_8 is as follows

$$H^1_{JO}(\overline{QO}(D_8))_{8k} = \begin{cases} [2^{4k+1}]^3 \oplus [2^{2k-1}] & \text{if } k \geq 2 \\ [32]^3 \oplus [2] & \text{if } k = 1 \\ [2]^2 & \text{if } k = 0 \\ 0 & \text{if } k \leq -1. \end{cases}$$

$$H^1_{JO}(\overline{QO}(D_8))_{8k+4} = \begin{cases} [2^{4k+4}]^2 \oplus [2^{4k+3}] \oplus [2^{2k+1}] & \text{if } k \geq 1 \\ [16]^2 \oplus [8] \oplus [2] & \text{if } k = 0 \\ [2] & \text{if } k = -1 \\ 0 & \text{if } k < -1. \end{cases}$$

Proof: Consider the part in degrees which are multiples of 8, and assume the order in degrees ≤ 0 has been established. Since p^* is injective on JU_4, we see that the order in degree $8k+8$ is the order in degree $8k$ times $|JU_4/p^*JU_4|$. Since p^* respects the decomposition $JU_4 = J_4' \oplus J_4''$, and the behaviour on J_4'' is just $\overline{\delta}^4$, we see the order is 2^8 times $|J_4'/p^*J_4'|$.

Now consider the diagram

$$\begin{array}{ccccccccc} 0 & \longrightarrow & J_4' & \longrightarrow & \mathcal{C} & \longrightarrow & \mathcal{C}/J_4' & \longrightarrow 0 \\ & & p^* \downarrow & & p^* \downarrow & & p^* \downarrow & \\ 0 & \longrightarrow & J_4' & \longrightarrow & \mathcal{C} & \longrightarrow & \mathcal{C}/J_4' & \longrightarrow 0. \end{array}$$

Note that $\mathcal{C}/J_4'$ is a finite group, so the kernel and cokernel of p^* have the same order (actually $J_4' = \mathcal{C}^2 + 2\mathcal{C}_*$ contains $p^*\mathcal{C}$, so p^* acts as zero), and we find

$$|J_4'/p^*J_4'| = |\mathcal{C}/p^*\mathcal{C}|.$$

Now $p^* = c(1)^2$ on $\mathcal{C}$, so that by Lemma 8.3.13 p^* has determinant 2^{2n+4} as required.

In degrees of the form $8k+4$ we assume the value is as given in degree 4, and deduce as above that stepping up 8 degrees multiplies the order by 2^{2n+4}. Stepping up from degree -4 to degree $+4$ is different because we have the diagram

$$\begin{array}{ccccccc} 0 & \longrightarrow & 2JU_1 & \xrightarrow{p^*} & 2JU_5 & \longrightarrow & H^1_{JO}(\overline{QO})_{-4} & \longrightarrow 0 \\ & & \downarrow & & p^* \downarrow & & \downarrow & \\ 0 & \longrightarrow & RSp(D)/\ker(p^*) & \xrightarrow{(p^*)^2} & 2JU_5 & \longrightarrow & H^1_{JO}(\overline{QO})_{+4} & \longrightarrow 0. \end{array}$$

Assuming, $H^1_{JO}(\overline{QO})_{-4}$ is finite, the stepping up is reduced by the order of the quotient $RSp(D)/(2JU_1 + \ker(p^*)) \cong RU(D)/(JU_1 + \mathrm{reg})$. This quotient is generated by the image of the unit, and the smallest multiple of it which is zero is the group order. The step up is thus two to the power $(2n+12) - (n+2) = n+10$. It therefore remains to verify

$$|JU_5/p^*JU_1| = 2 \text{ and } |JU_4/p^*RU(D)| = 2^2.$$

This could be done directly, but instead, in Remark 8.5.7 below, we use the local cohomology spectral sequence to deduce the answer from the calculation (Corollary 8.5.1) of $H^2_{\mathfrak{m}}(TO)$. Indeed, $ko_*(BD)$ is connective, so the differential

$$d^1 : H^1_{JO}(\overline{QO})_{4k-4} \longrightarrow H^2_{\mathfrak{m}}(TU)_{4k-4}$$

is an isomorphism for $k = 0$ and 1. □

8.5.E. The local cohomology spectral sequence. For definiteness, we display the local cohomology spectral sequence for D_8, in Figure 8.5.3, but it shows the typical pattern. Thus the values in the H^2 column will be the same for all dihedral groups, and the *positions* of the non-zero groups in the H^0 and H^1 columns will be the same for all dihedral groups. The entries in H^0 and H^1 in degrees divisible by 4 come from $\overline{QO}$ and the entries in H^0 and H^1 in degrees 1 and 2 mod 8 come from τ. All entries in H^2 come from TO. Since the only remaining d^1 differentials are from a connecting homomorphism from $\overline{QO}$ to T, the only ones which can be zero come from H^1 in the form $d^1 : H^1_{4k} \longrightarrow H^2_{4k}$.

REMARK 8.5.7. The local cohomology spectral sequence together with the connectivity of $ko_*(BD)$ shows (without calculation of H^1) that d^1 gives isomorphisms

$$H^1_{JO}(\overline{QO})_{-4} \cong 2$$

and

$$H^1_{JO}(\overline{QO})_0 \cong 2 \oplus 2.$$

Similarly, the fact that $ko_0(BD) \cong \mathbb{Z}$ shows that the differential d^2 from H^0 in degree 1 to H^2 in degree 2 is surjective.

LEMMA 8.5.8. *The differential*

$$d^1 : H^1_{JO}(\overline{QO})_{4k-4} \longrightarrow H^2_{\mathfrak{m}}(TU)_{4k-4}$$

has rank 1 if $k = 0$, 2 if $k = 1$ and 3 for $k \geq 2$.

Proof: The cases $k = 0, 1$ were dealt with in Remark 8.5.7. The proof of the upper bound is by comparison with the complex case. We have a commutative diagram

$$\begin{array}{ccc} H^1_{JO}(\overline{QO})_{4k-4} & \xrightarrow{d^1} & H^2_{\mathfrak{m}}(T0)_{4k-4} \\ \downarrow & & \downarrow \\ H^1_{JU}(QU)_{4k-4} & \xrightarrow{d^1} & H^2_{\mathfrak{m}}(TU)_{4k-4} \end{array}$$

It is easily checked that the map on H^2 is injective, so the rank of the complex d^1 is an upper bound. The upper bound of 3 is established in Lemma 8.4.4 above.

To establish that the bound is achieved, it suffices to look at degree 4, and to show that

$$d^1 : H^1_{JO}(\overline{QO})_4 \longrightarrow H^2_{\mathfrak{m}}(T0)_4 \cong 2^3$$

is surjective. The cokernel would contribute to $ko_2(BD)$, but $H^0_{\mathfrak{m}}(\tau) \cong 2^4$ necessarily survives, so it is sufficient to show $|ko_2(BD)| \leq 2^4$. This is immediate from the Bockstein spectral sequence, the relevant part of which is

0	0	0	$\mathbb{Z}$
0	0	$\mathbb{Z}$	2^2
0	$\mathbb{Z}$	2^2	$\mathbb{Z} \oplus 2$
$\mathbb{Z}$	2^2	$\mathbb{Z} \oplus 2$	?
0	1	2	3

The only relevant differentials are $d^1 : \mathbb{Z} \oplus 2 \longrightarrow \mathbb{Z}$, with image of index 2. This shows $|ko_2(BD)| \leq 2^4$ as required. □

COROLLARY 8.5.9. *The ko-homology of dihedral 2-groups is tabulated below, where an unbracketed* 2^k *means an elementary abelian group of order* 2^k, $[2^k]$ *means a cyclic group of order* 2^k *and* $\langle 2^k \rangle$ *means an undetermined abelian group of order* 2^k. *The lower bounds for* a_k, b_k *and* c_k *only apply for* $k \geq 2^{n-1}$.

$ko_n(BD)$	n		
$\langle 2^{g_k} \rangle$	$8k+7$	≥ 7	$g_k = (2n+12)(k+1)-1$
2^{f_k}	$8k+6$	≥ 6	$f_k = 2k+1$
0	$8k+5$	≥ 5	
$\mathbb{Z} \oplus 2^{e_k}$	$8k+4$	≥ 4	$e_k = 2k+2$
$\langle 2^{d_k} \rangle$	$8k+3$	≥ 3	$d_k = (2n+12)k+n+8$
2^{c_k}	$8k+2$	≥ 2	$k+4 \leq c_k \leq k+2^n+6$
2^{b_k}	$8k+1$	≥ 1	$2^n-1 \leq b_k \leq 2^{n+1}+2$
$\mathbb{Z} \oplus 2^{a_k}$	$8k+0$	≥ 0	$2k-2^n-2 \leq a_k \leq 2k+2^n$

REMARK 8.5.10. (i) The bounds on ranks in degrees 0,1,2 mod 8 are very crude, using only the fact that the rank of a differential cannot be bigger than the dimensions of the domain and codomain.

(ii) More precise information on the structure for $ko_*(BD_8)$ is tabulated below.

$ko_n(BD)$	n		
$\langle 2^{g_k} \rangle$	$8k+7$	≥ 7	$g_k = 14k+13$
2^{f_k}	$8k+6$	≥ 6	$f_k = 2k+1$
0	$8k+5$	≥ 5	
$\mathbb{Z} \oplus 2^{e_k}$	$8k+4$	≥ 4	$e_k = 2k+2$
$\langle 2^{d_k} \rangle$	$8k+3$	≥ 3	$d_k = 14k+9$
2^{c_k}	$8k+2$	≥ 2	$c_0 = 4, c_k = 2k+5$ for $k \geq 1$
2^{b_k}	$8k+1$	≥ 1	$b_0 = 3, b_k = 6$ for $k \geq 1$
$\mathbb{Z} \oplus 2^{a_k}$	$8k+0$	≥ 0	$a_0 = 0, a_k = 2k+2$ for $k \geq 1$

REMARK 8.5.11. Many detailed problems remain, especially in low degrees. Perhaps the first of them are the following.

(i) Identify the minimum number of generators of H^1_{4k} precisely.
(ii) Show for dihedral groups of order ≥ 16 that the only d_2 is from degree 1 (or find all the others).

8.5.F. Calculating $\mathcal{C}^k$. In order to determine the exact structure of the groups $H^1_{JO}(\overline{QO})_i$ we need to analyze the maps p^* in more detail. We begin the process here, but do not give a complete answer. Most readers will want to omit this subsection.

We first find a basis for $\mathcal{C}^k$ for the group $O(2)$ since it is uncluttered by relations, and then restrict to obtain bases for the finite dihedral groups.

The pattern is as follows. First $\mathcal{C}$ itself has basis consisting of $c(1)$, $c(2)$, $c(3)$, ... Next, we have seen in Lemma 8.3.8 that $\mathcal{C}^2$ has basis $\overline{c}(2) = -4c(1) + c(2)$, $\overline{c}(3) = -9c(1) + c(3)$, $\overline{c}(4) = -16c(1) + c(4), \ldots$. From this basis we can also read off the relation

$$\sum_{n \geq 1} n^2 \lambda_n = 0$$

which determines when $\sum_{n \geq 1} \lambda_n c(n)$ lies in $\mathcal{C}^2$.

FIGURE 8.5.3. The $E^{1\frac{1}{2}}$-page of the local cohomology spectral sequence for $ko_*(BD_8)$.

H^2_{JO}	H^1_{JO}	H^0_{JO}	
$2^7 \leftarrow$	$[2^{12}]^2 \oplus [2^{11}] \oplus [2^5]$	$\mathbb{Z}$	20
0	0	0	
2^5	2	2^5	
0	2	2^5	
$2^6 \leftarrow$	$[2^9]^2 \oplus [2^9] \oplus [2^3]$	$\mathbb{Z}$	16
0	0	0	
2^4	0	0	
0	0	0	
$2^5 \leftarrow$	$[2^8]^2 \oplus [2^7] \oplus [2^3]$	$\mathbb{Z}$	12
0	0	0	
2^3	2	2^5	
0	2	2^5	
$2^4 \leftarrow$	$[2^5]^2 \oplus [2^5] \oplus [2]$	$\mathbb{Z}$	8
0	0	0	
2^2	0	0	
0	0	0	
$2^3 \leftarrow$	$[2^4]^2 \oplus [2^3] \oplus [2]$	$\mathbb{Z}$	4
0	0	0	
2	0	2^4	
0	0	2^4	
$2^2 \leftarrow$	$[2]^2 \oplus [1] \oplus [1]$	$\mathbb{Z}$	0
0	0		
0	0		
0	0		
$2 \leftarrow$	2		-4

The pattern continues: at the kth stage we find $\mathcal{C}^k$ has basis $\bar{c}^k(k)$, $\bar{c}^k(k+1)$, $\bar{c}^k(k+2)$, $\bar{c}^k(k+3)$, ..., where

$$\bar{c}^k(n) = \kappa^k(n)_1 c(1) + \kappa^k(n)_2 c(2) + \cdots \kappa^k(n)_{n-1} c(n-1) + c(n)$$

for suitable integers $\kappa^k(n)_i$. Of course $\mathcal{C}^{k+1} \subseteq \mathcal{C}^k$, and we will prove below that for $n \geq k+1$,

$$\bar{c}^{k+1}(n) = -a^{k+1}(n)\bar{c}^k(k) + \bar{c}^k(n)$$

for suitable integers $a^{k+1}(n)$. It is convenient to let $a^{k+1}(k) = 1$, and $\bar{c}^{k+1}(k) = 0$. As above, this lets us read off the relation

$$\sum_{n \geq k} a^{k+1}(n)\lambda_n = 0$$

which determines when $\sum_{n \geq k} \lambda_n \bar{c}^k(n)$ lies in $\mathcal{C}^{k+1}$.

To actually determine these numbers it is convenient to consider the generating function

$$\alpha^{k+1}(t) = 1 + a^{k+1}(k+1)t + a^{k+1}(k+2)t^2 + a^{k+1}(k+3)t^3 + \cdots.$$

LEMMA 8.5.12. *The basis $\bar{c}^k(k)$, $\bar{c}^k(k+1)$, $\bar{c}^k(k+2)$, ... of $\mathcal{C}^k$ is determined as follows. For $k = 1$ we have $\bar{c}^1(n) = c(n)$, and the others are expressed recursively by*

$$\bar{c}^{k+1}(n) = -a^{k+1}(n)\bar{c}^k(k) + \bar{c}^k(n)$$

for $k \geq 1$. The integers $a^{k+1}(n)$ are determined by the generating functions

$$\alpha^{k+1}(t) = \sum_{i \geq 0} a^{k+1}(k+i)t^i,$$

where

$$\alpha^2(t) = 1 + 4t + 9t^2 + 16t^3 + 25t^4 + \cdots = \sum_{n=0}^{\infty}(n+1)^2 t^n = \frac{1+t}{(1-t)^3}$$

In general, for $k \geq 2$ we have

$$\alpha^k(t) = \frac{1+t}{(1-t)^{2k-1}}.$$

Proof: The proof is by induction on k. For $k = 1$ it is obvious. To avoid special treatment of the first case, we note that for $k = 2$ it was proved as part of Lemma 8.3.8 above. Now suppose $k \geq 2$ and that the results have been proved to apply up to $\mathcal{C}^k$. Observe that an element $\sum_{n \geq 1} \lambda_n c(n)$ of $\mathcal{C}^k$ is determined by $\lambda_k, \lambda_{k+1}, \lambda_{k+2}, \cdots$. This means that $-a^k(n)$ is the penultimate non-zero coefficient in the expression for $\bar{c}^k(n)$:

$$-a^k(n) = \kappa^k(n).$$

By Lemma 8.3.8, $\mathcal{C}^{k+1} = c(1)\mathcal{C}^k$, so $\mathcal{C}^{k+1}$ is spanned by the elements $c(1)\bar{c}^k(k)$, $c(1)\bar{c}^k(k+1)$, $c(1)\bar{c}^k(k+2)$, ... In tabular form, with a blank meaning zero, we have

	$\cdots$	$c(k-1)$	$c(k)$	$c(k+1)$	$c(k+2)$	$c(k+3)$	$c(k+4)$	$\cdots$
$\overline{c}^k(k)$	$\cdots$	$-a^k(k)$	1					
$\overline{c}^k(k+1)$	$\cdots$	$-a^k(k+1)$		1				
$\overline{c}^k(k+2)$	$\cdots$	$-a^k(k+2)$			1			
$\overline{c}^k(k+3)$	$\cdots$	$-a^k(k+3)$				1		
$\overline{c}^k(k+4)$	$\cdots$	$-a^k(k+4)$					1	

Multiplying by $c(1)$, we obtain

	$\cdots$	$c(k-1)$	$c(k)$	$c(k+1)$	$c(k+2)$	$c(k+3)$	$c(k+4)$	$\cdots$
$c(1)\overline{c}^k(k)$	$\cdots$	?	$a^k(k)+2$	-1				
$c(1)\overline{c}^k(k+1)$	$\cdots$	?	$a^k(k+1)-1$	2	-1			
$c(1)\overline{c}^k(k+2)$	$\cdots$	?	$a^k(k+2)$	-1	2	-1		
$c(1)\overline{c}^k(k+3)$	$\cdots$	?	$a^k(k+3)$		-1	2	-1	
$c(1)\overline{c}^k(k+4)$	$\cdots$	?	$a^k(k+4)$			-1	2	

These elements are obviously independent and therefore give a basis. We must use these to obtain the alternative basis

	$\cdots$	$c(k-1)$	$c(k)$	$c(k+1)$	$c(k+2)$	$c(k+3)$	$c(k+4)$	$\cdots$
$\overline{c}^{k+1}(k+1)$	$\cdots$	?	$-a^{k+1}(k+1)$	1				
$\overline{c}^{k+1}(k+2)$	$\cdots$	?	$-a^{k+1}(k+2)$		1			
$\overline{c}^{k+1}(k+3)$	$\cdots$	?	$-a^{k+1}(k+3)$			1		
$\overline{c}^{k+1}(k+4)$	$\cdots$	?	$-a^{k+1}(k+4)$				1	
$\overline{c}^{k+1}(k+5)$	$\cdots$	?	$-a^{k+1}(k+5)$					

This shows $a^{k+1}(k+1) = 2 + a^k(k)$, $a^{k+1}(k+2) = 3 + 2a^k(k) + a^k(k+1)$ $a^{k+1}(k+3) = 4 + 3a^k(k) + 2a^k(k+1) + a^k(k+2)$, or

$$\alpha^{k+1}(t) = (1 + 2t + 3t^2 + 4t^3 + \cdots)\alpha^k(t) = \frac{\alpha^k(t)}{(1-t)^2}$$

as required. □

REMARK 8.5.13. Reducing to the dihedral group D of order 2^{n+2} we see that for $k < 2^n$ the elements $\overline{c}^k(k), \cdots, \overline{c}^k(2^n)$ of $\mathcal{C}^k$ can be used as basis elements of $\mathcal{C}$, so that for the group D, the quotient $\mathcal{C}/\mathcal{C}^k$ requires at most $k-1$ generators. The relations $c(i) = c(2^{n+1}+i)$ and $c(2^n+i) = c(2^n-i)$ introduce additional relations. This gives a bound over the number of cyclic factors in $H^1_{JO}(\overline{QO})_{8k}$, which rises linearly to 2^n+2 as k increases (and similarly in degrees of the form $8k+4$). The number of cyclic factors does achieve this bound as one sees by comparison with periodic K-theory.

CHAPTER 9

The ko-Cohomology of Elementary Abelian 2-Groups

In this chapter we apply the Bockstein spectral sequence to deduce the coefficient ring ko_V^* from the ring ku_V^* calculated in [**28**, Chapter 4], where $V = V(r)$ is an elementary abelian 2-group of rank r.

The calculation involves a detailed understanding of the way that the Sq^2-action interacts with the truncated double Koszul resolution. Using the contents of Chapter 11 (which constructs an explicit resolution of TO) we give an explicit Hilbert series for the torsion part.

9.1. Summary of calculations

The conclusion of our calculations will be in terms of the short exact sequences

$$0 \longrightarrow T \longrightarrow ko_V^* \longrightarrow \overline{QO} \longrightarrow 0,$$

and

$$0 \longrightarrow \tau \longrightarrow T \longrightarrow TO \longrightarrow 0.$$

Here

$$\overline{QO} = \mathrm{im}(ko_V^* \longrightarrow KU_V^*)$$

and τ consists of the multiples of η.

In fact, we find that $TO = ZTU$, and we will give explicit structure results and Hilbert series in Subsection 9.12.B with full details in Chapter 11.

The crudest feature of τ is that it is bounded below for elementary abelian groups. It consists of $RU(V)/2$ in positive degrees $= 1$, $2 \bmod 8$, together with a few exceptional classes in negative degrees. We calculate its exact order in each degree.

Finally the image $\overline{QO}$ in KU_V^* is the subring consisting of the part in degrees ≥ 0 and the part generated by $JSp(V)$ in degree -4 and $JU(V)^4$ in degree -8. Thus it is concentrated in degrees divisible by 4, and there it is $RO(V) = RU(V)$ in degrees ≥ 0 divisible by 8 and $RSp(V) = 2RU(V)$ in positive degrees $\equiv 4 \bmod 8$, in degree $-8i$ it is $JU(V)^{4i}$ for $i \geq 1$ and in degree $-8i-4$ it is $JU(V)^{4i}JSp(V)$ for $i \geq 1$.

It is also notable that we find behaviour very regular in degrees -9 and above, starting with $ko_V^9 = 0$. The complex case might have suggested a non-zero answer in degree -9 and torsion in degree -8.

9.2. Recollections about ku^*BV

To retain as much information as possible, we work with ku_V^* rather than ku^*BV. However the practical difference is small, and the reader may replace ku_V^*

by ku^*BV throughout, if preferred. First recall that there is a short exact sequence

$$0 \longrightarrow TU \longrightarrow ku^*_V \longrightarrow QU \longrightarrow 0$$

where

$$QU = \mathrm{im}(ku^*_V \longrightarrow KU^*_V).$$

In fact QU is the Rees ring of $RU(V)$, which is $RU(V)$ in even degrees ≥ 0 and $JU(V)^n$ in degree $-2n$ with $n \geq 0$. This is generated by $RU(V)$ in degree 0, v in degree 2 and the Chern classes of simple representations in degree -2. For convenience, we choose generating simple representations $\alpha_1, \alpha_2, \ldots, \alpha_r$ and let $y_1, y_2, \ldots, y_r$ be the corresponding Euler classes (in degree -2). The only difference between ku^*_V and ku^*BV is that the part of QU in JU is 2-completed.

As for the v-power torsion TU, it is annihilated by 2 and by v. It is therefore a ku^*_V module via

$$ku^*_V \longrightarrow ku^*_V/(2, v) = \mathbb{F}_2[y_1, \ldots, y_r] = PC,$$

where PC stands for 'polynomial in the Chern classes'. Indeed TU is a PC-submodule of

$$H^*(BV; \mathbb{F}_2) = \mathbb{F}_2[x_1, x_2, \ldots, x_r] = PW$$

where PW stands for 'polynomial in the Stiefel-Whitney classes', and $y_i = x_i^2$. As a PC-module, we have a decomposition

$$TU = TU_2 \oplus TU_3 \oplus \cdots \oplus TU_r$$

where TU_i is the PC-submodule of PW generated by the elements

$$q_S = Q_0Q_1(\prod_{t \in S} x_t) = (\prod_{t \in S} x_t)(\sum_{\{l,m\}} (x_l x_m^3 + x_l^3 x_m))$$

where S runs through subsets of $\{1, 2, \ldots, r\}$ with exactly i elements. The sum is over 2-element subsets $\{l, m\}$. These modules TU_i have remarkable properties that will emerge later, but for the present we just note that TU_i is concentrated in degrees $\leq -4 - i$, and that TU_i has projective dimension i over PC [**28**, 4.6.3].

9.3. Deducing ko^*_V

To identify the E^2 term of the η Bockstein spectral sequence, we note that d^1 preserves the torsion submodule, so that

$$0 \longrightarrow TU \longrightarrow ku^*_V \longrightarrow QU \longrightarrow 0$$

is an exact sequence of chain complexes, and will therefore induce a long exact sequence of cohomology. The connecting homomomorphism is usually zero, but not always.

The first step is to identify the maps induced by d^1 on QU and on TU. Since the spectral sequence is compatible with complexification, the effect on QU is given by that on KU^*_V, which is $1 - \tau$ when the domain is a degree $\equiv 0$ mod 4 and $1 + \tau$ when the domain is a degree $\equiv 2$ mod 4 (see discussion in Chapter 4).

LEMMA 9.3.1. *The map d^1 induces $Sq^2 : TU \longrightarrow TU$.*

Proof: We saw in Chapter 4 that for any group G, the map

$$TU \longrightarrow ku^*_G \longrightarrow H^*(BG; \mathbb{F}_2)$$

takes d^1 to Sq^2. The result therefore follows since TU is detected in mod 2 cohomology when $G = V$ is elementary abelian [**28**, 4.2.7]. □

9.4. Homology of QU

We need to consider the homology of the complex

$$\cdots \longrightarrow RU(G) \xrightarrow{1-\tau} RU(G) \xrightarrow{1+\tau} RU(G) \xrightarrow{1-\tau} JU \xrightarrow{1+\tau} JU^2 \xrightarrow{1-\tau} JU^3 \xrightarrow{1+\tau} \cdots .$$

When G is elementary abelian, all representations are self-conjugate and therefore this becomes

$$\cdots \longrightarrow RU(V) \xrightarrow{0} RU(V) \xrightarrow{2} RU(V) \xrightarrow{0} JU \xrightarrow{2} JU^2 \xrightarrow{0} JU^3 \xrightarrow{2} \cdots .$$

From this, the space $Z(QU)$ of cycles is easily identified.

LEMMA 9.4.1. *The space of cycles $Z(QU)$ is the subring of KU_G^* consisting of $RU(V)$ in degree $4k$ for $k \geq 0$, and JU^{2n} in degree $-4n$ for $n \geq 0$.* □

Similarly, it is clear that the homology is zero in degrees $\equiv 2 \bmod 4$. It is also clear that it is $RU(V)/2$ in degrees ≥ 0 divisible by 4.

For the remaining degrees, we need a little combinatorics, described in Section 9.5. The homology is a mod 2 vector space and zero in degrees $< -2r$. The precise result is recorded in Corollary 9.5.2, noting that only half of the quotients $JU^i/2JU^{i-1}$ are relevant to the calculation of $H(QU, d^1)$.

9.5. The J-adic filtration for elementary abelian 2-groups

In this section we consider the filtration

$$\cdots \subseteq JU(V)^3 \subseteq JU(V)^2 \subseteq JU(V)^1 \subseteq RU(V)$$

where V is an elementary abelian 2-group of rank r. For brevity we write $J = JU(V)$ for the rest of the section. It turns out that the combinatorics is remarkably intricate: the subquotients J^n/J^{n+1} settle down to a single abelian group for large n, but for $n \leq r$ there is some exceptional behaviour. The second author is grateful to N.P.Dummigan for spotting the relevance of the Reed-Muller code.

PROPOSITION 9.5.1. *The subquotients J^n/J^{n+1} for $n \geq 1$ are elementary abelian 2-groups, with*

$$\dim_{\mathbb{F}_2}(J^n/J^{n+1}) = \begin{cases} \binom{r}{1} + \binom{r}{2} + \cdots + \binom{r}{n} & \text{for } n \leq r \\ 2^r - 1 & \text{for } n \geq r \end{cases}.$$

Indeed, the associated graded ring is

$$\mathrm{Gr}(RU(V), J) = \mathbb{Z}[x_1, \ldots, x_r]/(2x_i, x_i^2 x_j = x_i x_j^2 \mid 0 \leq i, j \leq r).$$

The ideal J is spanned as an abelian group by the Euler classes $e(\alpha)$ where α is a one dimensional simple representation. The character of $e(\alpha) = 1 - \alpha$ is 0 on elements of $\ker(\alpha)$ and 2 on the complement. Since

$$e(\alpha)^2 = (1-\alpha)^2 = 2(1-\alpha) = 2e(\alpha)$$

we see that

$$2 \cdot J^n \subseteq J^{n+1}.$$

This gives the inclusions

$$2J^{n+1} \subseteq 2J^n \subseteq J^{n+1}.$$

Noting that $J^{n+1}/2J^{n+1}$ is an elementary abelian 2-group of rank $2^r - 1$ and that $J^n/J^{n+1} \cong 2J^n/2J^{n+1}$ we have a short exact sequence

$$0 \longrightarrow J^n/J^{n+1} \longrightarrow \mathbb{F}_2^{2^r-1} \longrightarrow J^{n+1}/2J^n \longrightarrow 0,$$

showing in particular the fact that J^n/J^{n+1} is elementary abelian.

The result we will really need now follows from the proposition.

COROLLARY 9.5.2. *The quotients $J^{n+1}/2J^n$ are elementary abelian 2-groups and*

$$\dim_{\mathbb{F}_2}(J^{n+1}/2J^n) = \begin{cases} \binom{r}{n+1} + \binom{r}{n+2} + \cdots + \binom{r}{r} & \text{for } n \leq r-1 \\ 0 & \text{for } n \geq r \end{cases}.$$

Since the power of 2 in the character of every generator of J^n is 2^n, it is convenient to separate the combinatorics by writing $\overline{e}(\alpha)$ for the characteristic function of the complement of $\ker(\alpha)$, so that $e(\alpha) = 2\overline{e}(\alpha)$. We next consider the subgroup

$$I_n = \mathbb{Z}\{\overline{e}(\alpha_1)\cdots\overline{e}(\alpha_n) \mid \alpha_1, \ldots, \alpha_n \in V^*\}$$

of the free abelian group of class functions on V. We note that

$$J^n = 2^n I_n,$$

so that

$$J^{n+1}/2J^n \cong I_{n+1}/I_n.$$

These abelian groups are nested

$$I_1 \subseteq I_2 \subseteq I_3 \subseteq \cdots,$$

and in particular, I_1 is not an ideal.

The quotients I_{n+1}/I_n measure the deviation of the J-adic topology from its simplest possible behaviour, and Proposition 9.5.1 follows from the calculation of this deviation.

LEMMA 9.5.3. *The dimension of the deviation is*

$$\dim_{\mathbb{F}_2}(I_{n+1}/I_n) = \binom{r}{n+1} + \binom{r}{n+2} + \cdots + \binom{r}{r};$$

in particular, $I_n = I_r$ for $n \geq r$.

Proof: Since I_{n+1}/I_n is annihilated by 2, and tensor product is right exact, we see

$$I_{n+1}/I_n = (I_{n+1}/I_n) \otimes \mathbb{F}_2 = \operatorname{cok}(I_n \otimes \mathbb{F}_2 \longrightarrow I_{n+1} \otimes \mathbb{F}_2).$$

Now I_{n+1} is a free abelian group of rank $2^r - 1$, so that it suffices to show that the image of the map

$$I_n \otimes \mathbb{F}_2 \longrightarrow I_{n+1} \otimes \mathbb{F}_2$$

is of dimension

$$\binom{r}{1} + \binom{r}{2} + \cdots + \binom{r}{n}.$$

We claim that the image may be identified with the nth Reed-Muller vector space RM'_n of polynomial functions on V of degree $\leq n$ which vanish at the identity of V.

To see that RM'_n is the image, choose an isomorphism $V \cong \mathbb{F}_2^r$ and let $x_i : V \longrightarrow \mathbb{F}_2$ be projection onto the i-th factor. The corresponding simple representations α_i give a basis of V^*. Any simple representation α may be written as a product of the

α_i and $\overline{e}(\alpha)$ reduces mod 2 to the corresponding sum of the x_i. Hence a generator of I_n reduces to an n-fold product of linear forms, and the image is RM'_n.

To see that RM'_n has the required dimension, we argue as follows. Since $x_1, \ldots, x_r$ form a basis of V^* we see that RM'_1 is spanned by the homogeneous polynomials of degree 1, and it follows that RM'_n is spanned by the non-constant homogeneous polynomials of degree $\leq n$. Since two monomials give the same function if they involve the same variables, we obtain a spanning set consisting of products of $\leq n$ distinct variables x_i. Since the total space of functions on V is of dimension 2^r we see that this spanning set is also independent. □

REMARK 9.5.4. (i) In fact if the α_i are distinct, the generator $\overline{e}(\alpha_1) \cdots \overline{e}(\alpha_n)$ of I_n is non-zero on exactly 2^{r-n} elements of V, and I_r consists of all class functions vanishing at the identity of V.

(ii) The $\mathbb{F}_2$ vector space RM_n of polynomial functions on V of degree $\leq n$ is sometimes known as the Reed-Muller code of degree n.

9.6. The action of Sq^2 on PC

First, Sq^2 acts as a derivation on modules with $Sq^1 = 0$, and it is convenient to encode this behaviour using the action of a ring. We adjoin Sq^2 to PC to form a non-commutative ring $PC[Sq^2]$, suitable for acting on modules where Sq^2 acts as a derivation. To be more precise, we adjoin a generator σ of degree -2 with $\sigma^2 = 0$; all elements of the new ring can be uniquely expressed in the form $a + b\sigma$ with $a, b \in PC$ and with multiplication given by $\sigma a = a\sigma + Sq^2(a)$.

The key to the calculation of the Sq^2 cohomology is to note that Sq^2 is a PP-module map where $PP = \mathbb{F}_2[z_1, z_2, \ldots, z_r]$ is the polynomial on the Pontrjagin classes $z_i = y_i^2 = x_i^4$. Furthermore, as a PP-module

$$PC = PP \otimes \Lambda(y_1, y_2, \ldots, y_r).$$

For PC itself, the action of Sq^2 is especially simple. To describe it we will use Koszul complexes, so we briefly summarize what we need.

For any commutative ring R and any element $a \in R$ we have the Koszul complex

$$K_R(a) = (R \xrightarrow{a} R)$$

concentrated in degrees 0 and 1. For a sequence $a_1, a_2, \ldots, a_r$ of elements of R we may define the Koszul complex

$$K_R(a_1, a_2, \ldots, a_r) := K_R(a_1) \otimes_R K_R(a_2) \otimes_R \cdots \otimes_R K_R(a_r).$$

In particular it is easy to see that, for a polynomial ring $R = k[a_1, a_2, \ldots, a_r]$, the Koszul complex is acyclic:

$$H_*(K_R(a_1, a_2, \ldots, a_r)) = H_0(K_R(a_1, a_2, \ldots, a_r)) = k.$$

Accordingly, if K_i is the degree i piece, in homological degrees ≥ 1 the cycles are equal to the boundaries, and we have a resolution

$$0 \longleftarrow ZK_i \longleftarrow K_{i+1} \longleftarrow K_{i+2} \longleftarrow \cdots \longleftarrow K_r \longleftarrow 0$$

for these cycles. In homological degree 0, the free module K_0 consists entirely of cycles.

LEMMA 9.6.1. *As a PP-module chain complex, PC with Sq^2 as differential is isomorphic to the Koszul complex for $z_1, z_2, \ldots, z_r$*

$$(PC, Sq^2) = K_{PP}(z_1, z_2, \ldots, z_r).$$

Accordingly

$$H^*(PC; Sq^2) = \mathbb{F}_2.$$

Proof: A PP-basis of PC is given by the monomials $y_S = \prod_{t \in S} y_t$ as S runs through subsets of $\{1, 2, \ldots, r\}$. We then calculate

$$Sq^2 y_S = y_S \sum_{t \in S} y_t = \sum_{t \in S} z_t y_{S \setminus \{t\}},$$

which is the Koszul differential. The statement about cohomology follows since $z_1, z_2, \ldots, z_r$ is a regular sequence in PP. □

We have found it quite easy to get confused about degrees in the lemma, but the following explicit elements mitigate this difficulty.

NOTATION 9.6.2. For a sequence T of elements of $\{1, 2, \ldots, r\}$ we write

$$p_T := Sq^2(\prod_{s \in T} y_s) = \prod_{s \in T} y_s (\sum_{t \in T} y_t),$$

and we have

$$\deg(p_T) = -2|T| - 2,$$

where $|T|$ is the number of terms in the sequence T. This does not depend on the ordering of the sequence, but only on T as a subset with multiplicities. Often we are interested in the case when T is a subset of $\{1, \ldots, r\}$.

In particular $p_i = y_i^2 = z_i$ and $p_{ij} = q_{ij}$, but for subsets with more than two elements the p's and q's are different.

REMARK 9.6.3. Since the Koszul complex is exact, the Sq^2-cycles are generated by 1 and the p_T for subsets T with $|T| \geq 2$.

Furthermore, Sq^2 is PP-linear, and so the Sq^2-cycles are given by

$$Z(PC, Sq^2) = PP \oplus I_2 \oplus \cdots \oplus I_r,$$

where

$$I_j := PP\{p_T \mid |T| = j\} \text{ for } j \geq 1.$$

In particular, there are two free summands, PP (generated by 1) and the shift $I_r \cong PP(-2r-2)$ (generated by $p_{12\cdots r}$, where the parentheses indicate the module PP has been shifted down by $2r+2$ degrees). The Koszul complex itself gives a PP-projective resolution of I_j, showing that for $j \geq 1$ the PP-module I_j is of projective dimension $r - j$.

9.7. The action of Sq^2 on TU

The fun really starts with TU_i; once displayed, the following formulae are easily checked.

LEMMA 9.7.1. *The image $Sq^2(TU_i)$ of TU_i is contained in $TU_i \oplus TU_{i-2}$. More precisely, if q_T is a basis element of TU_i then*

$$Sq^2(q_T) = \sum_{i,j,k,l \in T \text{ distinct}} z_i y_j y_k y_l x_{T \setminus \{i,j,k,l\}},$$

which lies in TU_{i-2}. Indeed, this means

$$Sq^2(q_T) = \sum_{i,j \in T \text{ distinct}} y_i y_j q_{T \setminus \{i,j\}}.$$

Furthermore,

$$Sq^2(y_S q_T) = p_S q_T + y_S Sq^2 q_T,$$

where the first term lies in TU_i and the second term lies in TU_{i-2}. □

The fact that Sq^2 acts as a derivation shows that $TU = TU_{even} \oplus TU_{odd}$ as $PC[Sq^2]$-modules, and the lemma shows that both TU_{even} and TU_{odd} admit filtrations preserved by Sq^2, which allows us to calculate Sq^2 homology. We state the answer here, although the proof will occupy several sections.

PROPOSITION 9.7.2. *If $r = 1$ then $TU = 0$, and for larger r we have*

$$H(TU_{even}; Sq^2) = \binom{r}{2}\mathbb{F}_2(-6) \oplus \binom{r}{6}\mathbb{F}_2(-14) \oplus \binom{r}{10}\mathbb{F}_2(-22) \oplus \cdots$$

$$H(TU_{odd}; Sq^2) = \binom{r}{3}\mathbb{F}_2(-7) \oplus \binom{r}{7}\mathbb{F}_2(-15) \oplus \binom{r}{11}\mathbb{F}_2(-23) \oplus \cdots$$

Proof: For definiteness, we consider the even terms; the odd terms are precisely similar. We have the filtration

$$TU_2 \longrightarrow TU_{24} \longrightarrow TU_{246} \longrightarrow TU_{2468} \longrightarrow \cdots$$

of Sq^2 chain complexes, in which $TU_S = \bigoplus_{i \in S} TU_i$. There is an associated spectral sequence for calculating $H(TU_{even})$ from $H(TU_2)$, $H(TU_4)$, ... , where Sq^2 acts on TU_{2n} as the quotient $TU_{24\cdots(2n)}/TU_{24\cdots(2n-2)}$. If we carry on using homological grading, but halve degrees $H_{[n]} := H_{2n}$ (so as to get the differential $H_{[n]} \longrightarrow H_{[n-1]}$) the spectral sequence has

$$E^2_{p,q} = H_{[p+q]}(TU_{2p})$$

and standard homological differentials.

In the Section 9.8 we identify the E^2-term and then turn to the differentials in Section 9.10. □

9.8. The action of Sq^2 on the resolution of TU_n

We show that there is a Sq^2 action on the truncated Koszul PC-resolution of TU_{2p} given in [**28**, Section 4.6] making it into a complex of $PC[Sq^2]$-modules. In fact the $PC[Sq^2]$-modules which occur are all sums of copies of $PC[Sq^2]$ and PC.

We adapt the notation of [**28**, 4.6.3] in the interests of clarity and flexibility.

NOTATION 9.8.1. The resolution of TU_n by free PC-modules is of the form

$$0 \longleftarrow TU_n \longleftarrow \binom{r}{n} PC(-n-4) \longleftarrow \binom{r}{n+1} PC(-n-4) \oplus PC(-n-6)$$
$$\longleftarrow \binom{r}{n+2} PC(-n-6) \oplus PC(-n-8) \oplus PC(-n-10) \longleftarrow \cdots$$

Thus the hth syzygy is

$$F_h = \binom{r}{n+h} \left[PC(-n-4-2h) \oplus PC(-n-6-2h) \oplus \cdots \oplus PC(-n-4-4h) \right].$$

We name generators as follows. In homological degree h, for each set R with $n+h$ elements, there are generators $Q^0_R, Q^1_R, \ldots, Q^h_R$, with Q^i_R in degree $-(n+4+2h+2i)$.

In homological degree 0, the generator $Q_S = Q^0_S$ maps to q_S, and for higher syzygies we think of Q^i_R as the ith letter in a sequence, so we may write $Q^0_R = X_R, Q^1_R = Y_R$ in low homological degrees; we generally write $Q^h_R = Z_R$ for the generators of lowest degree in homological degree h.

In positive homological degree, the differential is defined as the sum of two differentials d_0 and d_1 as follows

$$dQ^i_R = d_0 Q^i_R + d_1 Q^i_R = \sum_{r \in R} y_r Q^i_{R \setminus \{r\}} + \sum_{r \in R} y_r^2 Q^{i-1}_{R \setminus \{r\}}.$$

In [**28**] the generators Q^i_R were written $x_j(R)$ with the relationship between i and j determined by examining degrees. If n is even the generators were denoted by $y_k(R)$ (of degree $-2k$). □

LEMMA 9.8.2. *The truncated Koszul resolution of TU_n may be made into a resolution of Sq^2 modules in a unique way.*

If h is odd, the hth syzygy is free over $PC[Sq^2]$ and hence is Sq^2-acyclic. More precisely,

$$F_h = \binom{r}{n+h} \left[PC[Sq^2](-n-4-2h) \oplus PC[Sq^2](-n-8-2h) \oplus \cdots \right.$$
$$\left. \cdots \oplus PC[Sq^2](-n-2-4h) \right];$$

a basis is given by $Q^0_R, Q^2_R, \ldots, Q^{h-1}_R$ for the subsets R with $n+h$ elements.

If h is even, the hth syzygy is the sum of a free module over $PC[Sq^2]$ and $\binom{r}{n+h} PC(-n-4-2h)$. The Sq^2-homology is thus $\binom{r}{n+h} \mathbb{F}_2(-n-4-2h)$. More precisely,

$$F_h = \binom{r}{n+h} \left[PC[Sq^2](-n-4-2h) \oplus PC[Sq^2](-n-8-2h) \oplus \cdots \right.$$
$$\left. \cdots \oplus PC[Sq^2](-n-4h) \oplus PC(-n-4-4h) \right];$$

a basis for the free summand is given by $Q^0_R, Q^2_R, \ldots, Q^{h-2}_R$ for the subsets R with $n+h$ elements, and the non-free summand has PC-basis Q^h_R for the subsets R with $n+h$ elements.

Proof: We restrict attention to TU_{2p}; the odd terms are precisely similar.

Starting in homological degree 0 with the generators Q_S^0 of degree $-2p-4$ with $|S|$ having $2p$ elements, we have $Sq^2 Q_S^0 = 0$ by Lemma 9.7.1. Now consider homological degree 1, and subsets R with $2p+1$ elements. Since

$$dQ_R^1 = d_1 Q_R^1 = \sum_s y_s^2 Q_{R\setminus\{s\}}^0$$

we see $Sq^2 Q_R^0 = 0$, and similarly, since

$$dQ_R^0 = d_0 Q_R^0 = \sum_s y_s Q_{R\setminus\{s\}}^0$$

we see $Sq^2 Q_R^0 = Q_R^1$.

We now deduce the general case by induction on the homological degree h. In fact we prove that

$$Sq^2 Q_R^i = \begin{cases} Q_R^{i+1} & \text{if } i \text{ is even} \\ 0 & \text{if } i \text{ is odd} \end{cases}$$

where Q_R^{i+1} is interpreted as 0 if $i = h$. Indeed, this is correct for $h = 0, 1$, and for the inductive step

$$dQ_R^i = \sum_s y_s Q_{R\setminus\{s\}}^i + \sum_s y_s^2 Q_{R\setminus\{s\}}^{i+1},$$

so that

$$dSq^2 Q_R^i = Sq^2 dQ_R^i = \sum_s y_s^2 Q_{R\setminus\{s\}}^i + \sum_s y_s^2 Sq^2 Q_{R\setminus\{s\}}^{i+1} + \sum_s y_s Sq^2 Q_{R\setminus\{s\}}^i.$$

It is then clear that the formula persists. $\square$

Splitting the resolution into short exact sequences, sparsity of the homology of the modules F_i lets us deduce the Sq^2-homology of TU_n.

COROLLARY 9.8.3. *The Sq^2-homology of TU_n is* $H(TU_n; Sq^2) =$

$$\binom{r}{n}\mathbb{F}_2(-n-4) \oplus \binom{r}{n+2}\mathbb{F}_2(-n-8) \oplus \binom{r}{n+4}\mathbb{F}_2(-n-12) \oplus \cdots. \quad \square$$

9.9. Module pictures

In [**28**] we introduced a pictorial representation of the resolutions of TU_k reflecting its nature as a truncated double Koszul complex. We have found this even more useful in describing and working with resolutions of TO_k, so we recall it here.

The principle is that a sum of n copies of $PC(-d)$ should be represented by a square

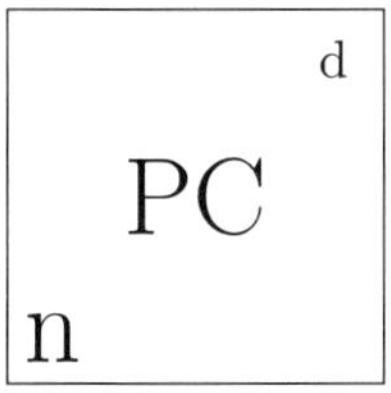

Considering rank 4 for definiteness, this means that the resolutions of TU_4, TU_3 and TU_2 are as follows, where we have indicated the names of the generators in the appropriate box.

$TU_4 =$

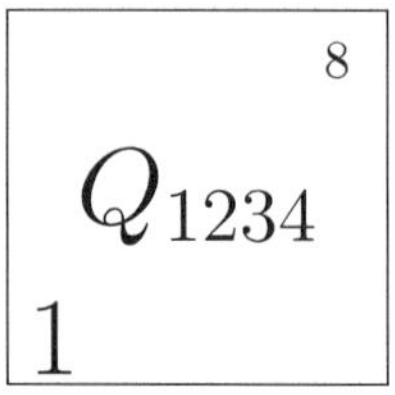

$TU_3 =$

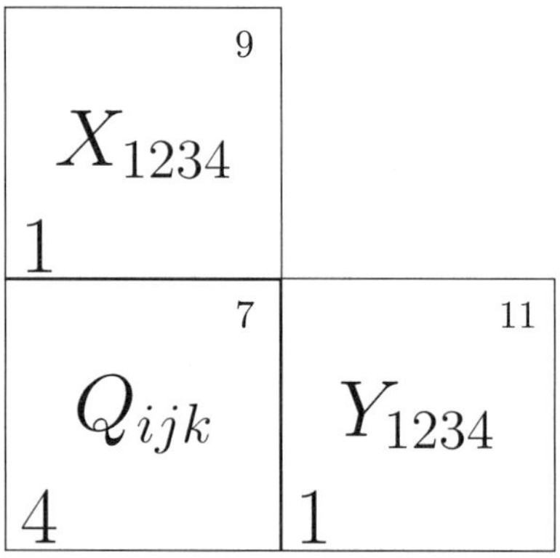

$TU_2 =$

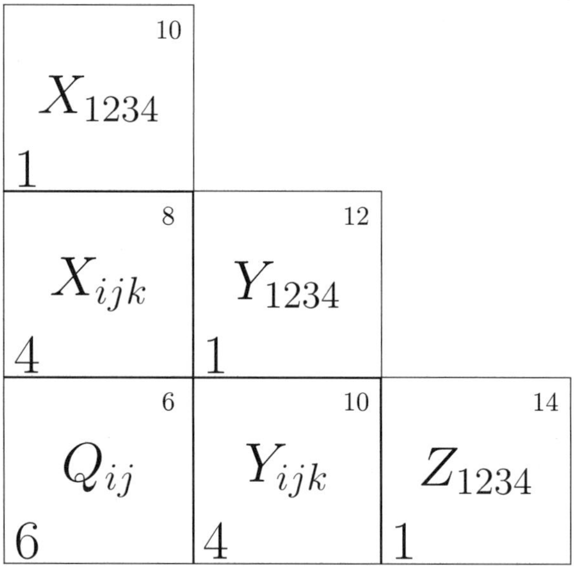

Note also that in these pictures, n copies of $PC[Sq^2](-d)$ appears in the form

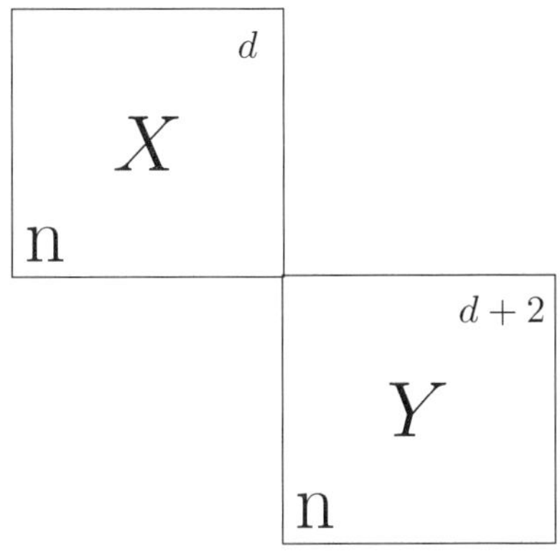

Along the diagonal representing each homological degree, the boxes are paired, starting from the top left. If there are an even number, all are paired into $PC[Sq^2]$s. If there are an odd number, all are paired except for a single lone PC at the bottom right.

9.10. The Sq^2-homology of TU

With the $PC[Sq^2]$-complex of Section 9.8 we can complete the proof of Proposition 9.7.2 giving the Sq^2 homology of TU. This shows that $H_*(TU_{even})$ has a vector space of dimension $\binom{r}{2}$, $\binom{r}{6}$, $\binom{r}{10}$, ... in degrees -6, -14, -22, ... Similarly $H_*(TU_{odd})$ has a vector space of dimension $\binom{r}{3}$, $\binom{r}{7}$, $\binom{r}{11}$, ... in degrees -7, -15, -23, ...

LEMMA 9.10.1. *The spectral sequence of Section 9.7 collapses at E^3.*

Before giving the proof, we illustrate it in low ranks. There is no room for differentials in ranks 1, 2 or 3. The first interesting case is in rank 4. We recommend that the reader identifies the generators of $H(TU_k)$ on the module diagrams of Section 9.9: following through the homology long exact sequences, the homology comes entirely from the unpaired copies of PC in the diagram. Each unpaired PC contributes homology consisting of a single $\mathbb{F}_2$ represented by its PC-generator in its top degree.

EXAMPLE 9.10.2. (Rank 4).

	TU_2	TU_4
$[-3]$	6	0
	0	0
$[-5]$	1	1

where an entry n means an $\mathbb{F}_2$ vector space of dimension n. The grading is such that in row $[-n]$ the degree in the TU_2 column is $[-n]$ and in the TU_4 column it is $[-(n-1)]$, and d^2 is horizontal. Thus the only possible differential is on the $[-5]$ row, where we have

$$1 \xleftarrow{d^2} 1.$$

The differential is non-zero since the formula of Lemma 9.7.1 shows that q_{1234} is not a cycle in TU_{24}.

EXAMPLE 9.10.3. (Rank 6).

	TU_2	TU_4	TU_6
[−3]	15	0	0
	0	0	0
[−5]	15	15	0
	0	0	0
[−7]	1	1	1

Thus on the [−5] row we have

$$15 \xleftarrow{d^2} 15$$

and on the [−7] row we have

$$1 \xleftarrow{d^2} 1 \xleftarrow{d^2} 1$$

Evidently, the 15's could cancel, but the three 1s cannot all cancel. We shall see that the right hand d^2 is always injective. For q_{123456} this is immediate.

EXAMPLE 9.10.4. (Rank 8).

	TU_2	TU_4	TU_6	TU_8
[−3]	28	0	0	0
	0	0	0	0
[−5]	70	70	0	0
	0	0	0	0
[−7]	28	28	28	0
	0	0	0	0
[−9]	1	1	1	1

Again the right hand d^2's are injective in each row. This leaves one possible d^2 at the left hand end of the [−9] row.

Proof of 9.10.1: Once again we consider TU_{even}, and TU_{odd} is precisely similar. From Lemma 9.8.2 we identify the E^2 term. The non-zero entries in the first column have dimensions equal to the binomial coefficients

$$\binom{r}{2}, \binom{r}{4}, \binom{r}{6}, \binom{r}{8}, \ldots$$

and the other rows are truncations.

To identify d^2 it is enough to work in the relevant $TU_{(2p)(2p-2)}$. The result is already suggested by the examples.

LEMMA 9.10.5. *The differentials d^2 between non-zero groups in each row alternate between being an isomorphisms and being zero, with the first being an isomorphism.*

It follows that the spectral sequence collapses at E^3 with the non-zero entries being in TU_2 column every fourth row (i.e., in degrees $-6, -14, -22, \ldots$), proving the proposition.

To prove the lemma we identify representative cycles for the homology of TU_{even}.

LEMMA 9.10.6. *Representative cycles for the basis of the Sq^2-homology group $H_{-2p-4m-4}(TU_{2p})$ are given by the $\binom{r}{2p+2m}$ elements*

$$z_{p+2m+2}(S) := \sum_{|T|=2m} y_T q_{S\backslash T},$$

where S runs through subsets of $\{1,2,\ldots,r\}$ with $2p+2m$ elements, and the sum is over subsets $T \subseteq S$ with $2m$ elements.

REMARK 9.10.7. The proof will show that the elements are indeed Sq^2-cycles, but the reader might be amused to give a direct verification with $p=2$ and $m=1$ in rank 6.

Proof: The Sq^2-homology is calculated using a PC-resolution of TU_{2p}. The homology group $H_{-2p-4-4m}(TU_{2p})$ comes from homological degree $2m$, where it is represented by the basis elements $q_{p+2+2m}(S)$ on the bottom edge, where S has $2p+2m$ elements. We need only follow this representative cycle through the spectral sequence.

The answer is clear for $m=0$ and if $m \geq 1$ we show how the process begins. Write

$$b_{2m} = Q_S^{2m}$$

with the subscript $2m$ indicating homological degree. Apply a differential to reduce homological degree

$$db_{2m}(S) = \sum_i y_i^2 Q_{S\backslash\{i\}}^{2m-1}$$

since the Sq^2-homology is zero in this degree (in homological degree $2m-2$) this is in the image of Sq^2. It is convenient to choose a representative on the bottom edge, namely

$$b_{2m-2} = \sum_i y_i Q_{S\backslash\{i\}}^{2m-1}$$

Applying the differential to reduce homological degree again, we obtain

$$db_{2m-2}(S) = \sum_{i\neq j} y_i y_j^2 Q_{S\backslash\{i,j\}}^{2m-2}$$

since the Sq^2-homology is zero in this degree (in homological degree $2m-4$) this is in the image of Sq^2. It is convenient to choose a representative on the bottom edge, namely

$$b_{2m-4} = \sum_i y_i y_j Q_{S\backslash\{i,j\}}^{2m-2}$$

This pattern continues until we find the representative

$$b_0 = \sum_{|T|=2m} y_T Q_{S\backslash T}^0$$

in homological degree 0. □

Proof of 9.10.5: Finally we may calculate d^2, since in $TU_{(2p-2)(2p)}$ we have

$$\begin{aligned} Sq^2 z_{p+2m+2}(S) &= \textstyle\sum_{|T|=2m}(Sq^2 y_T) q_{S\backslash T} + \sum_{|T|=2m} y_T Sq^2 q_{S\backslash T} \\ &= \textstyle\sum_{|T|=2m} y_T \sum_{i,j} y_i y_j q_{S\backslash(T\cup\{i,j\})} \end{aligned}$$

This is similar to $z_{(p-1)+2m+4}(S)$, but note that the term $q_{S\setminus U}$ with $|U| = 2m+2$ elements occurs $\binom{2m+2}{2}$ times. This is even when m is odd and odd when m is even, giving

$$Sq^2 z_{p+2m+2}(S) = \begin{cases} z_{(p-1)+2m+4}(S) & \text{if } m \text{ is even} \\ 0 & \text{if } m \text{ is odd,} \end{cases}$$

which immediately gives the stated formula in homology. □

This completes the proof of Lemma 9.10.1. □

9.11. Bockstein spectral sequence and the torsion part

The purpose of this section is to describe the process for calculating the precise value of ko^*BV in any given degree. To make it easier to follow the discussion, we have included charts of the Bockstein spectral sequence for elementary abelian groups of rank ≤ 8. For convenience we have put these together in Section 9.12 below, and we expect the reader will want to refer to these repeatedly during this section.

Step 1: The homology groups of the complex

$$\cdots \xrightarrow{0} RU \xrightarrow{2} RU \xrightarrow{0} JU \xrightarrow{2} JU^2 \xrightarrow{0} JU^3 \xrightarrow{2} JU^4 \xrightarrow{0} \cdots$$

(with $RU = RU(V)$ and $JU = JU(V)$) have been calculated. In effect we need to calculate half of the groups $JU^{k+1}/2JU^k$. This is an elementary abelian 2-group of rank $\binom{r}{k+1} + \binom{r}{k+2} + \cdots + \binom{r}{r}$ by Corollary 9.5.2.

Step 2: The homology of the complex $(TU; Sq^2)$ has been calculated in Section 9.10. Indeed, we saw in 9.7.2 that in even degrees we have

$$\binom{r}{2}\mathbb{F}_2(-6) \oplus \binom{r}{6}\mathbb{F}_2(-14) \oplus \binom{r}{10}\mathbb{F}_2(-22) \oplus \binom{r}{14}\mathbb{F}_2(-30) \oplus \cdots$$

and in odd degrees

$$\binom{r}{3}\mathbb{F}_2(-7) \oplus \binom{r}{7}\mathbb{F}_2(-15) \oplus \binom{r}{11}\mathbb{F}_2(-23) \oplus \binom{r}{15}\mathbb{F}_2(-31) \oplus \cdots$$

Step 3: Display this on a Bockstein spectral sequence chart as the $E^{1\frac{1}{2}}$ term. This is nearly the E^2-term, but the effect of the connecting homomorphism

$$H^n(QU) \longrightarrow H^{n+2}(TU)$$

has not been included, and it is treated as a late d^1. We make a number of observations.

First, note

- the homology of QU is zero below degree $-2r$ (and also in degree $-2r$ when r is odd).
- the homology of TU in even degrees is zero below degree $-6-8[(r-2)/4]$ and in odd degrees it is zero below degree $-7-8[(r-3)/4]$.

We conclude that, even at $E^{1\frac{1}{2}}$, the Bockstein spectral sequence is entirely on the zero line in large negative degrees (in any case from $-2r-2$ downwards).

Now consider the spectral sequence above the zero line. We refer to the line of slope 1 meeting the zero line in degree n as the n-diagonal. We know the entire behaviour in positive degrees, and note that, in this region, the only non-zero d^3 leaves the n-diagonal when $n \equiv 4 \bmod 8$. It follows that the same holds for the homology of QU: the only d^3 leaving the n-diagonal with $n \leq 0$ is for $n \equiv 4$ mod 8. In particular, no differential leaves the 0-diagonal, and we may consider only negative diagonals.

Next note that the situation in the n-diagonals above the zero line is as follows, according to the congruence class of n mod 4.

-0 mod 4: purely $H(QU)$
-1 mod 4: zero
-6,-7 mod 8: purely $H(TU)$
-2,-3 mod 8: zero

It now follows that, since d^3 is the longest differential, there are no differentials connecting elements of $H(TU)$. Since no QU-valued differentials leave n-diagonals with $n \equiv 0$ mod 8, it follows that all differentials come from diagonals with $n \equiv 4$ mod 8; there will typically be d^1, d^2 and d^3 and all must be surjective to ensure that the E^∞ page is on lines 3 and below.

Step 4: This gives the E^∞ term above the zero line (i.e., the η-multiples). It is zero except in degrees $-6, -7$ mod 8.

- In degree $d \equiv -6$ mod 8 it is $H_{d-1}(TU) \oplus H_{d-2}(QU)$ and
- In degree $n \equiv -7$ mod 8 it is $H_{d-1}(QU)$

We note that $H_*(TU_{even})$ therefore does not explicitly affect the answer above the 0 line, and that there are no η-multiples below degree $-2r$.

Step 5: Now consider the 0-line at E^∞.

Step 5a: The torsion is *exactly* $Z(TU) := \ker(Sq^2 : TU \longrightarrow TU)$ since no differentials leave the torsion submodule. This is immediate from the distribution of values above the zero line described in Step 3.

Step 5b: As for the torsion free quotient, we first note that the E^1 term is obvious: it is only non-zero in degrees divisible by 4, where it is $RU(V)$ in each positive degree and JU^{2k} in degree $-4k \leq 0$. At E^∞ we find JU^{4k} in degree $-8k$, and in degree $-8k-4$ we have $2JU^{4k+1}$ (for the latter, we are using the fact that the homology, $JU^{4k+4}/2JU^{4k+1}$ is exactly what is killed by the differentials above the 0-line). This is JU^{2n} in degree $-4n$ whenever n is large enough ($n \geq 2[(r-2)/4]+2$ suffices) but there are corrections in lower degrees not divisible by 8, starting with JSp in degree -4. In any case $\overline{QO}$ is generated by JSp in degree -4 and JU^4 in degree -8.

Step 6: We may calculate the Hilbert series of the kernel $Z(TU)$ of Sq^2 as follows.

Step 6a: The Hilbert series of $Z(PC)$ may be calculated from the fact that $(PC, Sq^2) = K_{PP}(z_1, \ldots, z_r)$. Indeed,

$$Z(PC) = PP \oplus I_2 \oplus \cdots \oplus I_r$$

where I_j is the part coming from Koszul degree j. Since the Koszul homology is entirely in degree 0, for $j \geq 2$

$$I_j = \ker(K_{j-1} \longrightarrow K_{j-2}) = \operatorname{im}(K_j \longrightarrow K_{j-1}),$$

and the Koszul complex itself gives a PP-projective resolution of Z_i, and hence a Hilbert series.

Step 6b: We have filtrations of TU_{even} and TU_{odd} with subquotients TU_i, and we have explicit resolutions of TU_i by free PC-modules F_{ij}. The modules F_{ij} have filtrations with subquotients suspensions of PC with Sq^2 being zero on the generator. Since the homology of all objects is known and small, we obtain short exact sequences for cycles (with small explicit error terms) and hence from Step 6a we can obtain the Hilbert series for $Z(PC)$.

9.12. Bockstein spectral sequence charts

We have collected together the Bockstein spectral sequence charts for elementary abelian 2-groups of rank ≤ 8 in the present section. This is designed to accompany the discussion in Section 9.11. Patterns emerge strongly when one flicks through them in order of increasing rank.

There is quite a bit of book keeping to implement the description from the previous section but to show it is practical we have made the structures completely explicit down to degree -32 in ranks ≤ 6. For example in rank 3, the only η-multiple in negative degrees is in degree -6, and the torsion on the zero line in degrees ≤ 0 is elementary abelian of rank

$$0, 0, 0, 0, 0, 0, 3, 1;\ \ 0, 0, 8, 3, 6, 3, 15, 7;\ \ 14, 8, 24, 13, 24, 15, 35, 21;\\ 36, 24, 48, 31, 50, 35, 63, 43;\ \ 66, \ldots$$

where the semicolons precede the degrees -8, -16, -24, ...

Before displaying the charts, we record some general facts to look out for.

9.12.A. η-multiples. The first striking thing is that below a certain degree, the $E^{1\frac{1}{2}}$-page is trivial above the 0-line. In particular, there are no η-multiples below a certain degree. This behaviour is probably unusual, and in any case for the dihedral groups the subgroup of η-multiples is not bounded below.

Next, it is interesting to look at the behaviour of the differentials at the bottom of the region which has groups above the 0-line. This is extremely regular as the rank varies, although it depends on the congruence class of the rank mod 4. The lowest degree for an η-multiple is as follows

PROPOSITION 9.12.1. *For $r \geq 1$, the lowest degree in $ko^*_{V(r)}$ from which η or η^2 is nonzero is as follows.*

Rank r	$4k$	$4k+1$	$4k+2$	$4k+3$
Lowest degree of x with $\eta x \neq 0$.	$-8k$	$-8k$	$-8k$	$-8k-7$
Lowest degree of x with $\eta^2 x \neq 0$.	$-8k$	$-8k$	$-8k$	$-8k$

Proof: First note that the result is correct for $r = 1$, and suppose that $r \geq 2$.

The only possible degrees in which there can be η-multiples, are of the form $8k+1$ or $8k+2$. Next, the pattern of differentials ensures that as soon as the $(8k+1)$-diagonal of the $E^{1\frac{1}{2}}$-page is non-trivial above the 0-line, this survives and there are non-zero η-multiples in degree $8k+2$ (and these are annihilated by η). Similarly, if the $8k$-diagonal is non-trivial above the 0-line, there is an η-multiple in degree $8k+1$ and an η^2-multiple in degree $8k+2$.

Entries above the 0-line come from $H(QU)$ or $H(TU; Sq^2)$. From Corollary 9.5.2 we see that if r is even, the lowest degree where the representation theory contributes entries above the 0-line is on the $2r$-diagonal. If r is odd, it is the $(2r-2)$-diagonal. From Proposition 9.7.2 lowest top even and odd homology of

$H(TU)$ is in degrees as follows.

r mod 4	0	1	2	3
Top even homology	$-2r+2$	$-2r+4$	$-2r-2$	$-2r$
Top odd homology	$-2r+1$	$-2r+3$	$-2r+5$	$-2r-1$

The lowest even homology is never in a degree of the form $8k$, but the lowest odd homology is always in a degree of the form $8k+1$. The result can now be easily verified. □

9.12.B. The torsion subgroup. The pattern of the ranks of the torsion part is very complicated, but using the methods of Chapter 11 one can write down its Hilbert series.

In the following, the Hilbert series are in terms of the codegree 1 variable t. It is natural to divide all modules into even and odd graded parts, and display the series separately. Furthermore, since the resolutions are in terms of the polynomial ring PP, it is convenient to introduce the variable $u = t^4$. All Hilbert series in rank r are expressed as a polynomial multiple of

$$[PP] = \frac{1}{(1-u)^r}.$$

The leading terms in each congruence class are expressed as powers of t for convenience in comparing with the charts.

EXAMPLE 9.12.2. *(Rank 2)* In this case $TO = TO_2$, concentrated in even degrees. The Hilbert series is given by

$$[TO_2] = \frac{t^6 + t^{12}}{(1-u)^2}$$

EXAMPLE 9.12.3. *(Rank 3)* In this case $TO = TO_2 \oplus TO_3$, with TO_2 in even degrees and TO_3 in odd degrees. The Hilbert series are given by

$$[TO_3] = \frac{t^7(1+3u) + t^9(3+u^2)}{(1-u)^3}$$

$$[TO_2] = \frac{t^6(3-u) + t^{12}(6-4u)}{(1-u)^3}.$$

EXAMPLE 9.12.4. *(Rank 4)* In this case $TU = TU_2 \oplus TU_3 \oplus TU_4$, with TU_2 and TU_4 in even degrees and TU_3 in odd degrees. In odd degrees, we find $TO = TO_3$, but in even degrees there is an exact sequence

$$0 \longrightarrow TO_2 \longrightarrow TO_{24} \longrightarrow TO_4 \longrightarrow \mathbb{F}_2(-8) \longrightarrow 0,$$

so that

$$[TO] = [TO_4] + [TO_3] + [TO_2] - t^8.$$

The resolutions of Chapter 11 give

$$[TO_4] = \frac{t^8(1+4u^2-u^3) + t^{14}(6-3u+u^2)}{(1-u)^4}$$

$$[TO_3] = \frac{t^7(4-u+10u^2-5u^3) + t^{13}(20-16u+4u^2)}{(1-u)^4}$$

$$[TO_2] = \frac{t^6(6-3u+u^2)+t^{12}(20-24u+7u^2+u^3)}{(1-u)^4}.$$

EXAMPLE 9.12.5. *(Rank 5)* In this case $TU = TU_2 \oplus TU_3 \oplus TU_4 \oplus TU_5$, with TU_2 and TU_4 in even degrees and TU_3 and TU_5 in odd degrees.

In even degrees there is an exact sequence

$$0 \longrightarrow TO_2 \longrightarrow TO_{24} \longrightarrow TO_4 \longrightarrow 5\mathbb{F}_2(-8) \longrightarrow 0,$$

so that

$$[TO_{24}] = [TO_4] + [TO_2] - 5t^8.$$

In odd degrees there is an exact sequence

$$0 \longrightarrow TO_3 \longrightarrow TO_{35} \longrightarrow TO_5 \longrightarrow \mathbb{F}_2(-9) \longrightarrow 0,$$

so that

$$[TO_{35}] = [TO_5] + [TO_3] - t^9.$$

The resolutions of Chapter 11 give

$$[TO_5] = \frac{t^9(1+10u^2-4u^3+u^4)+t^{15}(10-5u+4u^2+u^3)}{(1-u)^5}$$

$$[TO_4] = \frac{t^8(5-u+40u^2-25u^3+5u^4)+t^{14}(45-35u+19u^2+5u^3)}{(1-u)^5}$$

$$[TO_3] = \frac{t^7(10-4u+51u^2-45u^3+11u^4+u^5)}{(1-u)^5} + \frac{t^{13}(75-85u+40u^2+15u^3-u^4)}{(1-u)^5}$$

$$[TO_2] = \frac{t^6(10-5u+4u^2-u^3+50u^4+25u^5)}{(1-u)^5} + \frac{t^{12}(50-80u+40u^2-26u^3+19u^4)}{(1-u)^5}$$

EXAMPLE 9.12.6. *(Rank 6)* In this case $TU = TU_2 \oplus TU_3 \oplus TU_4 \oplus TU_5 \oplus TU_6$, with TU_2, TU_4 and TU_6 in even degrees and TU_3 and TU_5 in odd degrees.

In even degrees there is an exact sequence

$$0 \longrightarrow TO_2 \longrightarrow TO_{246} \longrightarrow TO_{46} \longrightarrow 15\mathbb{F}_2(-8) \longrightarrow 0,$$

$$0 \longrightarrow TO_4 \longrightarrow TO_{46} \longrightarrow TO_6 \longrightarrow \mathbb{F}_2(-12) \longrightarrow 0,$$

so that

$$[TO_{246}] = [TO_6] + [TO_4] + [TO_2] - 15t^8 - t^{12}.$$

In odd degrees there is an exact sequence

$$0 \longrightarrow TO_3 \longrightarrow TO_{35} \longrightarrow TO_5 \longrightarrow 6\mathbb{F}_2(-9) \longrightarrow 0,$$

so that

$$[TO_{35}] = [TO_5] + [TO_3] - 6t^9.$$

The resolutions of Chapter 11 give

$$[TO_6] = \frac{t^{10}(1+20u^2-9u^3+5u^4-u^5)+t^{16}(15-5u+10u^2-5u^3+u^4)}{(1-u)^6}$$

$$[TO_5] = \frac{t^9(6 - u + 105u^2 - 69u^3 + 29u^4 - 6u^5)}{(1-u)^6} + \frac{t^{15}(84 - 50u + 54u^2 - 30u^3 + 6u^4)}{(1-u)^6}$$

$$[TO_4] = \frac{t^8(15 - 5u + 211u^2 - 190u^3 + 75u^4 - 9u^5 - u^6)}{(1-u)^6} + \frac{t^{14}(189 - 179u + 129u^2 - 59u^3 + 10u^4 + u^5)}{(1-u)^6}$$

$$[TO_3] = \frac{t^7(20 - 9u + 175u^2 - 211u^3 - 9u^4 + 45u^5 - 32u^6 + 5u^7)}{(1-u)^6} + \frac{t^{13}(210 - 280u + 99u^2 - 5u^3 - 66u^4 + 31u^5 - 5u^6)}{(1-u)^6}$$

$$[TO_2] = \frac{t^6(15 - 5u + 10u^2 - 5u^3 + 281u^4 - 210u^5 + 84u^6 - 14u^7)}{(1-u)^6} + \frac{t^{12}(105 - 196u + 330u^2 - 250u^3 + 150u^4 - 68u^5 + 15u^6)}{(1-u)^6}$$

9.12.C. The first few negative groups. In Section 2.9 we have given a general, representation theoretic description of ko^G_i for $i \geq -7$. The complex case suggests that we should expect torsion to appear in all subsequent groups, and none of them to be zero. It is very striking that this does not happen for the elementary abelian groups.

PROPOSITION 9.12.7. *If $G = V(r)$ is an elementary abelian 2-group of rank r, we have*

$$ko^G_{-8} = JU^4 \text{ and } ko^G_{-9} = 0.$$

In particular, the group ko^G_{-8} is torsion free.

REMARK 9.12.8. This means that the groups in the first few negative degrees are purely representation theoretic, and have the following values.

n	-9	-8	-7	-6	-5	-4	-3	-2	-1	0
ko^G_n	0	JU^4	$2^{\binom{r}{3}}$	$2^{\binom{r}{2}}$	0	JU^2	0	0	0	RU

After this, the values are more complicated. For example, one may check by the same methods that in degree -10, the rank is

$$r\binom{r}{2} - \binom{r}{3} + \binom{r}{4} + \binom{r}{6}.$$

Proof: First, $H(TU; Sq^2)$ is zero in degrees -9, -10 and -11, the 0-line is the only contribution to ko^G_{-8}. Since no differentials leave the (-8)-diagonal, we have

$$ko^G_{-8} = JU^4 \oplus TO_{-8}.$$

Similarly,

$$ko^G_{-9} = TO_{-9},$$

so we need only show TO is zero in degrees -8 and -9.

This uses a little about the resolutions described in Chapter 11, but the essential ingredient is Proposition 9.7.2. For example, to understand degree 8 we need only consider TU_2 and TU_4, since the top of $TU_{67\cdots r}$ is in degree -10. The short exact sequence

$$0 \longrightarrow TU_2 \longrightarrow TU_{24} \longrightarrow TU_4 \longrightarrow 0$$

gives

$$0 \longrightarrow TO_2 \longrightarrow TO_{24} \longrightarrow TO_4 \longrightarrow H(TU_2) \longrightarrow H(TU_{24}) \longrightarrow \cdots$$

It is easy to check that TO_2 is zero in degree -8. Of course TO_4 is of rank $\binom{r}{4}$ in degree -8. However by Proposition 9.7.2, the map $H(TU_2) \longrightarrow H(TU_{24})$ has a kernel of precisely this size in degree -8, and TO_{24} is zero in degree -8 as required.

The argument for degree -9 is precisely similar, working from

$$0 \longrightarrow TU_3 \longrightarrow TU_{35} \longrightarrow TU_5 \longrightarrow 0.$$

□

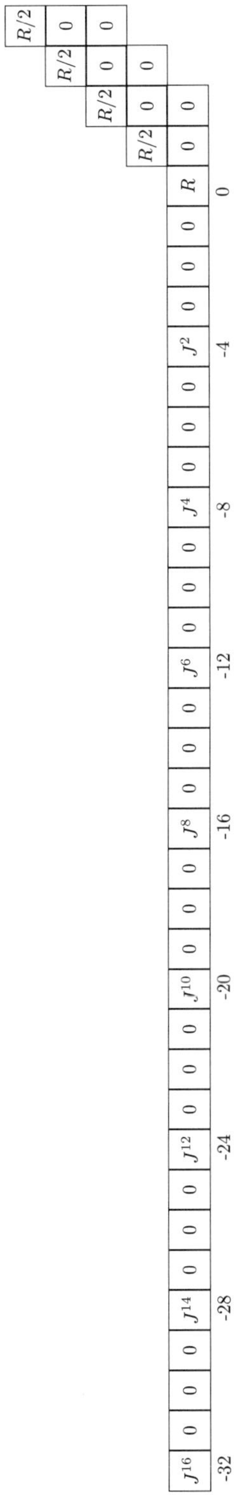

FIGURE 9.12.1. The $E^{1\frac{1}{2}}$-page for $V(1)$.

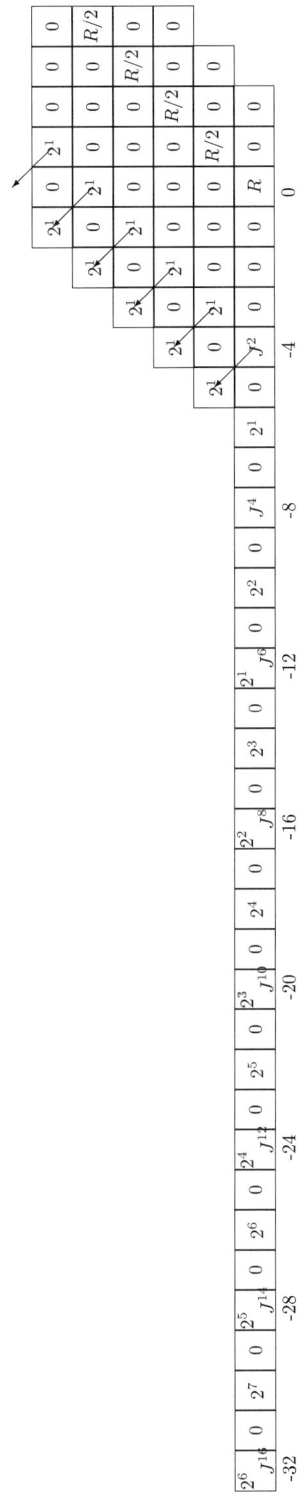

FIGURE 9.12.2. The $E^{1\frac{1}{2}}$-page for $V(2)$.

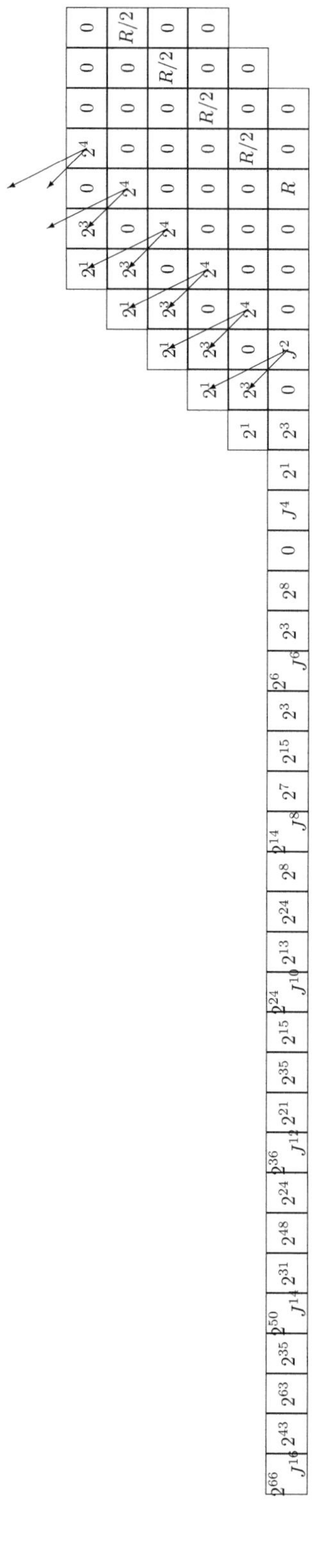

FIGURE 9.12.3. The $E^{1\frac{1}{2}}$-page for $V(3)$.

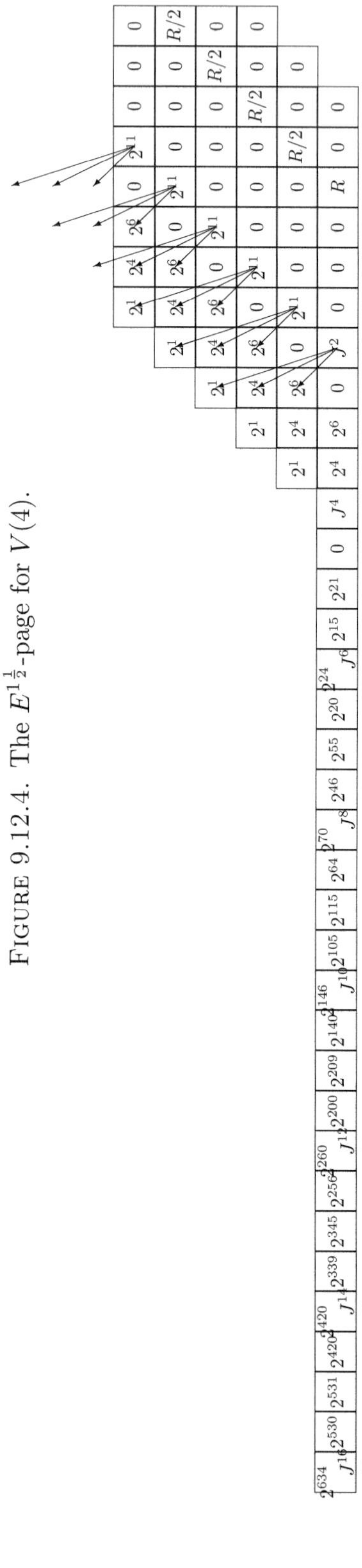

FIGURE 9.12.4. The $E^{1\frac{1}{2}}$-page for $V(4)$.

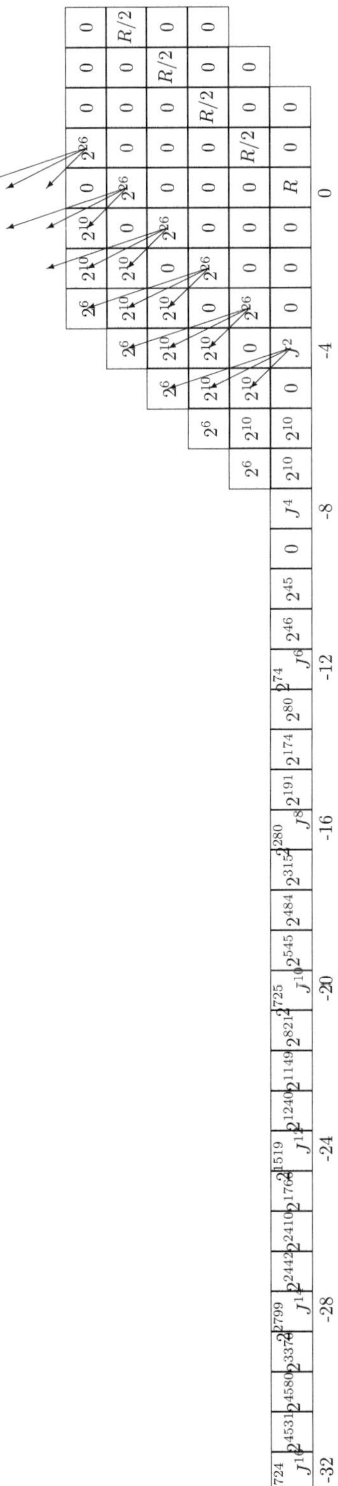

FIGURE 9.12.5. The $E^{1\frac{1}{2}}$-page for $V(5)$.

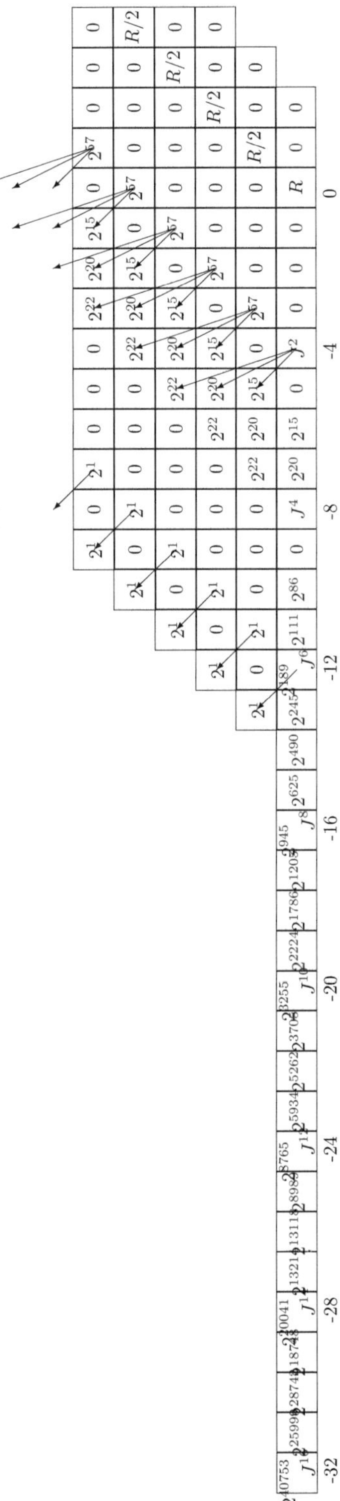

FIGURE 9.12.6. The $E^{1\frac{1}{2}}$-page for $V(6)$.

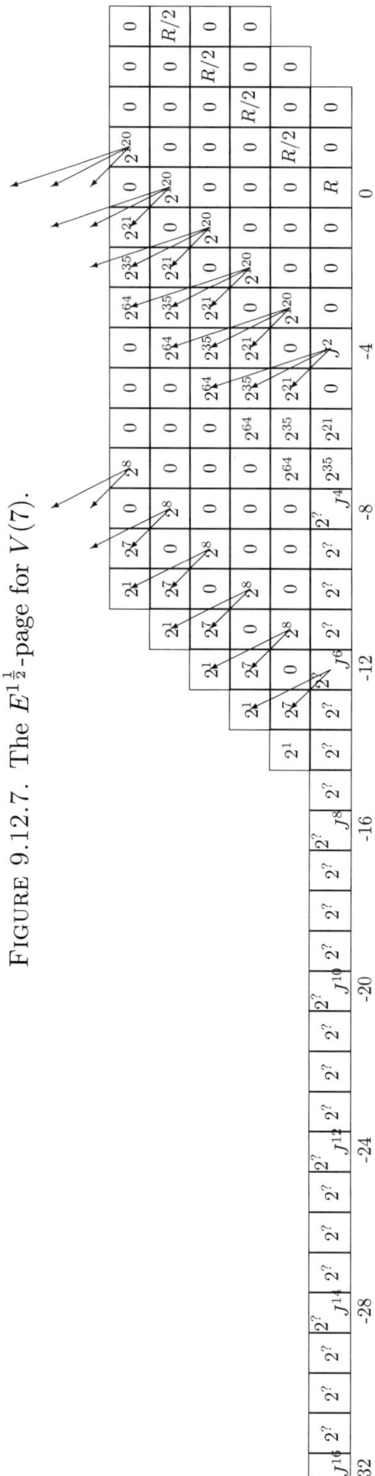

FIGURE 9.12.7. The $E^{1\frac{1}{2}}$-page for $V(7)$.

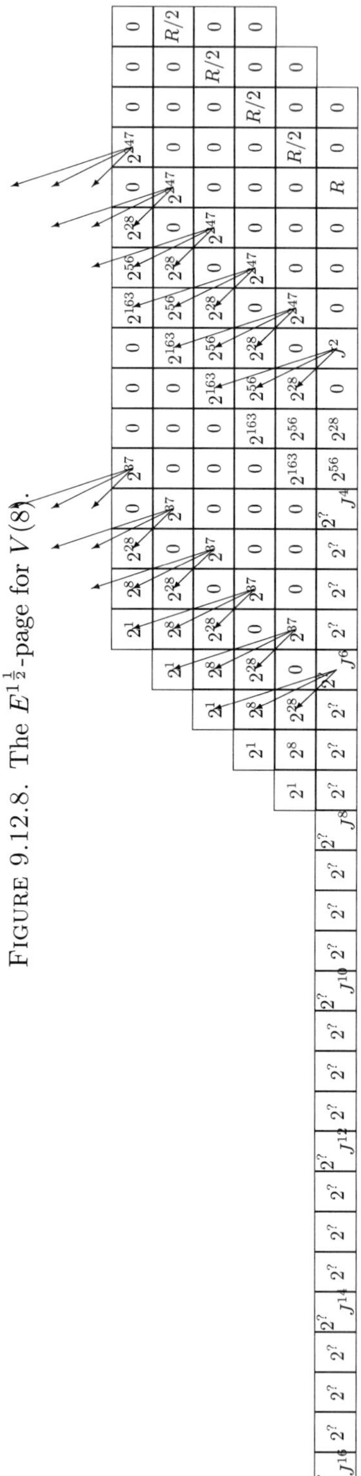

FIGURE 9.12.8. The $E^{1\frac{1}{2}}$-page for $V(8)$.

CHAPTER 10

The ko-Homology of Elementary Abelian Groups (BSS)

Chapter 9 described the calculation of cohomology, and we turn here to calculating the homology. By contrast with our usual practice of using the local cohomology spectral sequence to deduce $ko_*(BG)$ from $ko^*(BG)$, we use the homology η-Bockstein spectral sequence

$$ku_*(BV(r))[\tilde{\eta}] \Rightarrow ko_*(BV(r))$$

to deduce $ko_*(BV(r))$ from $ku_*(BV(r))$. We will consider the local cohomology spectral sequence in Chapter 12, but we can only obtain complete answers by that method in low rank.

With the exception of the additive extensions in degrees congruent to 3 mod 8, the analysis of the η-Bockstein spectral sequence is remarkably easy from what we already know. The main input is the homology TU-QU exact sequence

$$(6) \qquad 0 \longrightarrow \mathrm{Start}(2)TU^{\vee} \longrightarrow \widetilde{ku}_*(BV(r)) \longrightarrow \mathrm{Start}(1)2^{r-1}H^1_I(QU) \longrightarrow 0$$

from [**28**, 4.11.4]. These prior calculations give us a complete understanding of the E^1-term, and it remains only to find the differentials and resolves some additive extension problems.

10.1. The behaviour of the spectral sequence

To start with, we note that the homology TU-QU exact sequence is compatible with the Bockstein spectral sequence since $TU^{\vee}$ is the kernel of inverting the Bott class. Accordingly, we may view the TU-QU exact sequence as a short exact sequence of chain complexes.

As in cohomology, the differential on the TU part is given by Sq^2. Since we are working in homology, this means that the differential is the dual of $Sq^2 : TU \longrightarrow TU$. Similarly, the differential on the image of $ku_*(BV(r))$ in $KU_*(BV(r))$ is $1+\tau = 2$ from degrees 1 mod 4, and $1-\tau = 0$ from degrees 3 mod 4.

Using this, we may start by calculating the E^2-term above the 0-line. It turns out that this is rather sparse, and the fact that the spectral sequence must converge to 0 above the 2-line forces all remaining differentials.

Finally, we complete the analysis by returning to comment on the 0-line.

10.1.A. The Sq^2-homology of the TU-part. We gave a complete calculation of $H(TU; Sq^2)$ in Proposition 9.7.2. Of course $H(TU^{\vee}; Sq^2)$ is simply the vector space dual, so the purpose of this subsection is simply to record the answer in terms that permit direct transcription into the Bockstein spectral sequence.

Observe that, if we write $r = 4k+i$ for $i = 2$, 3, 4 or 5, then $H(TU; Sq^2)$ is zero for degrees $\leq -(8k+7)$ if $i = 2$ or $-(8k+8)$ otherwise. By dualization and shift, we see that $TU^\vee$ contributes nothing above the 0-line to the E^2-term in degrees $\geq 8k+3$ if $i=2$ or in degrees $\geq 8k+4$ otherwise. In low degrees there are some small contributions which may be transcribed directly from Proposition 9.7.2.

LEMMA 10.1.1. *The Sq^2-homology of $TU^\vee$ is concentrated in finitely many degrees. Its actual dimensions are as follows.*

Even degrees: *The homology group $H(\mathrm{Start}(2)TU^\vee_{ev}; Sq^2)$ is of dimension $\binom{r}{2}$, $\binom{r}{6}$, $\binom{r}{10}$, ... in degrees 2, 10, 18, 26, ...*

Odd degrees: *The homology group $H(\mathrm{Start}(2)TU^\vee_{od}; Sq^2)$ is of dimension $\binom{r}{3}$, $\binom{r}{7}$, $\binom{r}{11}$, ... in degrees 3, 11, 19, 27, ...*

□

10.1.B. The homology of the QU-part. The exact value of the E^1-term is given in [**28**, 4.4.7]. Indeed, in degrees ≥ 0

$$H^1_I(QU)_{2s} = \binom{r}{r-1}\mathbb{Z}/2^{s+r-1} \oplus \binom{r}{r-2}\mathbb{Z}/2^{s+r-2} \oplus \cdots \oplus \binom{r}{1}\mathbb{Z}/2^{s+1} \oplus \binom{r}{r}\mathbb{Z}/2^{s},$$

with the summands corresponding under multiplication by v as s varies. Accordingly,

$$2^{r-1}H^1_I(QU)_{2s} = \binom{r}{r-1}\mathbb{Z}/2^{s} \oplus \binom{r}{r-2}\mathbb{Z}/2^{s-1} \oplus \cdots$$
$$\oplus \binom{r}{1}\mathbb{Z}/2^{s-r+2} \oplus \binom{r}{r}\mathbb{Z}/2^{s-r+1},$$

where terms with negative exponent are treated as zero.

LEMMA 10.1.2. *The contribution of the QU term to the E^2 term above the 0-line is entirely concentrated on n-diagonals where n is of the form $4k+1$, where it is elementary abelian. For $4k+1 \geq 2r-1$ this is of dimension 2^r-1, and in lower degrees the dimension is the sum $\binom{r}{1} + \binom{r}{2} + \cdots + \binom{r}{2k+1}$.*

Proof: We note that the map

$$2 : 2^{r-1}H^1_I(QU)_{2s} \longrightarrow 2^{r-1}H^1_I(QU)_{2s-2}$$

is an epimorphism for all s, and the kernel consists of the elements of order 2. The rank can be read off from the values quoted before the statement of the lemma. □

10.1.C. Higher differentials. We have now calculated the E^2-term above the 0-line, and before proceeding, the reader may find it helpful to refer to the displays in ranks ≤ 6 given in Section 10.3 below. For each rank we have displayed the reduced as well as the unreduced form to emphasize the patterns.

First we consider large degrees. The identification of d^1 on the TU and QU parts shows that in large degrees there are only non-zero entries on $(4k+1)$-diagonals. In particular, there are no further d^1 or d^2-differentials, and these remaining groups in large degrees must be cancelled by the d^3 differentials in the sense that the complex they give in high degrees is exact. To find out the ranks of successive d^3s we work up from low degrees.

Turning to low degrees, close inspection of the dimensions shows that the differentials are forced by the requirement that E^4 is zero above the 3-line. Working from low degrees we see that the differentials must come from the $(8k+5)$-diagonal, but now there is a d^1, a d^2 and a d^3.

In more detail, we argue by induction on k that these three differentials are epimorphisms, and that the intersection of their kernels is zero. This is trivial for $k=-1$. If this has been proved in lower degrees we argue first that the $(8k+1)$- , $(8k+2)$- and $(8k+3)$-diagonals consist of infinite cycles. On the $(8k+1)$-diagonal $d^1=0$ and $d^2=0$ by sparsity and $d^3=0$ from the inductive hypothesis. All three differentials are zero on the $(8k+2)$-diagonal by sparsity. On the $(8k+3)$-diagonal $d^2=0$ and $d^3=0$ by sparsity and $d^1=0$ since the QU-homology is in degrees congruent to 1 mod 4. Given that the $(8k+3)$-, $(8k+2)$- and $(8k+1)$-diagonals are infinite cycles it follows that they must be boundaries, so that d^1, d^2 and d^3 from the $(8k+5)$-diagonal must be epimorphic to cancel the homology. Adding up dimensions, we see that this also precisely cancels the groups in their domain (i.e. on the $(8k+5)$-diagonal itself) and the intersections of the kernels of these three differentials is zero. This completes the inductive step.

Once the exceptional lower degrees have been dealt with we see that the high dimensional d^3 again goes from the $(8k+5)$-diagonal to the $(8k+1)$-diagonal of rank 2^r-1.

10.1.D. The 0-line. The discussion of the previous subsection identified all differentials in high degrees. From the structure of the Bockstein spectral sequence, this determines all differentials. We may use this to identify the 0-line.

The QU part is straightforward.

LEMMA 10.1.3. *The contribution of the QU part to the 0-line is the same as the homology on the $(4k+1)$-diagonal, and it is the entire $M_{4k-1}=2^{r-1}H^1_I(QU)_{4k}$ on the $(4k-1)$-diagonal.*

Proof: It suffices to recall that the differential is $1-\tau=0$ from degrees 3 mod 4. □

The contribution of TU can easily be expressed in terms of known quantities.

LEMMA 10.1.4. *The dual $TU^\vee$ contributes* $\mathrm{Start}(2)(TU/Sq^2TU)^\vee$ *to the 0-line.* □

However, it is less easy to make this explicit. To approach TU/Sq^2TU we use the exact sequence

$$0 \longrightarrow TO \longrightarrow TU \xrightarrow{Sq^2} \Sigma^2 TU \longrightarrow \Sigma^2 TU/Sq^2 \longrightarrow 0.$$

From Chapter 11 one can deduce structural information, but for our present purposes we will be satisfied with determining the Hilbert series of the module. The exact sequence gives

$$[TU/Sq^2]=(1-t^2)[TU]+t^2[TO].$$

10.2. The calculation

We have now assembled the information required to make the additive calculation. The statement of the theorem is rather complicated, and to understand it, it is necessary to be familiar with the form of the η-Bockstein spectral sequence. In particular, looking at the region near degree $8k$ we have the following pattern at the E^∞ page of the spectral sequence.

$$\begin{array}{ccccc}
0 & 0 & 8k_2 & (8k+1)_2 & 0 \\
0 & 8k_1 & (8k+1)_1 & (8k+2)_1 & 0 \\
8k_0 & (8k+1)_0 & (8k+2)_0 & ? & ? \\
\hline
8k & 8k+1 & 8k+2 & 8k+3 & 8k+4
\end{array}$$

First, note that all entries are zero above the 2-line, so that we have shown all non-zero rows in the region. We will justify the entries in due course, but the point is that the entries $8k_i$ form the $8k$-diagonal, and contribute to the groups in degrees $8k, 8k+1$ and $8k+2$. Similarly the entries $(8k+1)_i$ form the $(8k+1)$-diagonal, and contribute to the groups in degrees $8k+1, 8k+2$ and $8k+3$. Finally, the entries $(8k+2)_i$ form the $(8k+2)$-diagonal, and contribute to the groups in degrees $8k+2$ and $8k+3$ (since the entry $(8k+2)_2$ is zero).

THEOREM 10.2.1. *The connective real K-homology of an elementary abelian group of rank r, $ko_*(BV(r))$ may be described as follows. We have*

$$ko_*(BV(r))/\eta \cong ku_{\equiv 0(4)} \oplus D_{\equiv 1(8)} \oplus D_{\equiv 2(8)} \oplus \text{Start}(2)(TU/Sq^2)^\vee \oplus \Sigma^{-1} 2^{r-1} H_I^1(QU)_{\equiv 0(4)}.$$

The first summand just consists of $\mathbb{Z}$ in degrees 0 mod 4. The second summand $D_{\equiv 1(8)}$ is zero except in degrees of the form $8k+1$, where it is elementary abelian of rank $\binom{r}{1}+\binom{r}{2}+\cdots+\binom{r}{4k+1}$ (which is to say 2^r-1 for $r \geq 4k+1$). The third summand $D_{\equiv 2(8)}$ is a group of rank $\binom{r}{4k+2}$ (which is to say 0 when $r < 4k+2$). The fourth summand is a complicated but self-explanatory elementary abelian group. The fifth summand is the part of of $2^{r-1}H_I^1(QU)$ in degrees which are multiples of 4, shifted down by one degree: the contribution in degree $4k-1$ is

$$2^{r-1}H_I^1(QU)_{4k} = \binom{r}{r-1}\mathbb{Z}/2^{2k} \oplus \binom{r}{r-2}\mathbb{Z}/2^{2k-1} \oplus \cdots \oplus \binom{r}{1}\mathbb{Z}/2^{2k-r+2} \oplus \binom{r}{r}\mathbb{Z}/2^{2k-r+1},$$

where summands with negative exponent are treated as zero. The submodule of η-multiples is elementary abelian, and concentrated on the n-diagonals where $n = 8k, 8k+1$ and $8k+2$. On the $8k$-diagonal we only have the η-multiples of ko_*, which is to say it is of rank 1 on the 1- and 2-line. On the $(8k+1)$-diagonal we have D_{8k+1} on the 1- and 2-line. On the $(8k+2)$-diagonal we have D_{8k+2} on the 1-line, and nothing on the 2-line.

In degrees n not of the form $8k-1$, $ko_n(BV(r))$ is the direct sum of the η-multiple part and the quotient by the η-multiples. When $n = 8k-1$ the extension is highly non-split; in large degrees, the elementary abelian group of η-multiples (of rank 2^r-1) increases the size of all cyclic factors in the QU part.

Writing

$$h_i = \dim_{\mathbb{F}_2}(\text{Start}(2)(TU/Sq^2)^\vee)_i,$$

the groups in each degree are given in tabular form below.

$n \geq 0$	$ko_n(BV(r))$
$8k+7$	$[2^{4k+4}]^{\binom{r}{r-1}} \oplus [2^{4k+3}]^{\binom{r}{r-2}} \oplus \cdots \oplus [2^{4k-r+6}]^{\binom{r}{1}} \oplus [2^{4k-r+5}]^{\binom{r}{0}} \oplus 2^{h_{8k+7}}$
$8k+6$	$2^{h_{8k+6}}$
$8k+5$	$2^{h_{8k+5}}$
$8k+4$	$\mathbb{Z} \oplus 2^{h_{8k+4}}$
$8k+3$	$[2^{4k+3}]^{\binom{r}{r-1}} \oplus [2^{4k+2}]^{\binom{r}{r-2}} \oplus \cdots \oplus [2^{4k-r+5}]^{\binom{r}{1}} \oplus [2^{4k-r+4}]^{\binom{r}{0}} \oplus 2^{h_{8k+3}-\binom{r}{4k+3}}$
$8k+2$	$2 \oplus 2^{\binom{r}{1}+\cdots+\binom{r}{4k+1}} \oplus 2^{h_{8k+2}}$
$8k+1$	$2 \oplus 2^{\binom{r}{1}+\cdots+\binom{r}{4k+1}} \oplus 2^{h_{8k+1}}$
$8k+0$	$\mathbb{Z} \oplus 2^{h_{8k}}$

Note that for $n = 8k+3$ the subtraction of $\binom{r}{4k+3}$ is present precisely when the last of the cyclic terms $[2^a]$ has $a = 1$.

Proof: We have described the behaviour of the Bockstein spectral sequence, so the conclusion follows up to extensions. It is classical that ko_* is a summand. We must argue that the η-multiples split off in dimensions $8k+1$ and $8k+2$. This follows since $(TU/Sq^2)^\vee$ is generated by its elements in degrees not congruent to 1 and 2 mod 8, where there are no η-multiples.

Finally, we must explain the very intriguing extension in degrees of the form $8k+3$. This comes from the local cohomology spectral sequence discussed in Chapter 12. On the E^2-page, the (-1)-column is $H^1_I(\overline{QO})$. We may calculate local cohomology of $\overline{QO}$ from the known local cohomology of QU [**28**, 4.4.7] as follows. First note that the map $\overline{QO} \longrightarrow QU$ is a monomorphism, with image in degrees which are multiples of 4. In calculating the local cohomology it suffices to work with the ring $\mathbb{Z}[(p^*)^2]$, and we have a short exact sequence of modules

$$0 \longrightarrow \overline{QO} \longrightarrow QU^{\equiv 0(4)} \longrightarrow C \longrightarrow 0.$$

This is especially convenient since C is zero in sufficiently negative degrees so that $H^*_{(p^*)}(C) = H^0_{(p^*)}(C) = C$. Thus we have the exact sequence

$$0 \longrightarrow H^0_{JO}(\overline{QO}) \longrightarrow H^0_{JU}(QU)^{\equiv 0} \longrightarrow C \longrightarrow H^1_{JO}(\overline{QO}) \longrightarrow H^1_{JU}(QU)^{\equiv 0} \longrightarrow 0$$

It is easy to evaluate C in degrees ≥ 0: it is zero in degree divisible by 8, it is $RU/RSp \cong 2^{2^r}$ in degrees of the form $8k+4 \geq 4$, It is easy to see that the map $\overline{QO} \longrightarrow QU^{\equiv 0(4)}$ in 0th local cohomology has cokernel $\mathbb{Z}/2$ in degrees $8k+4$. Since we know the rank is at most $2^r - 1$, this gives the answer in positive degrees, and in particular

$$H^1_{JO}(\overline{QO})_{8k+4} = [2^{4k+r+2}]^{\binom{r}{r-1}} \oplus [2^{4k+r+1}]^{\binom{r}{r-2}} \oplus \cdots \oplus [2^{4k+4}]^{\binom{r}{1}} \oplus [2^{4k+3}]$$

Now, there are $r-1$ differentials in the local cohomology spectral sequence, each with codomain annihilated by 2, so the smallest possible group at E^∞ on the (-1)-column is

$$[2^{4k+3}]^{\binom{r}{r-1}} \oplus [2^{4k+2}]^{\binom{r}{r-2}} \oplus \cdots \oplus [2^{4k-r+5}]^{\binom{r}{1}} \oplus [2^{4k-r+4}]$$

Since this occurs as a subquotient, the extension in the Bockstein spectral sequence must also have this group as a quotient. This requires an extension by an elementary abelian 2-group of rank

$$\left[\binom{r}{r-1}+\binom{r}{r-2}+\cdots+\binom{r}{r-(4k+2)}\right]+\binom{r}{r-(4k+3)}.$$

We have bracketed this to make visible that the bracket comes from QU (see Lemma 10.1.2), and the final term from TU (see Lemma 10.1.1). This completes the proof. □

10.3. Bockstein spectral sequence charts

We display the Bockstein spectral sequence in low degrees for ranks ≤ 6. We give both the full version and the reduced version, since the latter is less cluttered and has a clearer pattern of differentials.

Most of the entries are either elementary abelian 2^a or the integers. The only group which involves cyclic groups of other orders is

$$M_{4k-1} = 2^{r-1}H^1_I(QU)_{4k}.$$

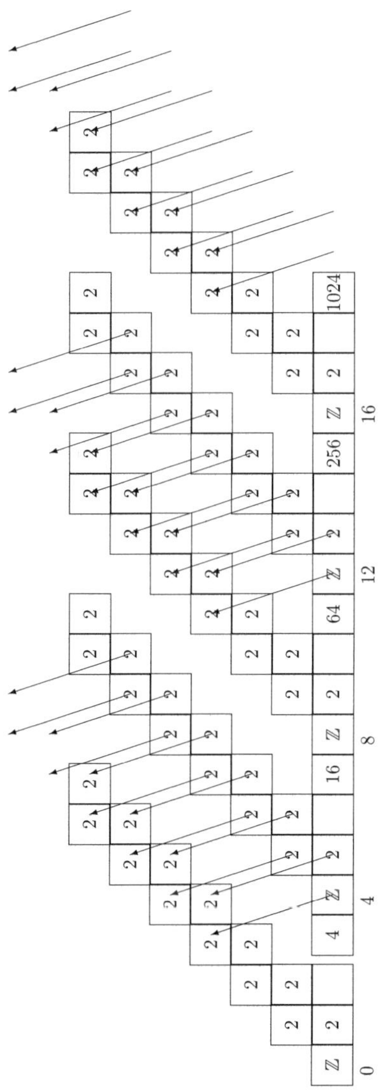

FIGURE 10.3.1. The E^2-page for $V(1)$

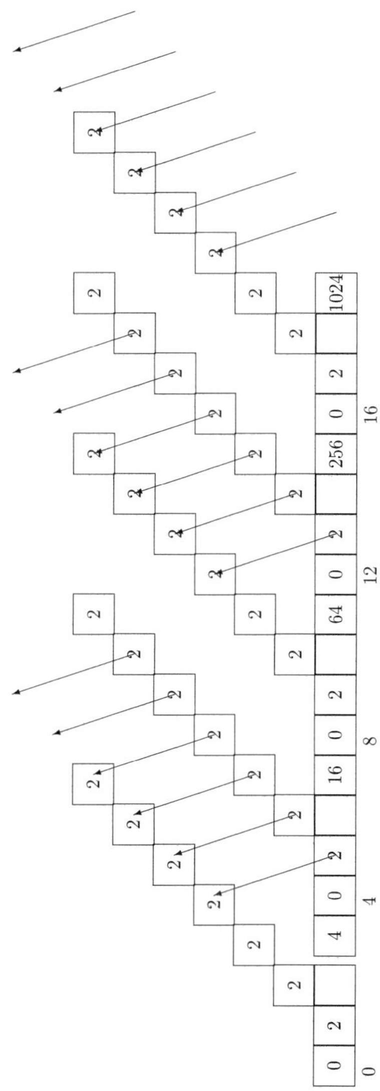

FIGURE 10.3.2. The E^2-page for $V(1)$ (reduced)

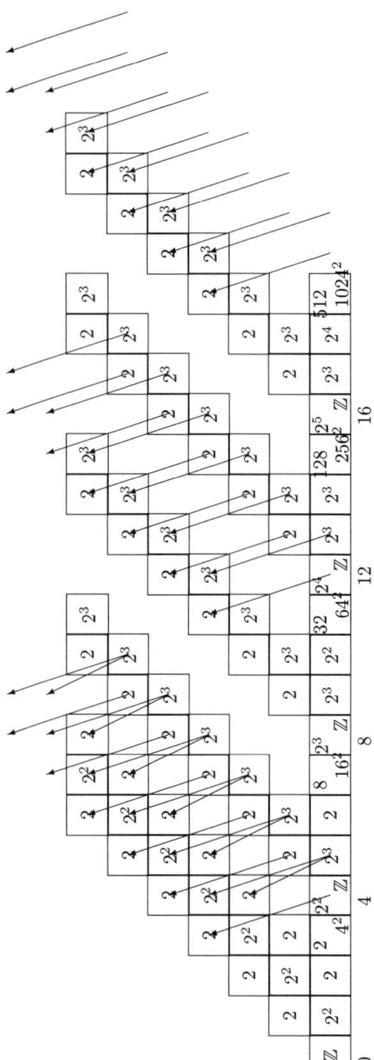

FIGURE 10.3.3. The E^2-page for $V(2)$

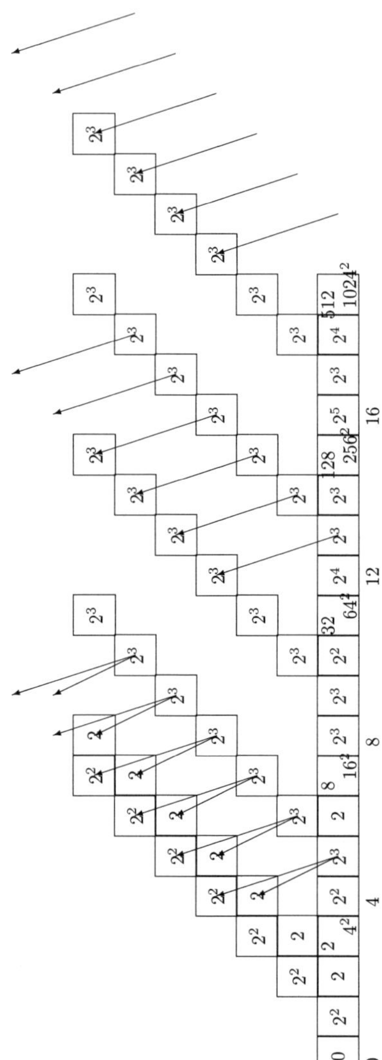

FIGURE 10.3.4. The E^2-page for $V(2)$ (reduced)

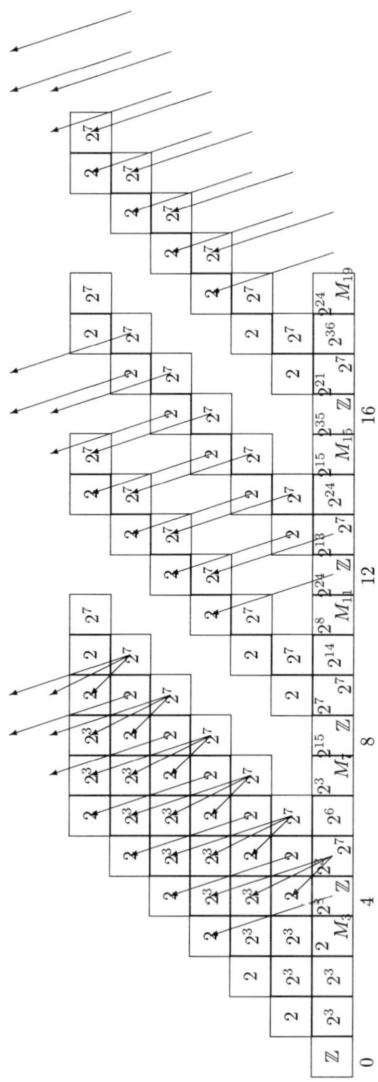

FIGURE 10.3.5. The E^2-page for $V(3)$

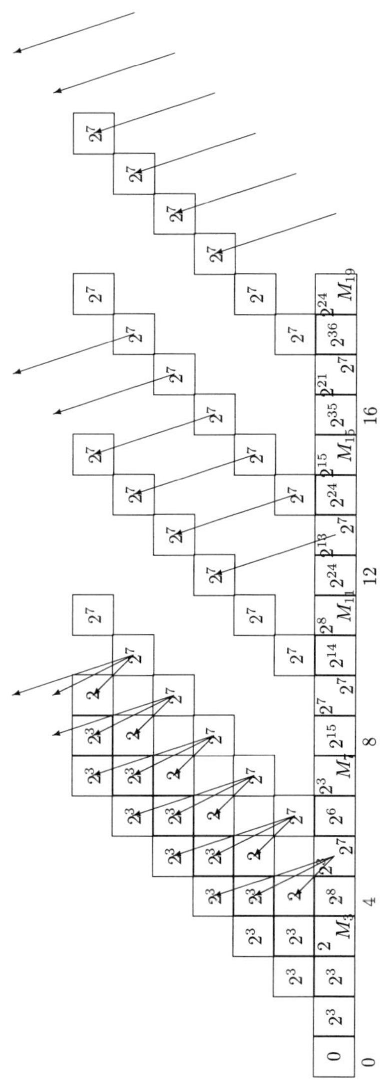

FIGURE 10.3.6. The E^2-page for $V(3)$ (reduced)

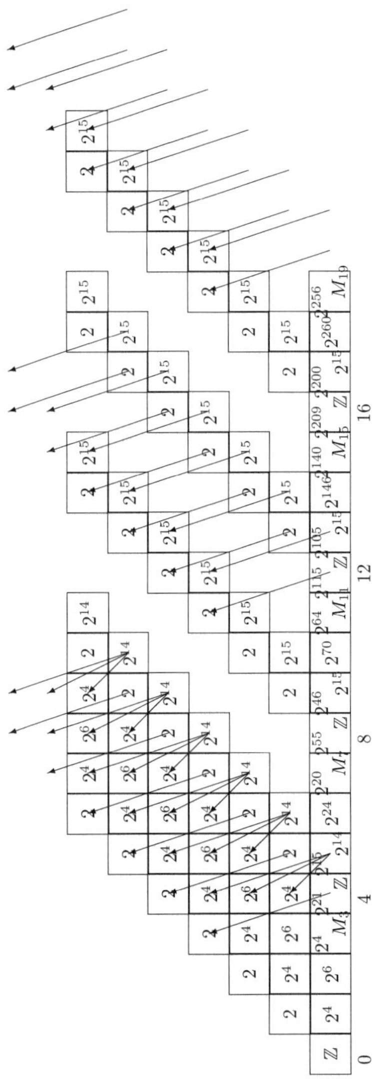

FIGURE 10.3.7. The E^2-page for $V(4)$

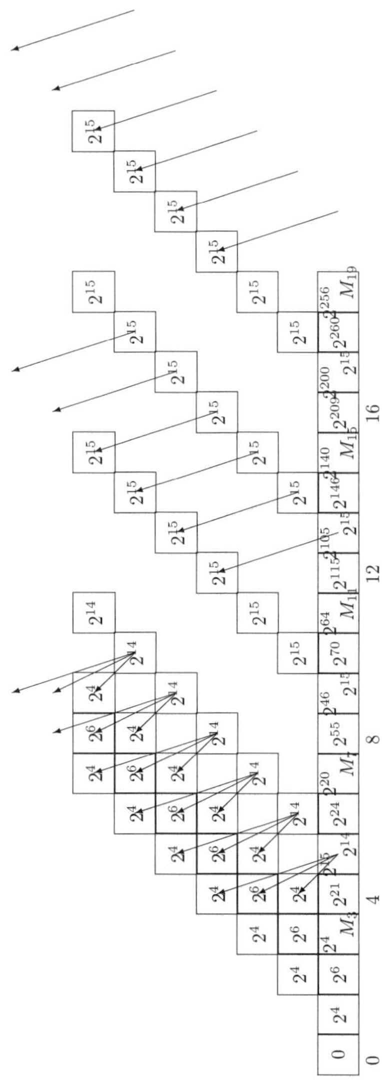

FIGURE 10.3.8. The E^2-page for $V(4)$ (reduced)

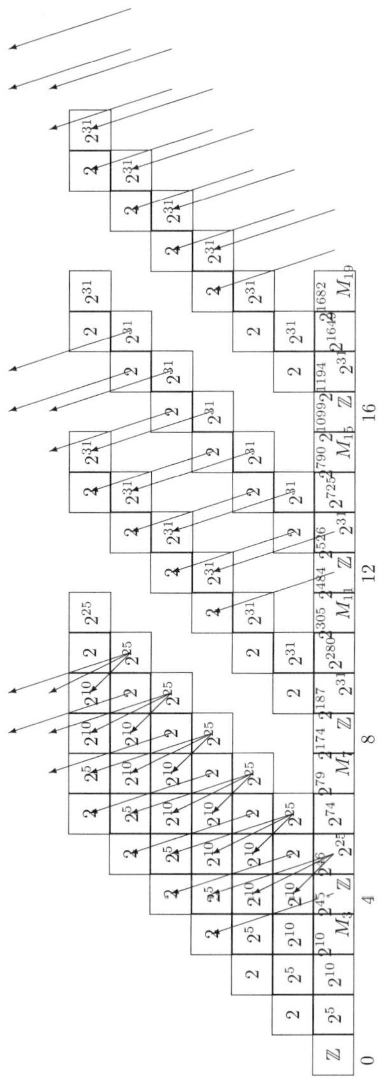

FIGURE 10.3.9. The E^2-page for $V(5)$

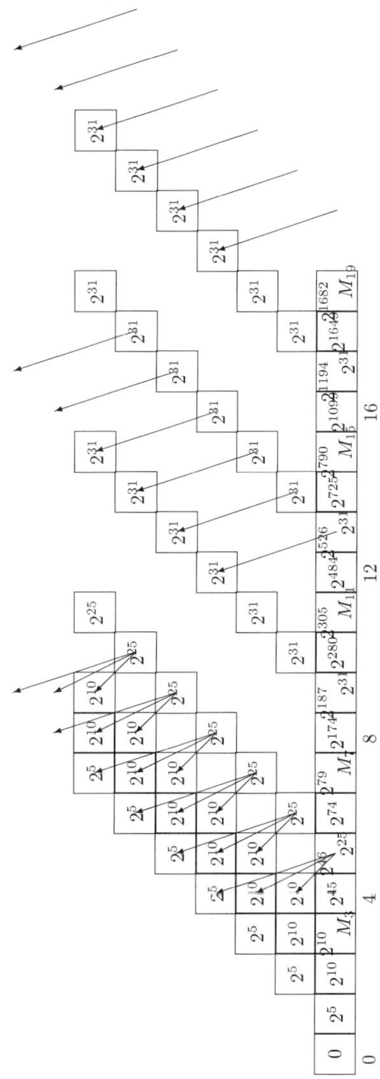

FIGURE 10.3.10. The E^2-page for $V(5)$ (reduced)

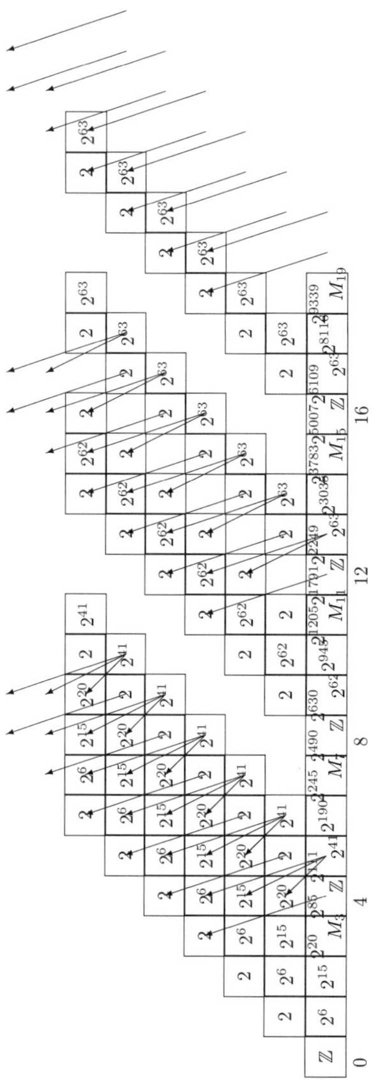

FIGURE 10.3.11. The E^2-page for $V(6)$

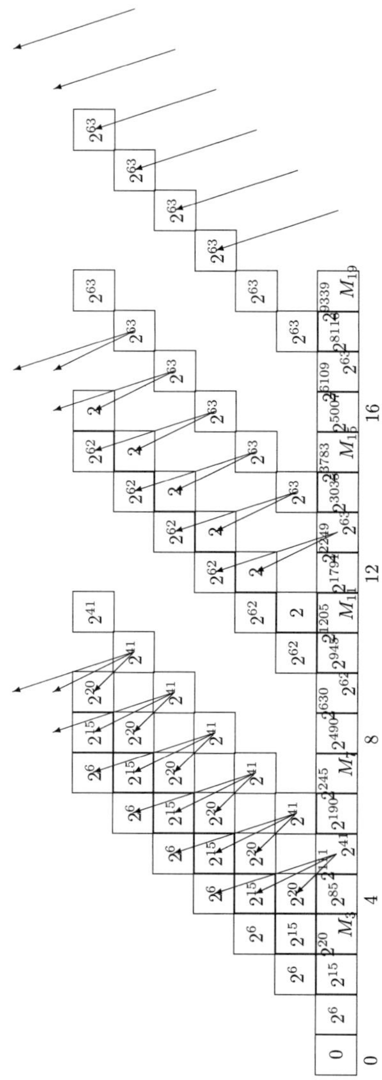

FIGURE 10.3.12. The E^2-page for $V(6)$ (reduced)

CHAPTER 11

The Structure of TO

For an elementary abelian 2-group $V = V(r)$ of rank r, we have shown that the module TO of v-torsion in ko^*_V detected in mod 2 cohomology consists precisely of the Sq^2-cycles. However, to understand ko^*_V more explicitly, we may want to know the Hilbert series of TO or its structure as a module over PP. Accordingly we spend this chapter in the remarkably intricate construction of a resolution of TO. This has the added benefit that we can use the resolution to calculate local cohomology, which provides the input to the local cohomology spectral sequence for $ko_*(BV)$ considered in Chapter 12.

We have seen that TO_i consists exactly of the Sq^2-cycles of TU_i. Our strategy for constructing a PP-projective resolution of TO_i is to work from the PC-resolution of TU_i, which Lemma 9.8.2 shows admits an action of Sq^2. Since passage to Sq^2-cycles is not exact, the process is fairly intricate, but we have not found another method except for TO_2. We are grateful to M.Katzman who made Macaulay calculations in ranks ≤ 4, which were critical in guiding our investigations at an early stage.

We showed in Section 9.8 that the resolution admits the structure of a $PC[Sq^2]$-complex and that the terms are sums of copies of $PC[Sq^2]$ and and PC. We obtain a new complex by taking Sq^2-cycles, this is very nearly exact, and the deviation is given by Sq^2-homology.

We begin with two sections considering the cycles in the constituents of the resolution.

11.1. Cycles in $PC[Sq^2]$

The Sq^2-cycles in $PC[Sq^2]$ are surprisingly simple.

LEMMA 11.1.1. *The PP-module of Sq^2-cycles in $PC[Sq^2]$ is free, and given by*

$$Z(PC[Sq^2]) = \binom{r}{0}PP(-2)\oplus\binom{r}{1}PP(-4)\oplus\binom{r}{2}PP(-6)\oplus\cdots\oplus\binom{r}{r}PP(-2r-2).$$

If we write X for the non-zero element of degree 0 and $Y = Sq^2X$ for the non-zero element of degree -2, then the generators are

$$m_B := Sq^2(y_BX) = p_BX + y_BY,$$

for subsets $B \subset \{1,\ldots,r\}$, so that in particular $m_\emptyset = Y$, and $\deg(m_B) = -(2|B|+2)$.

REMARK 11.1.2. (a) One may restate the conclusion neatly by saying that, as a *right* PC-submodule, $Z(PC[Sq^2])$ is free on the single generator $Y = Sq^2$.

(b) The Hilbert series of $ZPC[Sq^2]$ is

$$t^2(1+t^2)^r \cdot [PP].$$

Proof: Since Sq^2 is exact, cycles are boundaries. On the other hand PC is free over PP on the generators y_B. $\square$

It is worth making explicit how to express the elements $p_B Y$ in terms of the basis.

LEMMA 11.1.3. *Inside* $ZPC[Sq^2]$ *we have the relation*

$$p_B Y = \sum_{b \in B} z_b m_{B \setminus \{b\}}.$$

Proof: First note $p_B Y = \sum_b z_b y_{B\setminus\{b\}} Y$. On the other hand,

$$z_b m_{B\setminus\{b\}} = z_b (p_{B\setminus\{b\}} X + y_{B\setminus\{b\}} Y),$$

so the result is immediate from the fact that $\sum_b z_b p_{B\setminus\{b\}} = 0$. $\square$

In diagrammatic terms, sticking with the example of rank 4, the Sq^2-cycles in a pair

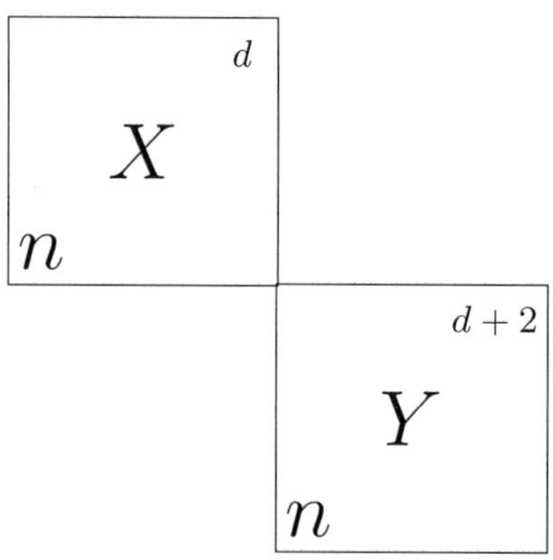

can be depicted

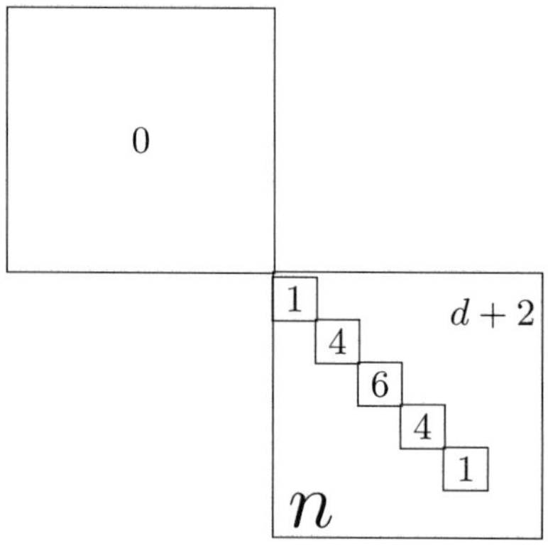

Here each entry $\boxed{s}$ represents s copies of PP (since there is one of these for each of the n original copies of $PC[Sq^2]$ it gives rise to ns copies altogether). The top left little box is in the same degree as the big box (namely $-d-2$), and each term in the succession of little boxes is 2 degrees lower than the previous one. The reason there is a gap for a sixth box at the end will appear when we consider ZPC. Of course the actual generators are not multiples of Y, so they do not come from the lower big box: it is just that their leading terms are from there.

11.2. Cycles in PC

We analyzed the Sq^2-action on PC in Section 9.6, and we record the answer for reference.

LEMMA 11.2.1. *The PP-module of Sq^2-cycles in PC is given by*

$$ZPC = PP \oplus I_2 \oplus \cdots \oplus I_r,$$

where I_j is the PP-module generated by the p_B for subsets B with $|B| = j$, with $\deg(p_B) = -(2j+2)$. The index j refers to the fact that I_j is the image of the module in Koszul homological dimension j.

The PP-module I_j is of projective dimension $r-j$, and has a minimal resolution

$$0 \longleftarrow I_j \longleftarrow \binom{r}{j} PP(-2j-2) \longleftarrow \binom{r}{j+1} PP(-2j-6) \longleftarrow \binom{r}{j+2} PP(-2j-10) \longleftarrow \cdots \longleftarrow \binom{r}{r} PP(2j-4r-2) \longleftarrow 0$$

obtained by truncating the Koszul complex. The generators in homological degree h are P_U with $|U| = j+h$, so that P_B maps to p_B and $dP_U = \sum_t z_t P_{U\setminus\{t\}}$ in positive homological degree.

In diagrammatic terms, sticking with the example of rank 4, the Sq^2-cycles in

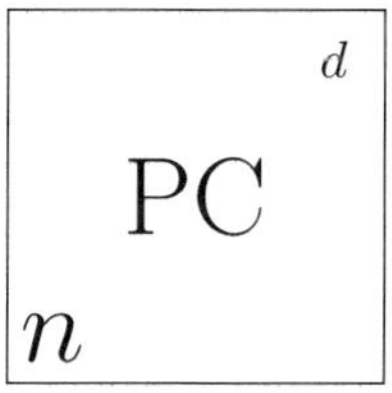

can be depicted

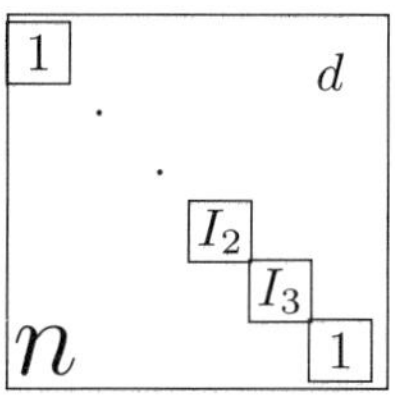

As before, an entry $\boxed{s}$ represents a single free PP-module of rank s (since the entire PC-box has multiplicity n, this gives ns copies of PP altogether). The entries I_2 and I_3 refer to the modules of those names. The module in the top left hand box has top degree the same as the ambient box (namely $-d$), and each successive box (or dot) has a module with top degree 2 less: thus I_2 has top 6 degrees below the ambient top and I_3 has top 8 degrees below the ambient top. The final entry 1 (with top 10 degrees below) could equally be called I_4.

The truncated Koszul resolution of a copy of I_k with top in degree 0 shows that it has Hilbert series

$$[I_k] = \binom{r}{k} - \binom{r}{k+1}u + \binom{r}{k+2}u^2 - \cdots \pm \binom{r}{r}u^{r-k} = \frac{(1-u)^r_{[k]}}{(-u)^k},$$

where we have used the polynomial $(1-u)^r_{[k]}$ consisting of terms of degree $\geq k$ in $(1-u)^r$. We can read off the Hilbert series of ZPC.

LEMMA 11.2.2. *The Hilbert series of ZPC is given by the formula*

$$[ZPC] = \frac{v(1+v)^{r-1} + (1-v)(1-v^2)^{r-1} - 1}{(1-u)^r}$$

where $v = t^2$ and $u = t^4$ where t is of degree -1.

REMARK 11.2.3. The pattern of coefficients in the numerator becomes much simpler above codegree $2r$.

Proof: From the earlier discussion, it is clear that we have

$$[ZPC] = (1+t^6\frac{(1-u)^r_{[2]}}{(-u)^2}+t^8\frac{(1-u)^r_{[3]}}{(-u)^3}+\cdots+t^{2r-2}\frac{(1-u)^r_{[r-1]}}{(-u)^{r-1}}+t^{2r}\frac{(1-u)^r_{[r]}}{(-u)^r})[PP].$$

A short calculation shows that if we sum the coefficients in each degree we have the following totals in codegrees ≥ 6

codegree	$8n$	$8n+2$	$8n+4$	$8n+6$
coefficient	$\binom{r-1}{4n-1}+\binom{r-1}{2n}$	$\binom{r-1}{4n}-\binom{r-1}{2n}$	$\binom{r-1}{4n+1}-\binom{r-1}{2n+1}$	$\binom{r-1}{4n+2}+\binom{r-1}{2n+1}$

It is then easy to verify the given formula. □

11.3. Minimal generating sets for TO_n

To find a minimal generating set for TO_n, we need only consider a presentation of TU_n:

$$0 \longleftarrow TU_n \longleftarrow \binom{r}{n}PC(-n-4) \stackrel{d}{\longleftarrow} \binom{r}{n+1}PC[Sq^2](-r-6).$$

We write Q_S for the generators in homological degree 0 and $X_R = Q^0_R$, $Y_R = Q^1_R$ for the generators in homological degree 1, where $|S| = n$ and $|R| = n+1$.

Thus TO_n is generated by the images of the PP-generators of $\binom{r}{n}ZPC(-n-4)$, namely Q_S and p_AQ_S for subsets A, together with a small number of additional generators to account for the Sq^2-homology. In this section, we identify exactly which of these are redundant. Indeed, the images of the generators

$$m_{B,R} = Sq^2(y_BX_R) = p_BX_R + y_BY_R$$

for subsets B are relations, sometimes showing generators are redundant.

Note that if i occurs twice in the sequence B, $p_B = z_i p_{B\setminus\{i,i\}}$, so that if B has repetitions, p_B is PP-decomposable.

LEMMA 11.3.1.

$$d(m_{B,R}) = d(p_B X_R + y_B Y_R) = \sum_{i\in R\cap B} z_i p_{B\setminus\{i\}} Q_{R\setminus\{i\}} + \sum_{i\in R\setminus B} p_{B\cup\{i\}} Q_{R\setminus\{i\}}$$

Thus

$$d(m_{B,R}) = d(p_B X_R + y_B Y_R) = \sum_{i\in R\setminus B} p_{B\cup\{i\}} Q_{R\setminus\{i\}} \text{ mod decomposables.}$$

Proof: We make a short calculation

$$d(p_B X_R + y_B Y_R) = p_B \sum_{i\in R} y_i Q_{R\setminus\{i\}} + y_B \sum_{i\in R} z_i Q_{R\setminus\{i\}} = \sum_{i\in R} (p_B y_i + y_B z_i) Q_{R\setminus\{i\}}.$$

Now note that $y_i p_B = p_{B\cup\{i\}}$ if $i \notin B$ and $z_i p_{B\setminus\{i\}}$ if $i \in B$. □

Now consider the combinatorics of this. First, all terms have the same

$$(B \cup \{i\}) \cup (R \setminus \{i\}) = B \cup R =: U$$

and

$$(B \cup \{i\}) \cap (R \setminus \{i\}) = B \cap R =: I.$$

We refer to U as the *universe* of the relation and to I as its *intersection*.

Now fix U and I, with $|U| = u$, $|I| = i$, and consider the set of generators

$$\mathcal{G}_{U,I} := \{p_A Q_S \mid |S| = n, A \cup S = U \text{ and } A \cap S = I\}$$

with this universe and intersection. All these generators have degree $-(2u-n+6)$. Notice that an element $p_A Q_S$ of $\mathcal{G}_{U,I}$ is uniquely specified by $S' := S \setminus I$, since $S = S' \cup I$ and $A = U \setminus S'$. We therefore write $g_{S'} := p_A Q_S$, and there are precisely $\binom{u-i}{n-i}$ such generators.

The corresponding relations come from

$$\mathcal{R}_{U,I} := \{p_B X_R + y_B Y_R \mid |R| = n+1, B \cup R = U \text{ and } B \cap R = I\}$$

Again, an element $p_B X_R + y_B Y_R$ of $\mathcal{R}_{U,I}$ is uniquely specified by $R' := R \setminus I$, since $R = R' \cup I$ and $B = U \setminus R'$. We therefore write $\rho_{R'} := p_B X_R + y_B Y_R$, and there are precisely $\binom{u-i}{n+1-i}$ such generators.

Finally, the relation is

$$\rho_{R'} = p_B X_R + y_B Y_R \longmapsto \sum_{i\in R\setminus I} p_{B\cup\{i\}} Q_{R\setminus\{i\}} = \sum_{i\in R'} g_{R'\setminus\{i\}},$$

which is the differential

$$C_{n+1-i}(\Delta(U \setminus I)) \longrightarrow C_{n-i}(\Delta(U \setminus I))$$

from the simplex with vertex set $U \setminus I$. The number of redundant generators is the rank of this map, and since the simplex is acyclic this is the kernel of the next differential.

LEMMA 11.3.2. *The number of redundant generators of TO_n in $\mathcal{R}_{U,I}$ is exactly*

$$\binom{u-i-1}{n-i}.$$

The total number of redundant generators of TO_n for universes with u elements and intersections with i elements is

$$\binom{r}{u}\binom{u}{i}\binom{u-i-1}{n-i}. \quad \square$$

REMARK 11.3.3. Notice that for any n, r all the generators $p_{12\cdots r}Q_S$ are always redundant.

EXAMPLE 11.3.4. It is natural in applications to be discussing a particular module TO_n, so n is fixed. It is then natural to list the possibilities for $i = 0$, 1, 2, ..., n in succession. Since A always has at least two elements, we start with $u = n + 2 - i$. In each case we name a sample generator of the relevant type.

(a) TO_2 in rank 4, so that $n = 2, r = 4$.

i=0	u=4	$p_{12}Q_{34}$	Redundancy	$1 \times 1 \times 3 = 3$
i=1	u=3	$p_{12}Q_{23}$	Redundancy	$4 \times 3 \times 1 = 12$
	u=4	$p_{123}Q_{34}$	Redundancy	$1 \times 4 \times 2 = 8$
i=2	u=2	$p_{12}Q_{12}$	Redundancy	$6 \times 1 \times 0 = 0$
	u=3	$p_{123}Q_{34}$	Redundancy	$4 \times 3 \times 1 = 12$
	u=4	$p_{1234}Q_{34}$	Redundancy	$1 \times 6 \times 1 = 6$

(b) TO_3 in rank 5, so that $n = 3, r = 5$.

i=0	u=5	$p_{12}Q_{345}$	Redundancy	$1 \times 1 \times 3 = 3$
i=1	u=4	$p_{12}Q_{234}$	Redundancy	$5 \times 4 \times 1 = 20$
	u=5	$p_{123}Q_{345}$	Redundancy	$1 \times 5 \times 3 = 15$
i=2	u=3	$p_{123}Q_{123}$	Redundancy	$10 \times 3 \times 0 = 0$
	u=4	$p_{123}Q_{234}$	Redundancy	$5 \times 6 \times 1 = 30$
	u=5	$p_{1234}Q_{345}$	Redundancy	$1 \times 10 \times 2 = 20$
i=3	u=3	$p_{123}Q_{123}$	Redundancy	$10 \times 1 \times 0 = 0$
	u=4	$p_{1234}Q_{234}$	Redundancy	$5 \times 4 \times 1 = 20$
	u=5	$p_{12345}Q_{345}$	Redundancy	$1 \times 10 \times 1 = 10$

11.4. The resolution of TO_{r-1}

Because of the intricacy of the answer it is worth making the first few cases completely explicit. Since $TU_r = PC(-r-4)$, the case of TO_r is implicit in 11.2.1.

The resolution of TU_{r-1} is the short exact sequence

$$0 \longleftarrow TU_{r-1} \longleftarrow rPC(-r-3) \longleftarrow PC[Sq^2](-r-5) \longleftarrow 0.$$

We write $Q_{\hat{i}}$ for the generator in homological degree 0 corresponding to the subset omitting i and X for the generator of degree $-r-5$ and Y for the generator of degree $-r-7$ in homological degree 1.

Since $PC[Sq^2]$ is Sq^2-acyclic, taking Sq^2-cycles gives the exact sequence

$$0 \longleftarrow TO_{r-1} \longleftarrow rZPC(-r-3) \longleftarrow ZPC[Sq^2](-r-5) \longleftarrow 0.$$

or explicitly

$$0 \longleftarrow TO_{r-1} \longleftarrow r\left[PP \oplus I_2 \oplus \cdots \oplus I_r\right](-r-3) \longleftarrow$$
$$\left[\binom{r}{0}PP(0) \oplus \binom{r}{1}PP(-2) \oplus \cdots \oplus \binom{r}{r}PP(-2r)\right](-r-7) \longleftarrow 0.$$

In diagrammatic terms we may illustrate the answer with TO_3 in rank 4:

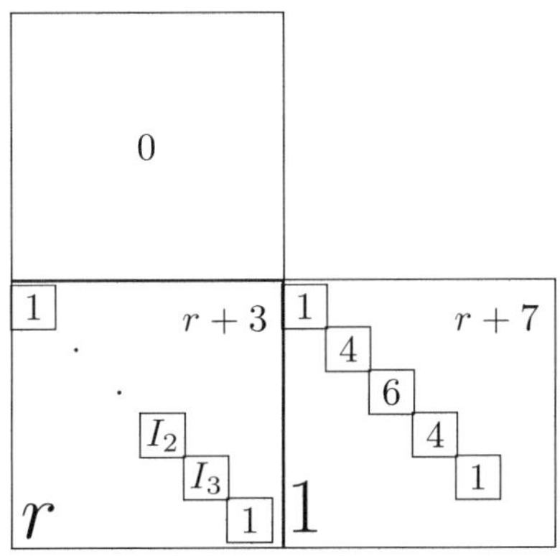

In the general case, the strings of little boxes are longer, and those in homological degree 1 filled with binomial coefficients $\binom{r}{i}$.

At this stage we may write down a free resolution, replacing each I_j by its resolution. However this is not minimal.

We consider redundancy. Since $n = r - 1$ the universe of a generator either has $r - 1$ or r elements. By Lemma 11.3.2 there is no redundancy from universes with $r - 1$ elements and a redundancy of $\binom{r}{i}$ from the universe with r elements and intersection i. This happens to be all of the generators in the relevant degrees. In fact only the two extreme generators $Y \cdot 1$ and $Y \cdot y_{12\cdots r}$ remain.

In the following displays the columns each give one homological degree. The first column is homological degree 0, and the first non-zero entry is in degree $\alpha := -r-3$, and each row is two degrees lower. We describe the general behaviour after the examples.

EXAMPLE 11.4.1. TO_2 in rank 3

3	
.	
.	1
6	
.	
	4

EXAMPLE 11.4.2. TO_3 in rank 4

4		
.		
.	1	
20		
10		
.	16	
	5	
		4

EXAMPLE 11.4.3. TO_4 in rank 5

5			
.			
.	1		
45			
40			
15	50		
.	25		
	6	25	
		5	
			5

EXAMPLE 11.4.4. TO_5 in rank 6

6				
.				
.	1			
84				
105				
70	120			
21	90			
.	36	90		
	7	36		
		6	36	
			6	
				6

In fact the entries in the first column are

$$\binom{r}{r-1}, 0, 0, r\binom{r}{2} - \binom{r}{1}, r\binom{r}{3} - \binom{r}{2}, \dots, r\binom{r}{r-1} - \binom{r}{r-2},$$

in the second column

$$0, 0, 1, 0, 0, r\binom{r}{3}, r\binom{r}{4}, \dots, r\binom{r}{r-1}, r\binom{r}{r} + 1.$$

After this the pattern is regular, with the hth column starting with $2h+3$ zero entries, followed by

$$r\binom{r}{h+2}, r\binom{r}{h+3}, \dots, r\binom{r}{r}.$$

The last non-zero column (and the projective dimension of TO_{r-1}) is $h-2$.

The resolution has a filtration with subquotients obtained from Koszul complex. The subcomplex consists of the tail of a Koszul complex, the next $r-3$ pieces are truncations of r copies of the Koszul complex with some generators removed in homological degree 0, and the final quotient is like this but has one addtional generator in homological degree 1.

11.5. The resolution of TO_{r-2}

We make one further case completely explicit, and we use notation consistent with the general case. We remind the reader that we use the notation p_B even when the sequence B has multiplicity; this simplifies formulae enormously.

The resolution of TU_{r-2} is the exact sequence

$$0 \longleftarrow TU_{r-2} \longleftarrow \binom{r}{r-2} PC(-r-2) \longleftarrow \binom{r}{r-1} PC[Sq^2](-r-4) \\ \longleftarrow PC[Sq^2](-r-6) \oplus PC(-r-10) \longleftarrow 0.$$

We write $Q_{\hat{i}\hat{j}}$ for the generator in homological degree 0 corresponding to the subset omitting $\{i,j\}$ and $X_{\hat{i}}$ and $Y_{\hat{i}}$ for the generators in homological degree 1 omitting i and X, Y, Z for the generators of homological degree 2.

We will construct an exact sequence which may be expanded to a resolution.

PROPOSITION 11.5.1. *There is an exact sequence*

$$0 \longleftarrow TO_{r-2} \longleftarrow ZF_0 \oplus PP(-r-6) \longleftarrow ZF_1 \oplus PP(-r-8) \oplus rPP(-r-10) \\ \longleftarrow ZF_2/(I_2' \oplus I_3') \oplus rPP(-r-12) \oplus \binom{r}{2} PP(-r-14) \longleftarrow 0$$

where the F_i are the free PC-modules mentioned in the resolution of TU_{r-2}, and where I_2' is generated by elements $p_{ab}Z$ and I_3' is generated by elements $p_{abc}Z$.

We may illustrate this with the case of TO_2 in rank 4. As a PP-module, this is the sum of a modified version of what we have already seen and a second complex of free modules.

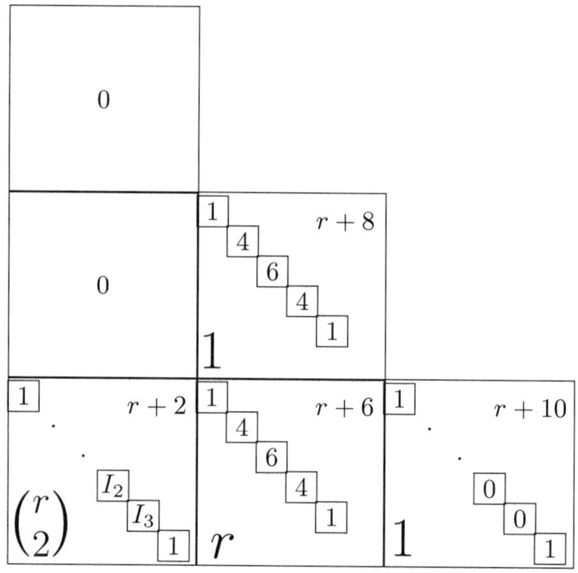

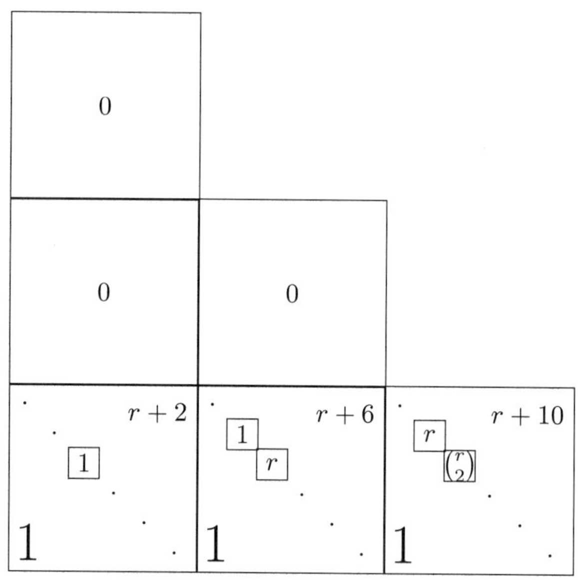

Note that in the first summand, the two boxes in the third column that would have I_2 and I_3 in ZPC now have entries 0.

In the general cases, the strings of little boxes in the first summand are longer, and those in the second column are filled with binomial coefficients $\binom{r}{i}$. There are still only two zeroes in the string of boxes in the third column. The second summand is correct for all r.

We break the resolution of TU_{r-2} into two short exact sequences

$$0 \longleftarrow TU_{r-2} \longleftarrow F_0 \longleftarrow L_0 \longleftarrow 0$$

and

$$0 \longleftarrow L_0 \longleftarrow F_1 \longleftarrow F_2 \longleftarrow 0.$$

We define L_0 to be a submodule of F_0 and name elements accordingly. The known homology of F_i shows that $H(L_0) = \mathbb{F}_2(-r-8)$, with a representative cycle

$$\overline{h} := \sum_{\{i,j\}} p_{ij} Q_{\hat{i}\hat{j}} = Sq^2 \sum_{\{i,j\}} y_i y_j Q_{\hat{i}\hat{j}}.$$

Similarly, $H(TU_{r-2}) = \binom{r}{2}\mathbb{F}_2(-r-2) \oplus \mathbb{F}_2(-r-6)$, and a representative cycle for the homology in degree $-r-6$ is

$$h = \sum_{\{i,j\}} y_i y_j q_{\hat{i}\hat{j}}.$$

Taking Sq^2-cycles gives the exact sequences

$$TO_{r-2} \longleftarrow ZF_0 \longleftarrow ZL_0 \longleftarrow 0$$

and

$$ZL_0 \longleftarrow ZF_1 \longleftarrow ZF_2 \longleftarrow 0.$$

Our next task is to adjoin generators to make the left hand maps surjective. The case of ZL_0 is easier, so we take it first.

LEMMA 11.5.2. *There is a short exact sequence*

$$0 \longleftarrow ZL_0 \longleftarrow ZF_1 \oplus PP(-r-8) \longleftarrow ZF_2/I_2' \oplus rPP(-r-12) \longleftarrow 0,$$

where I_2' is the I_2-summand of ZF_2 beginning in degree $-r-8$ and generated by the elements $p_{ij}Z$.

Proof: We adjoin a free generator $\overline{H}$ to map to the cycle $\overline{h}$, and define $\overline{B}$ by the exact sequence

$$0 \longleftarrow ZL_0 \longleftarrow ZF_1 \oplus PP(-r-8) \longleftarrow \overline{B} \longleftarrow 0.$$

It remains to show $\overline{B} = ZF_2/I_2' \oplus rPP(-r-12)$, and to specify generators $\overline{H}_a$ for the free summand.

The idea is that for each a the element $z_a\overline{h}$ is already in the image of ZF_1, so we may choose suitable $\overline{\Gamma}_i \in ZF_1$ with degree $-r-12$ mapping to $z_a\overline{h}$. Now we take

$$\overline{u}_a = (\overline{\Gamma}_a, z_a\overline{H}).$$

We write $\overline{U}_a$ for the corresponding element of $\overline{B}$, so that $d\overline{U}_a = \overline{u}_a$, and let F' be the submodule generated by $\overline{U}_1, \ldots, \overline{U}_r$. We then show that F' is free (on the elements $\overline{U}_a$) and a summand of $\overline{B}$ with complement ZF_2/I_2'. To fit in with the general pattern, we write $\overline{H}_a = \overline{U}_a$ for the free generators.

Now I_2' is generated by elements $\overline{\Gamma}_{ab} = p_{ab}Z$, and we write $\overline{\gamma}_{ab} = d\overline{\Gamma}_{ab}$ for their images in ZF_1. It suffices to choose $\overline{\Gamma}_a$ so that

$$z_b\overline{\Gamma}_a + z_a\overline{\Gamma}_b = \overline{\gamma}_{ab}.$$

Since $z_b\overline{u}_a + z_a\overline{u}_b = (z_b\overline{\Gamma}_a + z_a\overline{\Gamma}_b, 0)$, this shows that I_2' is in the submodule F', and there is an exact sequence

$$0 \longrightarrow I_2' \longrightarrow F' \longrightarrow PP(-r-8) \longrightarrow \mathbb{F}_2(-r-8) \longrightarrow 0.$$

We calculate

$$\overline{\gamma}_{ab} = dp_{ab}Z = p_{ab}\sum_s z_s Y_{\hat{s}} = z_a\overline{\Gamma}'_b + z_b\overline{\Gamma}'_a,$$

where

$$\overline{\Gamma}'_a = y_a \sum_s z_s Y_{\hat{s}}.$$

However, $\overline{\Gamma}'_a$ is not a Sq^2-cycle, so we add on a correction term to obtain

$$\overline{\Gamma}_a = y_a \sum_s z_s Y_{\hat{s}} + z_a \sum_s y_s Y_{\hat{s}} = \sum_s p_{as} Y_{\hat{s}} = \sum_s z_s m_{a,\hat{s}} + z_a m_{s,\hat{s}}.$$

Evidently the correction terms for $\overline{\Gamma}_a$ and $\overline{\Gamma}_b$ cancel to give

$$\overline{\gamma}_{ab} = d\overline{\Gamma}_{ab} = z_a\overline{\Gamma}_b + z_b\overline{\Gamma}_a,$$

as required.

Finally, we may calculate

$$\overline{\gamma}_a = d\overline{\Gamma}_a = \sum_s p_{as} \sum_{t\neq s} z_t Q_{\hat{s}\hat{t}},$$

and

$$z_s p_{at} + z_t p_{as} = z_s(y_a z_t + z_a y_t) + z_t(y_a z_s + z_a y_s) = z_a(y_s z_t + z_s y_t) = z_a p_{st},$$

so that

$$\overline{\gamma}_a = d\overline{\Gamma}_a = z_a\overline{h}$$

as required. □

This resolution is not minimal. An argument just like that of Lemma 11.3.2 shows that all generators of $ZPC[Sq^2](-r-6)$ give redundancy except for the one of lowest degree. More precisely each of the $\binom{r}{b}$ generators $p_B X + y_B Y$ eliminate a generator unless $b = |B| = r$. Finally, the images $(\overline{\Gamma}_a, z_a\overline{H})$ of the generators $\overline{U}_a$ are decomposable.

LEMMA 11.5.3. *There is an exact sequence*

$$0 \longleftarrow ZTU_{r-2} \longleftarrow ZF_0 \oplus PP(-r-6) \longleftarrow$$
$$ZF'_1 \oplus rPP(-r-10) \longleftarrow ZF'_2/I'_3 \oplus \binom{r}{2} PP(-r-14) \longleftarrow 0,$$

where $ZF'_1 = ZF_1 \oplus PP(-r-8)$, $ZF'_2 = ZF_2/I'_2 \oplus rPP(-r-12)$ *as in Lemma 11.5.2 and* I'_3 *is generated by elements* $p_{abc}Z$.

Proof: We adjoin a free generator H of degree $-r-6$ to map to the cycle h, and define B by the exact sequence

$$0 \longleftarrow ZTU_{r-2} \longleftarrow ZF_0 \oplus PP(-r-6) \longleftarrow B \longleftarrow 0.$$

It remains to construct a resolution of B. By definition, there is a short exact sequence

$$0 \longleftarrow I_1 \longleftarrow B \longleftarrow ZL_0 \longleftarrow 0,$$

and Lemma 11.5.2 gives

$$0 \longleftarrow ZL_0 \longleftarrow ZF'_1 \longleftarrow ZF'_2 \longleftarrow 0.$$

In fact the changes for ZL_0 are in degrees $\equiv r \bmod 4$, whilst the changes for ZTU_{r-2} are in degrees $\equiv r+2 \bmod 4$, so $ZF_i = ZF_i'$ in the degrees of interest. Now we adjoin r new generators H_a of degree $-r-10$ to map to the generators of I_1, and define C by

$$0 \longleftarrow B \longleftarrow ZF_1 \oplus rPP(-r-10) \longleftarrow C \longleftarrow 0.$$

We show that $C = ZF_2/I_3' \oplus \binom{r}{2} PP(-r-14)$, and denote the $\binom{r}{2}$ generators of degree $-r-14$ by H_{ab}.

We first describe how to choose the generators H, H_a and H_{ab} and then make the necessary checks. We name some elements $\Gamma_{abc} \in ZF_2$, $\Gamma_{ab} \in ZF_1$ and $\Gamma_a \in ZF_0$ and then use these to describe new generators H, H_a, and H_{ab}.

The I_3-module I_3' in ZF_2 is generated by elements

$$\Gamma_{abc} = p_{abc}Z \in ZF_2,$$

and their images in ZF_1 may be written in the form

$$\gamma_{abc} = p_{abc}dZ = z_a\Gamma_{bc} + z_b\Gamma_{ac} + z_c\Gamma_{ab};$$

indeed, we may take (check below)

$$\Gamma_{ab} := \sum_s p_{abs}Y_{\hat{s}} \in ZF_1.$$

Now the image of Γ_{ab} in ZF_0 may be written in the form

$$\gamma_{ab} = d\Gamma_{ab} = z_a\Gamma_b + z_b\Gamma_a.$$

In fact we may take (check below)

$$\Gamma_a = \sum_{\{s,t\}} p_{ast}Q_{\hat{st}} \in ZF_0.$$

Now we calculate $d\Gamma_a = z_a h$ (check below), and we may proceed as in 11.5.2, taking

$$u_a = (\Gamma_a, z_aH) \in ZF_0 \oplus PP(-r-6).$$

Since $du_a = 0$ there is an element $U_a \in B$ with $dU_a = u_a$. Now in homological degree 1 adjoin new generators H_a in degree $-r-10$ mapping to U_i, so that combined with ZF_1 we map onto the first syzygy. Now

$$z_bH_a + z_aH_b \longmapsto z_b(\Gamma_a, z_aH) + z_a(\Gamma_b, z_bH) = (\Gamma_{ab}, 0),$$

and the element

$$u_{ab} = (\Gamma_{ab}, z_bH_a + z_aH_b) \in ZF_1 \oplus rPP(-r-10)$$

maps to zero, and therefore there are elements $U_{ab} \in C$ mapping to u_{ab}. We claim that the elements U_{ab} generate a free summand, and therefore we write $H_{ab} = U_{ab}$. Indeed,

$$\begin{aligned} &z_aU_{bc} + z_bU_{ac} + z_cU_{ab} \\ &\quad = z_a(\Gamma_{bc}, z_cH_b + z_bH_c) + z_b(\Gamma_{ac}, z_cH_a + z_aH_c) + z_c(\Gamma_{ab}, z_bH_a + z_aH_b) \\ &\quad = (\Gamma_{abc}, 0), \end{aligned}$$

and since we know $\Gamma_{abc} = p_{abc}Z$ generate an I_3-summand, we conclude the module generated by the H_{ab} is free.

To see the formula for Γ_{ab} is properly related to Γ_{abc}, we calculate

$$\gamma_{abc} = d\Gamma_{abc} = p_{abc}\sum_s z_sY_{\hat{s}} = z_a\Gamma_{bc}' + z_a\Gamma_{bc}' + z_a\Gamma_{bc}',$$

where

$$\Gamma'_{ab} = y_a y_b \sum_s z_s Y_{\hat{s}}.$$

However this is not a Sq^2-cycle, so we add a correction term to form

$$\Gamma_{ab} = y_a y_b \sum_s z_s Y_{\hat{s}} + p_{ab} \sum_s y_s Y_{\hat{s}} = \sum_s p_{abs} Y_{\hat{s}},$$

where the final equality is $p_{abs} = y_a y_b z_s + p_{ab} y_s$. The fact that the correction terms from Γ_{ab}, Γ_{ac} and Γ_{bc} cancel to give

$$\gamma_{abc} = d\Gamma_{abc} = p_{abc} \sum_s z_s Y_{\hat{s}} = z_a \Gamma_{bc} + z_a \Gamma_{bc} + z_a \Gamma_{bc}$$

comes from $z_a p_{bc} + z_b p_{ac} + z_c p_{ab} = 0$.

Next

$$\gamma_{ab} = d\Gamma_{ab} = \sum_s p_{abs} \sum_{t \neq s} z_t Q_{\hat{st}} = p_a b \sum_{\{s,t\}} p_{st} Q_{\hat{st}} = z_b \Gamma'_a + z_a \Gamma'_b,$$

where

$$\Gamma'_a = y_a \sum_{\{s,t\}} p_{st} Q_{\hat{st}},$$

and the penultimate equality comes from $z_t p_{abs} + z_s p_{abt} = p_{ab} p_{st}$. However this is not a Sq^2-cycle, so we add a correction term to form

$$\Gamma_a = y_a \sum_{\{s,t\}} p_{st} Q_{\hat{st}} + z_a \sum_{\{s,t\}} y_{st} Q_{\hat{st}} = \sum_{\{s,t\}} p_{ast} Q_{\hat{st}}.$$

Finally, we check that

$$\gamma_a = \sum_{\{s,t\}} p_{ast} q_{\hat{st}}$$

is equal to

$$z_a h = \sum_{\{s,t\}} z_a p_{st} q_{\hat{st}}.$$

Indeed, $p_{ast} = z_a p_{st} + y_a p_{st}$, so that it suffices to show $\sum_{\{s,t\}} p_{st} q_{\hat{st}} = 0$. Now

$$q_{\hat{st}} = \prod_{i \notin \{s,t\}} x_i \sum_{\{l,m\} \subseteq \hat{st}} x_l x_m (y_l + y_m),$$

so

$$\sum_{\{s,t\}} p_{st} q_{\hat{st}} = (\prod_i x_i) \sum_{\{s,t\}} x_s x_t (y_s + y_t) \sum_{\{l,m\} \subseteq \hat{st}} x_l x_m (y_l + y_m).$$

Each term $x_i x_j (y_i + y_j) x_p x_q (y_p + y_q)$ occurs twice, once with $\{s,t\} = \{i,j\}, \{l,m\} = \{p,q\}$ and once with $\{s,t\} = \{p,q\}, \{l,m\} = \{i,j\}$, so the sum vanishes as required. □

At this stage we may write down a free resolution, replacing each I_j by its resolution. However this is not minimal.

We consider redundancy. Since $n = r - 2$ the universe of a generator either has $r-2$, $r-1$ or r elements. By Lemma 11.3.2 there is no redundancy if $u = r-2$, the redundancy if $u = r-1$ is $r\binom{r-1}{i}$, redundancy if $u = r$ is $\binom{r}{i}(r-1-i)$.

It is also clear that the generator H removes the redundant generator h. The generator H_a maps to $u_a = (\Gamma_a, z_a H)$, and since Γ_a is a sum of generators $p_{ast} Q_{\hat{st}}$, and since a can be inferred from the indices, these r generators remove r redundant generators. The generators H_a necessarily remove redundant generators since they are in degree $-r-12$ (and homological degree 2), whilst the only generators in homological degree 1 in higher degree (namely $-r-8$) have now all been removed killing redundant generators in homological degree 0. More specifically, H_a maps to $u_a = (\Gamma_a, z_a H)$, but Γ_a is decomposable, being the sum of terms $z_s m_{a,\hat{s}}$; when we factor out the generators $m_{a,\hat{s}}$ and H we are obliged to factor out the H_a as well.

In the following displays the columns each give one homological degree. The first column is homological degree 0, and the first non-zero entry is in degree $-r-2$, and each row is two degrees lower.

EXAMPLE 11.5.4. TO_2 in rank 4 (obtained by M.Katzman using Macaulay).

6		
.		
1	4	
20	.	
.	.	1
	24	.
		.
		7
		.
		1

EXAMPLE 11.5.5. TO_3 in rank 5

10			
.			
1	5		
75	.		
50	.	1	
10	95	.	
.	45	.	
	10	50	
		11	
		5	10
		1	.
			1

EXAMPLE 11.5.6. TO_4 in rank 6

15				
.				
1	6			
189	.			
210	.	1		
120	294	.		
20	210	.		
.	96	225		
	15	90		
		30	90	
		6	15	
		1	6	15
			1	.
				1

In fact the entries in the first column are

$$\binom{r}{r-2},\ 0,\ 1,\ \binom{r}{r-2}\binom{r}{2}-\binom{r}{3}\binom{r}{1},\ \binom{r}{r-2}\binom{r}{3}-\binom{r}{3}\binom{r}{2}-r,$$
$$\binom{r}{r-2}\binom{r}{4}-\binom{r}{3}\binom{r}{3},\ \ldots,\ \binom{r}{r-2}\binom{r}{r-1}-\binom{r}{3}\binom{r}{r-2},$$

so the first three, the fifth and the last entry are exceptions to a simple pattern. In the second column

$$0,\ 0,\ r,\ 0,\ 0,\ \binom{r}{2}\binom{r}{3}-r,\ \binom{r}{2}\binom{r}{4}-\binom{r}{2},$$
$$\binom{r}{2}\binom{r}{5},\ \binom{r}{2}\binom{r}{6},\ \ldots,\ \binom{r}{2}\binom{r}{r-1}+r,\ \binom{r}{2}\binom{r}{r},$$

so that the first seven and the penultimate non-zero entry are exceptions to a simple pattern. In the third column the entries are

$$0,0,0,0,1,0,0,\binom{r}{2}\binom{r}{4},\ \binom{r}{2}\binom{r}{5},$$
$$\binom{r}{2}\binom{r}{6}+\binom{r}{4},\ \binom{r}{2}\binom{r}{7}+\binom{r}{5},\ \ldots,\ \binom{r}{2}\binom{r}{r}+\binom{r}{r-2},\ \binom{r}{r-1},\ \binom{r}{r},$$

so that after the ninth entry, there is a simple pattern. After this the pattern is regular, with the hth column starting with $2h+3$ zero entries, followed by

$$\binom{r}{2}\binom{r}{h+2},\ \binom{r}{2}\binom{r}{h+3},$$
$$\binom{r}{2}\binom{r}{h+4}+\binom{r}{h+2},\ \binom{r}{2}\binom{r}{h+5}+\binom{r}{h+3},\ \ldots,$$
$$\binom{r}{2}\binom{r}{r}+\binom{r}{r-2},\ \binom{r}{r-1},\ \binom{r}{r}.$$

The last non-zero column is the $(r-1)$st, so the projective dimension of TO_{r-2} is $r-1$.

11.6. An exact sequence for TO_{r-k}

We are going to describe the resolution of TO_{r-k} as a modification of the known modules ZF_h where F_h is the hth term in the resolution of TU_{r-k}. We recall that ZF_h is a free PP-module if h is odd and the sum of a free module and $\binom{r}{k-h}(I_2 \oplus I_3 \oplus \cdots \oplus I_r)$ if h is even. Finally, we write $\alpha = r - k - 4$ for the top degree of TO_{r-k}.

We define certain free PP-modules $A(m,i)_s$ where m, i, s are integers. The number m refers to the homological degree where the family starts, i indexes the families starting in homological degree m (and varies between $m/2$ and $k/2$), and the subscript s refers to homological degree; the family $A(m,i)_s$ runs from homological degree $s = m$ to $s = 2i$. More precisely, the modules starting in even homological degree are entirely in degrees $\equiv \alpha$ mod 4 and given by

$$A(2p,i)_s = \begin{cases} \binom{r}{k-2i}\binom{r}{s-2p} PP(\alpha - 4(i+s-p)) & p+1 \le i \le k/2,\ 2p \le s \le 2i \\ 0 & \text{otherwise} \end{cases}$$

and the modules starting in odd homological degree are entirely in degrees $\equiv \alpha + 2$ mod 4

$$A(2p+1,i)_s = \begin{cases} \binom{r}{k-2i}\binom{r}{2p+1-s} PP(\alpha - 4(i+s-p)+2) & p+1 \le i \le k/2,\ 2p+1 \le s \le 2i \\ 0 & \text{otherwise} \end{cases}$$

In either case the generators of $A(m,i)_s$ are $K(m,i)_{S,A}$ where $|S| = r - k + 2i$ and $|A| = h$.

We are now ready to state a general result.

THEOREM 11.6.1. *There is an exact sequence*

$$0 \longleftarrow TO_{r-k} \longleftarrow F'_0 \longleftarrow F'_1 \longleftarrow \cdots \longleftarrow F'_k \longleftarrow 0,$$

where

$$F'_{2a} = ZF_{2a}/\binom{r}{k-2a}(I_2 \oplus I_3 \oplus \cdots \oplus I_{2a+1}) \oplus \bigoplus_{m,i} A(m,i)_{2a}$$

(the nonzero $A(m,i)_{2a}$ are $m \le 2a$, $a+1 \le i \le k/2$) and

$$F'_{2a+1} = ZF_{2a} \oplus \bigoplus_{m,i} A(m,i)_{2a+1}$$

(the nonzero $A(m,i)_{2a+1}$ are $m \le 2a+1$, $a+1 \le i \le k/2$). The maps in the resolution will be described in Section 11.7.

The statement is a little indigestible, so we make the first few terms explicit when k is even (when k is odd, the formula for the suspension in the final term is slightly different).

$$F'_0 = ZF_0 \oplus \binom{r}{k-2}\binom{r}{0}PP(\alpha-4) \oplus \binom{r}{k-4}\binom{r}{0}PP(\alpha-8) \oplus \cdots$$
$$\oplus \binom{r}{0}\binom{r}{0}PP(\alpha-2k),$$

$$F_1' = ZF_1 \oplus$$
$$\binom{r}{k-2}\binom{r}{1}PP(\alpha-8)\oplus\binom{r}{k-4}\binom{r}{1}PP(\alpha-12)\oplus\cdots\oplus\binom{r}{0}\binom{r}{1}PP(\alpha-2k-4)\oplus$$
$$\binom{r}{k-2}\binom{r}{0}PP(\alpha-6)\oplus\binom{r}{k-4}\binom{r}{0}PP(\alpha-10)\oplus\cdots\oplus\binom{r}{0}\binom{r}{0}PP(\alpha-2k-2)$$

$$F_2' = ZF_2/\binom{r}{k-2}(I_2\oplus I_3)\oplus$$
$$\binom{r}{k-2}\binom{r}{2}PP(\alpha-12)\oplus\binom{r}{k-4}\binom{r}{2}PP(\alpha-16)\oplus\cdots\oplus\binom{r}{0}\binom{r}{2}PP(\alpha-2k-8)\oplus$$
$$\binom{r}{k-2}\binom{r}{1}PP(\alpha-10)\oplus\binom{r}{k-4}\binom{r}{1}PP(\alpha-14)\oplus\cdots\oplus\binom{r}{1}\binom{r}{0}PP(\alpha-2k-6)\oplus$$
$$\binom{r}{k-4}\binom{r}{0}PP(\alpha-12)\oplus\binom{r}{k-4}\binom{r}{0}PP(\alpha-16)\oplus\cdots\oplus\binom{r}{0}\binom{r}{0}PP(\alpha-2k-8).$$

In a different direction, we make explicit the resolution for small values of k.

The correction terms $A(m,i)_s$ are all zero for TO_r and TO_{r-1}, so the exact sequences are simply $TO_r = ZF_0$ and $0 \longleftarrow TO_{r-1} \longleftarrow ZF_0 \longleftarrow ZF_1 \longleftarrow 0$ respectively.

We make the exact sequences explicit for TO_{r-2}, TO_{r-3} and TO_{r-4}. With luck this should make the general pattern clear.

EXAMPLE 11.6.2. TO_{r-2}
Here the only two non-zero families are $A(0,1)$ and $A(1,1)$, where we have

i	0	1	2
$A(0,1)_i$	$\binom{r}{0}PP(\alpha-4)$	$\binom{r}{1}PP(\alpha-8)$	$\binom{r}{2}PP(\alpha-12)$
$A(1,1)_i$	0	$\binom{r}{0}PP(\alpha-6)$	$\binom{r}{1}PP(\alpha-10)$

In Section 11.5, the generators $H(0,1)_{all,A}$ of $A(0,1)$ were called H_A and the generators $H(1,1)_{all,A}$ of $A(1,1)$ were called $\overline{H}_A$

The resolution is

$$0 \longleftarrow TO_{r-2} \longleftarrow ZF_0 \oplus A(0,1)_0 \longleftarrow ZF_1 \oplus A(0,1)_1 \oplus A(1,1)_1 \longleftarrow$$
$$ZF_2/(I_2\oplus I_3) \oplus A(0,1)_2 \oplus A(1,1)_2 \longleftarrow 0.$$

EXAMPLE 11.6.3. TO_{r-3}
Again the only two non-zero families are $A(0,1)$ and $A(1,1)$, where we have

i	0	1	2
$A(0,1)_i$	$\binom{r}{1}\binom{r}{0}PP(\alpha-4)$	$\binom{r}{1}\binom{r}{1}PP(\alpha-8)$	$\binom{r}{1}\binom{r}{2}PP(\alpha-12)$
$A(1,1)_i$	0	$\binom{r}{1}\binom{r}{0}PP(\alpha-6)$	$\binom{r}{1}\binom{r}{1}PP(\alpha-10)$

Thus the resolution is

$$0 \longleftarrow TO_{r-3} \longleftarrow ZF_0 \oplus A(0,1)_0 \longleftarrow ZF_1 \oplus A(0,1)_1 \oplus A(1,1)_1 \longleftarrow$$
$$ZF_2/\binom{r}{1}(I_2 \oplus I_3) \oplus A(0,1)_2 \oplus A(1,1)_2 \longleftarrow ZF_3 \longleftarrow 0$$

EXAMPLE 11.6.4. TO_{r-4}

The non-zero families are $A(0,1)$, $A(0,2)$ and $A(1,1)$, $A(1,2)$, $A(2,2)$ and $A(3,2)$. For typographical reasons, we have written P for PP in the following table.

i	0	1	2	3	4
$A(0,1)_i$	$\binom{r}{2}\binom{r}{0}P(\alpha-4)$	$\binom{r}{2}\binom{r}{1}P(\alpha-8)$	$\binom{r}{2}\binom{r}{2}P(\alpha-12)$	0	0
$A(0,2)_i$	$\binom{r}{0}\binom{r}{0}P(\alpha-8)$	$\binom{r}{0}\binom{r}{1}P(\alpha-12)$	$\binom{r}{0}\binom{r}{2}P(\alpha-16)$	$\binom{r}{0}\binom{r}{3}P(\alpha-20)$	$\binom{r}{0}\binom{r}{4}P(\alpha-24)$
$A(1,1)_i$	0	$\binom{r}{2}\binom{r}{0}P(\alpha-6)$	$\binom{r}{2}\binom{r}{1}P(\alpha-10)$	0	0
$A(1,2)_i$	0	$\binom{r}{0}\binom{r}{0}P(\alpha-10)$	$\binom{r}{0}\binom{r}{1}P(\alpha-14)$	$\binom{r}{0}\binom{r}{2}P(\alpha-18)$	$\binom{r}{0}\binom{r}{3}P(\alpha-22)$
$A(2,2)_i$	0	0	$\binom{r}{0}\binom{r}{0}P(\alpha-12)$	$\binom{r}{0}\binom{r}{1}P(\alpha-16)$	$\binom{r}{0}\binom{r}{2}P(\alpha-20)$
$A(3,2)_i$	0	0	0	$\binom{r}{0}\binom{r}{0}P(\alpha-14)$	$\binom{r}{0}\binom{r}{1}P(\alpha-18)$

Thus the resolution is

$$0 \longleftarrow TO_{r-4} \longleftarrow ZF_0 \oplus A(0,1)_0 \oplus A(0,2)_1 \longleftarrow$$
$$ZF_1 \oplus A(0,1)_1 \oplus A(0,2)_1 \oplus A(1,1)_1 \oplus A(1,2)_1 \longleftarrow$$
$$ZF_2/\binom{r}{2}(I_2 \oplus I_3) \oplus A(0,1)_2 \oplus A(0,2)_2 \oplus A(1,1)_2 \oplus A(1,2)_2 \oplus A(2,2)_2 \longleftarrow$$
$$ZF_3 \oplus A(0,2)_3 \oplus A(1,2)_3 \oplus A(2,2)_3 \oplus A(3,2)_3 \longleftarrow$$
$$ZF_4/\binom{r}{0}(I_2 \oplus I_3 \oplus I_4 \oplus I_5) \oplus A(0,2)_4 \oplus A(1,2)_4 \oplus A(2,2)_4 \oplus A(3,2)_4 \longleftarrow 0$$

11.7. The maps in the resolution

We begin with some notation. As before F_h is the hth term in the PC-resolution of TO_{r-k}, and we will modify the complex

$$0 \longleftarrow TO_{r-k} \longleftarrow ZF_0 \longleftarrow ZF_1 \longleftarrow \cdots \longleftarrow ZF_k \longleftarrow 0$$

to make it exact, as in the case of TO_{r-2}. We recall that Sq^2 is exact on F_h if h is odd and $H(F_{2s}) = \binom{r}{r-k-2s}\mathbb{F}_2(\alpha - 4s)$.

We let

$$L_h := \operatorname{im}(F_{h+1} \longrightarrow F_h), L_{-1} := TO_{r-k}$$

so that we have short exact sequences

$$0 \longleftarrow L_{h-1} \longleftarrow F_h \longleftarrow L_h \longleftarrow 0$$

for $h \geq 0$. This allows us to calculate the Sq^2-homology of all the terms, finding

$$\Sigma^2 H(L_{2p}) = \bigoplus_{i=p+1}^{k/2} \binom{r}{k-2i}\mathbb{F}_2(\alpha - 4(i+p))$$

and

$$\Sigma^2 H(L_{2p+1}) = \bigoplus_{i=p+1}^{k/2} \binom{r}{k-2i} \mathbb{F}_2(\alpha - 4(i+p) - 2).$$

For degree reasons the map $H(F_h) \longleftarrow H(L_h)$ is zero so we have an exact sequence

$$0 \longleftarrow \Sigma^2 H(L_h) \longleftarrow ZL_{h-1} \longleftarrow ZF_h \longleftarrow ZL_h \longleftarrow 0.$$

We now begin to modify the complex with terms ZF_h, starting at the top end where we have the exact sequence

$$0 \longleftarrow \Sigma^2 H(L_{k-1}) \longleftarrow ZL_{k-2} \longleftarrow ZF_{k-1} \longleftarrow ZF_k \longleftarrow 0.$$

More generally, we suppose successively that all families $A(n,i)$ with $n \geq m+1$ have been added to form certain PP-modules

$$F''_{m+1}, F''_m \dots F''_r,$$

where F''_h is obtained from ZF_h by adding free modules, and if h is even, factoring out $\binom{r}{k-h}[I_2 \oplus \cdots I_{h-m}]$. These modules depend on m, but this is not displayed in the notation. The modules will fit into an exact sequence

$$0 \longleftarrow ZL_m \longleftarrow F''_{m+1} \longleftarrow F''_{m+2} \longleftarrow \cdots \longleftarrow F''_r \longleftarrow 0.$$

Now we splice this with the exact sequence

$$0 \longleftarrow \Sigma^2 H(L_m) \longleftarrow ZL_{m-1} \longleftarrow ZF_m \longleftarrow ZL_m \longleftarrow 0$$

to obtain

$$0 \longleftarrow \Sigma^2 H(L_m) \longleftarrow ZL_{m-1} \longleftarrow ZF_m \longleftarrow F''_{m+1} \longleftarrow F''_{m+2} \longleftarrow \cdots \longleftarrow F''_r \longleftarrow 0.$$

Now assume for definiteness that $m = 2p$ is even, so that

$$\Sigma^2 H(L_{2p}) = \bigoplus_{i=p+1}^{k/2} \binom{r}{k-2i} \mathbb{F}_2(\alpha - 4(i+p)).$$

We add generators to ZF_m to map onto this, and the new generators force us to modify the rest of the resolution. Let us fix attention to a particular value of i; this will give the sequence $A(m,i)$. The summands in $\binom{r}{k-2i}\mathbb{F}_2(\alpha - 4(i+p))$ are naturally indexed by sets S with $r-k+2i$ elements, and we fix one of these. Now for $m \geq 1$ the resulting family comes from the representative cycle

$$h(m,i)_S = \sum_{|T|=2i-m} p_T Z_{S\setminus T},$$

where, in any particular homological degree, we write Z_U for the generator Q_U in lowest degree. For $m = 0$ we find

$$h(0,i)_S = \sum_{|T|=2i} y_T q_{S\setminus T}.$$

for the Sq^2-homology of L_m as in 9.10.6. We adjoin a new generator $H(m,i)_S$ in homological degree m to map to $h(m,i)_S$ and then $H(m,i)_{S,B}$ in homological degree $m+|B|$ for $|B| \leq e$, where $e := 2i - m$ is the homological excess of the top of the family $A(m,i)$.

To construct the resolution we need some auxiliary elements, and to make these, we begin with the module I_j that is responsible for finishing the sequence. Since $m = 2p$, the last term in the sequence is

$$A(2p,i)_{2i} = \binom{r}{k-2i}\binom{r}{2i-2p} PP(\alpha - 4(3i-p)),$$

and we find $\binom{r}{k-2i} I_{e+1}$ with top in degree $\alpha - 4(3i-p) - 4$ as a summand in ZF_{2i}. Again the summands I_{e+1} are indexed by subsets S with $r-k+2i$ elements. Since S, m, i are fixed in the discussion, we omit them from the notation where possible, writing $H = H(m,i)_S$ and so forth.

Now for $B \subseteq S$ with $1 \leq |B| \leq e+1$, let

$$\Gamma_B := \sum_{|T|=e-|B|+1} p_{BT} Z_{S \setminus T}$$

(notice that the sequence BT always has $e+1$ elements, independent of B). These elements have the properties that

$$d\Gamma_B = \sum_{b \in B} z_b \Gamma_{B \setminus \{b\}}$$

and

$$d\Gamma_b = z_b h.$$

We check these properties below, and proceed to define the generators $H(2p,i)_{S,A}$. Indeed,

$$u_b = (\Gamma_b, z_b H) \in ZF_{2p} \oplus A(2p,i)_{2p}$$

now maps to zero and we adjoin a generator H_b to map to u_b. However

$$z_b U_a + z_a U_b \longmapsto z_b(\Gamma_a, z_a H) + z_a(\Gamma_b, z_b H) = (\Gamma_{ab}, 0)$$

and so the element

$$u_{ab} = (\Gamma_{ab}, z_b H_a + z_a H_b) \in ZF_{2p+1} \oplus A(2p,i)_{2p+1}$$

maps to zero and we adjoin a new generator H_{ab} to map to it. We continue in this way until we reach $|B| = e+1$ where we have $\Gamma_B = p_B Z_S$, and we know that as B varies, these elements generate the I_{e+1} summand indexed by S. We conclude that the elements u_B with $|B| = e$ generate a free summand, finishing the sequence.

The case with m odd is exactly analogous, taking into account the slightly different formula for $\Sigma^2 H(L_{2p+1})$, and the fact that the excess is now odd.

It remains to make the verifications. As usual we permit p_B where B is a sequence with repetitions, and we write BT for the concatenation of the sequences B and T. Similarly, we write $B \setminus T$ for the sequence in which terms in T which occur in B are deleted in turn.

We suppose first that $m \geq 1$; the calculations for $m = 0$ look a little different at the final step. First suppose $|B| \geq 1$ and calculate

$$\begin{aligned} \gamma_B = d\Gamma_B &= \sum_{|T|=e-|B|+1} p_{BT} dZ_{S\setminus T} \\ &= \sum_{|T|=e-|B|+1} p_{BT} \sum_{u \in S\setminus T} z_u Z_{(S\setminus T)\setminus\{u\}} \end{aligned}$$

Thus if $|U| = e - |B| + 2$, the coefficient of $Z_{T\setminus U}$ is

$$\sum_{u\in U} z_u p_{BU\setminus\{u\}} = y_{BU}\left(\sum_{u\in U} y_u(\sum_{b\in B} y_b + \sum_{v\neq u} y_v)\right) = p_B p_U,$$

since the term $y_u y_v$ cancels with $y_v y_u$. Now, first note that if $B = \{b\}$ has one element, $p_b = z_b$ and

$$\gamma_b = d\Gamma_b = z_b \sum_U p_U Z_{S\setminus U} = z_b h.$$

Now suppose $|B| \geq 2$; since $p_B = \sum_{b\in B} z_b y_{B\setminus\{b\}}$, we find

$$\gamma_B = d\Gamma_B = p_B \sum_U p_U Z_{S\setminus U} = \sum_{b\in B} z_b \Gamma'_{B\setminus\{b\}},$$

where $\Gamma'_{B\setminus\{b\}} = y_{B\setminus\{b\}} \sum_U p_U Z_{S\setminus U}$. Unfortunately $\Gamma'_{B\setminus\{b\}}$ is not a Sq^2-cycle but we may add a correction term to obtain

$$y_{B\setminus\{b\}} \sum_U p_U Z_{S\setminus U} + p_{B\setminus\{b\}} \sum_U y_U Z_{S\setminus U} = \sum_U p_{(B\setminus\{b\})U} Z_{T\setminus U} = \Gamma_{B\setminus\{b\}};$$

the additional terms cancel in the sum for $d\Gamma_B$ since $\sum_{b\in B} z_b p_{B\setminus b} = 0$.

The only difference when $m = 0$ is the verification that $d\Gamma_b = z_b h$. Indeed,

$$\gamma_b = d\Gamma_b = \sum_{|T|=e} p_{Tb} q_{S\setminus T},$$

whereas

$$z_b h = \sum_{|T|=e} z_b y_T q_{S\setminus T}.$$

Since $p_{Tb} = z_b y_T + y_b p_T$, we need

$$\sum_{|T|=e} \sum_{t\in T} y_t y_T y_b q_{S\setminus T} = 0.$$

Expanding $q_{S\setminus T}$ and dividing by $x_b x_S$ we must show

$$\sum_{|T|=e} \sum_{t\in T} y_t x_T \sum_{\{i,j\}\subseteq(S\setminus T)} (x_i x_j^3 + x_i^3 x_j) = 0.$$

Now, for a fixed subset T and element t, we have $e-1$ exponents 1 and one exponent 3 in T and the exponents in $S\setminus T$ are all zero except for one 1 and one 3. Altogether we have e exponents 1 and 2 exponents 3. On the other hand, if we specify a set of e elements to have exponent 1 and 2 elements to have exponent 3, this comes from $2e$ different pairs (T, t) and the contributions therefore cancel.

CHAPTER 12

The ko-Homology of Elementary Abelian Groups (LCSS)

In this chapter we collect together the information to calculate $ko_*(BV)$ for elementary abelian 2-groups $V = V(r)$ of rank ≤ 4 via the local cohomology spectral sequence. It is revealing to comment that the cases of rank 1, 2, 3 and 4 take roughly 2 pages, 3 pages, 8 pages, and 32 pages respectively. Not only does this show the increase of complexity, but it also compares very unfavourably with the Bockstein spectral sequence method of Chapter 10. On the other hand, the local cohomology spectral sequence does not just calculate the groups: it displays a filtration which can be significant in applications (for example to the Gromov-Lawson-Rosenberg conjecture, as described in Section 2.7).

The chapter ends in Section 12.6 with a discussion of the calculations and some conjectures on the general behaviour.

It is worth reminding the reader of some conventions: we write 2^a for an elementary abelian group of rank a and $[n]$ for a cyclic group of order n. In writing Hilbert series, t is of codegree 1 (i.e., of degree -1), and $u = t^4$.

12.1. The local cohomology of ko_V^*

Our experience suggests that the best approach to this calculation is to consider the exact sequences

$$0 \longrightarrow T \longrightarrow ko_V^* \longrightarrow \overline{QO} \longrightarrow 0,$$

and

$$0 \longrightarrow \tau \longrightarrow T \longrightarrow TO \longrightarrow 0.$$

Here

$$\overline{QO} = \mathrm{im}(ko_V^* \longrightarrow KU_V^*)$$

and τ consists of the multiples of η.

Our calculations with the cohomology η-Bockstein spectral sequence in Chapter 9 show that for elementary abelian groups $TO = ZTU$, and give an exact description of $\overline{QO}$.

In fact $\overline{QO}$ is the subring of KU_V^* consisting of the part in degrees ≥ 0 and the part generated by $JSp(V)$ in degree -4 and $JU(V)^4$ in degree -8. Thus it is concentrated in degrees divisible by 4, and there it is $RO(V) = RU(V)$ in degrees ≥ 0 divisible by 8 and $RSp(V) = 2RU(V)$ in positive degrees $\equiv 4 \bmod 8$, in degree $-8i$ it is $JU(V)^{4i}$ for $i \geq 1$ and in degree $-8i-4$ it is $JU(V)^{4i}JSp(V)$ for $i \geq 1$ The module τ consists of $RU(V)/2$ in positive degrees $\equiv 1, 2 \bmod 8$, together with a few exceptional classes in negative degrees.

12.2. Calculation of $ko_*(BV)$ in rank 1

12.2.A. Overview. This case is rather trivial, and the reader may like to make the calculation as an exercise before reading on. We record it to be able to make comparison with higher ranks.

We will apply the local cohomology spectral sequence to the known value of ko_V^*. To calculate the E^2-term we need to calculate local cohomology.

In rank 1, $TO = 0$, and we have the short exact sequence

$$0 \longrightarrow \tau \longrightarrow ko_V^* \longrightarrow \overline{QO} \longrightarrow 0$$

where $\overline{QO}$ is the image in periodic complex K-theory, and τ consists of the η-multiples.

First of all, the module τ is bounded below, so torsion, and

$$H_I^*(\tau) = H_I^0(\tau) = \tau.$$

Indeed, τ is 2^2 in positive degrees congruent to 1 or 2 mod 8, and otherwise it is zero.

Thus $H_I^0(ko_V^*) = H_I^0(\overline{QO}) \oplus \tau$ and $H_I^1(ko_V^*) = H_I^1(\overline{QO})$.

12.2.B. The local cohomology of $\overline{QO}$. From the Bockstein spectral sequence we find that $\overline{QO}$ is $RO(G) = RU(G)$ in degrees 0, 8, 16, ..., and $RSp(G) = 2RU(G)$ in degrees 4, 12, 20, ... In negative degrees it is JU^2, JU^4, JU^6, JU^8, ... in degrees -4, -8, -12, -16, ... We identify elements of $RU(G)$ with their characters, displayed as a vector starting with the value at the identity. The augmentation ideal is the principal ideal generated by

$$p^* = (04) \in JU^2$$

so the local cohomology is given by the exact sequence

$$0 \longrightarrow H_{JO}^0(\overline{QO}) \longrightarrow \overline{QO} \longrightarrow \overline{QO}[1/p^*] \longrightarrow H_{JO}^1(\overline{QO}) \longrightarrow 0.$$

Evidently the kernel of inverting p^* consists of characters supported at the identity, so $H_{JO}^0(\overline{QO})$ is just $\mathbb{Z}$ in degrees 0, 4, 8, 12, 16, 20, 24, ...

Turning to H^1, JU^{2k} is generated by the element (04^k) for $k \geq 1$, so that p^* is an isomorphism from degree -4 downwards. Thus H^1 is in degrees 0, 4, 8, 12, 16, ... A little linear algebra gives the following conclusion, where $[a]$ means a cyclic group of order a.

$$H_{JO}^1(\overline{QO})_i = \begin{cases} [2^{4k}] & i = 8k \geq 0 \\ [2^{4k+3}] & i = 8k+4 \geq 4 \\ 0 & \text{otherwise} \end{cases}$$

12.2.C. The local cohomology spectral sequence. We have now calculated the E^2 page of the local cohomology spectral sequence. We display the spectral sequence in low degrees (≤ 20) in Figure 12.2.1.

Since ko_* is a summand of $ko_*(BV)$, the copies of $\mathbb{Z}$ on the H^0 column survive. There are no other possible differentials.

This gives the E^∞ term of the spectral sequence, and there are no extension problems.

FIGURE 12.2.1. The E^2-page of the local cohomology spectral sequence for $ko_*(BV(1))$.

$[2^{11}]$	$\mathbb{Z}$	20
0	0	
0	2^2	
0	2^2	
$[2^8]$	$\mathbb{Z}$	16
0	0	
0	0	
0	0	
$[2^7]$	$\mathbb{Z}$	12
0	0	
0	2^2	
0	2^2	
$[2^4]$	$\mathbb{Z}$	8
0	0	
0	0	
0	0	
$[2^3]$	$\mathbb{Z}$	4
	0	
	2^2	
	2^2	
	$\mathbb{Z}$	0
		-4
		-8
H^1_{JO}	H^0_{JO}	

12.2.D. Conclusion: the ko-homology groups. The values for the hololgy groups $ko_*(BV(1))$ can be read off from the local cohomology spectral sequence. By sparsity there are no additive extensions. In degrees ≥ 0 they are given by the following formulae.

ϵ	$ko_{8k+\epsilon}(BV(1))$
7	$[2^{4k+4}]$
6	0
5	0
4	$\mathbb{Z}$
3	$[2^{4k+3}]$
2	2^2
1	2^2
0	$\mathbb{Z}$

12.3. Calculation of $ko_*(BV)$ in rank 2

12.3.A. Overview. We will apply the local cohomology spectral sequence to the known value of ko^*_V. To calculate the E^2-term we need to calculate local cohomology.

We have the short exact sequences

$$0 \longrightarrow T \longrightarrow ko^*_V \longrightarrow \overline{QO} \longrightarrow 0$$

where $\overline{QO}$ is the image in periodic complex K-theory, and

$$0 \longrightarrow \tau \longrightarrow T \longrightarrow TO \longrightarrow 0$$

where τ consists of the η-multiples.

First of all, the module τ is bounded below, so torsion, and

$$H^*_I(\tau) = H^0_I(\tau) = \tau.$$

Indeed, τ is 2^4 in positive degrees congruent to 1 or 2 mod 8, and otherwise it is zero.

12.3.B. The local cohomology of $\overline{QO}$. From the Bockstein spectral sequence we find that $\overline{QO}$ is $RO(G) = RU(G)$ in degrees 0, 8, 16, ..., and $RSp(G) = 2RU(G)$ in degrees 4, 12, 20, ... In negative degrees it is $2JU$, JU^4, JU^6, JU^8, ... in degrees -4, -8, -12, -16, ... We identify elements of $RU(G)$ with their characters, displayed as a vector starting with the value at the identity. The augmentation ideal is the radical of the principal ideal generated by

$$(p^*)^2 \in JU^4, \text{ where } p^* = (0444),$$

so the local cohomology is given by the exact sequence

$$0 \longrightarrow H^0_{JO}(\overline{QO}) \longrightarrow \overline{QO} \longrightarrow \overline{QO}[1/p^*] \longrightarrow H^1_{JO}(\overline{QO}) \longrightarrow 0.$$

Evidently the kernel of inverting p^* consists of characters supported at the identity, so $H^0_{JO}(\overline{QO})$ is just $\mathbb{Z}$ in degrees 0, 4, 8, 12, 16, 20, 24, ...

Turning to H^1, it is easy to check that JU^{2k} is generated by the diagonal elements (04^k00), (004^k0), ..., (0004^k) for $k \geq 1$, so that p^* is an isomorphism from degree -8 downwards. Thus H^1 is in degrees -4, 0, 4, 8, 12, 16, ... A little linear algebra gives the following conclusion, where $[a]$ means a cyclic group of order a.

$$H^1_{JO}(\overline{QO})_i = \begin{cases} [2] & i=-4 \\ [2^{4k+1}]^2 \oplus [2^{4k}] & i = 8k \geq 0 \\ [2^{4k+4}]^2 \oplus [2^{4k+3}] & i = 8k+4 \geq 0 \\ 0 & \text{otherwise} \end{cases}$$

12.3.C. The local cohomology of TO. In rank 2, $TU = TU_2 = PC(-6)$, with generator q_{12}. The Sq^2 action makes TU into a Koszul complex for the maximal ideal of PP. Thus $TO = TO_2 = ZPC(-6) = PP(-6) \oplus PP(-12)$, with generators q_{12} and $p_{12}q_{12}$. Accordingly,

$$H^*_I(TO) = H^2_I(TO) = PP^\vee(2) \oplus PP^\vee(-4).$$

12.3.D. The local cohomology spectral sequence. We have now calculated the $E^{1\frac{1}{2}}$ page of the local cohomology spectral sequence. This is nearly the E^2-page, but the local cohomology connecting homomorphism $d^1 : H^1_{JO}(\overline{QO}) \longrightarrow H^2_{JO}(TO)$ is treated as a late differential, which has not yet been taken into account. We display the spectral sequence in low degrees (≤ 20) in Figure 12.3.2.

First note that each column is bounded below. The H^0 column has lowest entry in degree 0, the H^1 column in degree -4, and the H^2 column in degree -4. However we will argue below that after the action of d^1, The H^0 column has lowest entry in degree 0, the H^1 column in degree 4, and the H^2 column in degree 2, so that regularity of $ko^*_{V(2)}$ is 0.

Now consider differentials. The principal fact is that the spectral sequence converges to $ko_*(BV(2))$, which is zero in negative degrees. The reader should note the slope -1 line below which the E^∞-page must be zero. Similarly, we know $ko_0(BG) \xrightarrow{\cong} KO_0(BG)$ is an isomorphism so that the only non-zero entry $E^\infty_{p,-p}$ has $p=0$, where it is $E^2_{0,0}$. This forces certain differentials. Specifically,

- the d^1 leaving H^1_{-4} must be an isomorphism.
- the d^1 leaving H^1_0 must be an isomorphism.
- the d^2 leaving H^0_1 must be surjective, and therefore has kernel 2^3

It is not hard to deal with differentials leaving the H^0 column.

LEMMA 12.3.1. *There are no further differentials leaving the H^0-column.*

Proof: We know that ko_* is a summand, so it remains only to discuss the classes in degrees 1 and 2 mod 8. Those in degrees 2 mod 8 survive because H^2 is zero in odd degrees. Since $ko_*(BG)[1/\beta] = KO_*(BG)$, we know that all of the 2^4 survives in sufficiently large degrees. Since the image in KO^G_* is a ko_*-module, it suffices to deal with degree 9. For this we use the fact that the differentials are maps of ko^*_V-modules. Every element of H^2_{10} is detected by some map $H^2_{10} \longrightarrow H^2_6$ coming from multiplication by an element λ of ko^4_V, so if $d^2(x)$ is non-zero in H^2_{10} and detected by λ there would have to be a non-zero $d^2(\lambda x)$. However λx lies in the zero group. □

The other differentials also fall into a simple pattern.

LEMMA 12.3.2. *The differential $d^1 : H^1_{JO}(\overline{QO}) \longrightarrow H^2_{JO}(TO)$ is of maximal rank. Thus it is of rank 3 in degree $4k$ for $k \geq 1$, and an isomorphism in degrees* 0

FIGURE 12.3.2. The $E^{1\frac{1}{2}}$-page of the local cohomology spectral sequence for $ko_*(BV(2))$.

H^2_{JO}	H^1_{JO}	H^0_{JO}	
$2^7 \leftarrow$	$[2^{12}]^2 \oplus [2^{11}]$	$\mathbb{Z}$	20
0	0	0	
2^5	0	2^4	
0	0	2^4	
$2^6 \leftarrow$	$[2^9]^2 \oplus [2^8]$	$\mathbb{Z}$	16
0	0	0	
2^4	0	0	
0	0	0	
$2^5 \leftarrow$	$[2^8]^2 \oplus [2^7]$	$\mathbb{Z}$	12
0	0	0	
2^3	0	2^4	
0	0	2^4	
$2^4 \leftarrow$	$[2^5]^2 \oplus [2^4]$	$\mathbb{Z}$	8
0	0	0	
2^2	0	0	
0	0	0	
$2^3 \leftarrow$	$[2^4]^2 \oplus [2^3]$	$\mathbb{Z}$	4
0	0	0	
2	0	2^4	
0	0	2^4	
$2^2 \leftarrow$	$[2]^2 \oplus [1]$	$\mathbb{Z}$	0
0	0		
0	0		
0	0		
$2 \leftarrow$	2		−4
			−8

and -4. *Accordingly*

$$\ker(d^1 : H^1_{JO}(\overline{QO}) \longrightarrow H^2_{JO}(TO)) = 2H^1_{JO}(\overline{QO}).$$

Proof: The codomain is annihilated by 2, so d^1 factors through

$$\overline{d}^1 : H^1_{JO}(\overline{QO})/2 \longrightarrow H^2_{JO}(TO).$$

We consider the dual

$$(H^1_{JO}(\overline{QO})/2)^\vee \stackrel{(\overline{d}^1)^\vee}{\longleftarrow} H^2_{JO}(TO)^\vee.$$

We already know it is surjective in degrees -4 and 0, so it suffices to show that $(H^1_{JO}(\overline{QO})/2)^\vee$ is generated in these degrees. Here generation refers to generation over ko^*_V, since we began with extension of modules over this ring. In degree -8 we have $ko^8_V = JU^4$, which is diagonal in the sense that it consists of all characters zero at the identity whose values are multiples of 16.

If we identify all local cohomology in degrees which are multiples of 4 with a quotient of JU^4 then generators are given by characters (02^400), (002^40), (0002^4), and elements of ko^8_V act by 1/16 of their values. Thus we can act by any integer function zero at the identity, and evidently all generators can be obtained in this way. Similarly for local cohomology in degrees 2 mod 4, using JU^6 rather than JU^4. □

This gives the E^∞ term of the spectral sequence.

12.3.E. Conclusion: the *ko*-homology groups. The values for the homology groups $ko_*(BV(2))$ can be read off from the local cohomology spectral sequence and are given in Figure 12.3.3. The only place where there is room for an additive extension is in degree 2 mod 8; to see this splits we could argue that the elements in the H^0 column in this degree are η-multiples (an alternative argument is given in higher rank).

REMARK 12.3.3. We note that there is a misprint in the table on Page 15 of [**45**], where the value of $ku_7(BV(2))$ is given incorrectly.

12.4. Calculation of $ko_*(BV)$ in rank 3

12.4.A. Overview. We will apply the local cohomology spectral sequence to the known value of ko^*_V. To calculate the E^2-term we need to calculate local cohomology. This will occupy several subsections, and we will display part of the local cohomology spectral sequence in Subsection 12.4.F when this is done.

We have the short exact sequences

$$0 \longrightarrow T \longrightarrow ko^*_V \longrightarrow \overline{QO} \longrightarrow 0$$

where $\overline{QO}$ is the image in periodic complex K-theory, and

$$0 \longrightarrow \tau \longrightarrow T \longrightarrow TO \longrightarrow 0$$

where τ consists of the η-multiples.

First of all, the module τ is bounded below, so torsion, and

$$H^*_I(\tau) = H^0_I(\tau) = \tau.$$

FIGURE 12.3.3. The ko homology of $BV(2)$

n	$ko_n(BV(2))$
$8k+7 \geq 7$	$[2^{4k+4}]^2 \oplus [2^{4k+3}]$
$8k+6 \geq 6$	2^{2k+1}
$8k+5 \geq 5$	0
$8k+4 \geq 4$	$\mathbb{Z} \oplus 2^{2k+2}$
$8k+3 \geq 3$	$[2^{4k+3}]^2 \oplus [2^{4k+2}]$
$8k+2 \geq 2$	2^{2k+4}
$8k+1 \geq 9$	2^4
$8k+0 \geq 8$	$\mathbb{Z} \oplus 2^{2k+1}$
15	$[2^8]^2 \oplus [2^7]$
14	2^3
13	0
12	$\mathbb{Z} \oplus 2^4$
11	$[2^7]^2 \oplus [2^6]$
10	2^6
9	2^4
8	$\mathbb{Z} \oplus 2^3$
7	$[2^4]^2 \oplus [2^3]$
6	2
5	0
4	$\mathbb{Z} \oplus 2^2$
3	$[2^3]^2 \oplus [2^2]$
2	2^4
1	2^3
0	$\mathbb{Z}$

Indeed, τ is 2^8 in positive degrees congruent to 1 or 2 mod 8, and 2 in degree -6, and otherwise it is zero. In the following section we will calculate the local cohomology of $\overline{QO}$, and then turn to that of TO.

12.4.B. The local cohomology of $\overline{QO}$. From the Bockstein spectral sequence we find that $\overline{QO}$ is $RO(G) = RU(G)$ in degrees $0, 8, 16, \ldots$, and $RSp(G) = 2RU(G)$ in degrees $4, 12, 20, \ldots$. In negative degrees it is $2JU$, JU^4, JU^6, $JU^8, \ldots$ in degrees $-4, -8, -12, -16, \ldots$. We identify elements of $RU(G)$ with their characters, displayed as a vector starting with the value at the identity. The augmentation ideal is the radical of the principal ideal generated by

$$(p^*)^2 \in JU^4, \text{ where } p^* = (04444444),$$

so the local cohomology is given by the exact sequence

$$0 \longrightarrow H^0_{JO}(\overline{QO}) \longrightarrow \overline{QO} \longrightarrow \overline{QO}[1/p^*] \longrightarrow H^1_{JO}(\overline{QO}) \longrightarrow 0.$$

Evidently the kernel of inverting p^* consists of characters supported at the identity, so $H^0_{JO}(\overline{QO})$ is just $\mathbb{Z}$ in degrees $0, 8, 16, 24, \ldots$.

Turning to H^1, it is easy to check that JU^{2k} is generated by the diagonal elements $(04^k000000)$, (004^k00000), $\ldots$, (00000004^k) for $k \geq 2$, so that p^* is an isomorphism from degree -8 downwards. Thus H^1 is in degrees -4, 0, 4, 8, 12, 16, ... A little linear algebra gives the following conclusion, where $[a]$ means a cyclic group of order a.

$$H^1_{JO}(\overline{QO})_i = \begin{cases} [2]^3 \oplus [1]^3 \oplus [4] & i = -4 \\ [2^{4k+2}]^3 \oplus [2^{4k+1}]^3 \oplus [2^{4k}] & i = 8k \geq 0 \\ [2^{4k+1}]^3 \oplus [2^{4k}]^3 \oplus [2^{4k-1}] & i = 8k+4 \geq 0 \\ 0 & \text{otherwise} \end{cases}$$

12.4.C. Resolutions of TO. To calculate the local cohomology of TO we use the resolutions we constructed in Chapter 11. Recapitulating, in rank 3 we have $TU = TU_2 \oplus TU_3$ as PC-modules with Sq^2 action. Here $TU_3 = PC(-7)$ with generator q_{123}, and the Sq^2 action makes it a Koszul complex for the maximal ideal of PP. There is a short exact sequence

$$0 \longleftarrow TU_2 \longleftarrow 3PC(-6) \longleftarrow PC(-8) \oplus PC(-10) \longleftarrow 0.$$

The generators of TU_2 are the elements q_{ij}, and elements Q_{ij} generate the corresponding free PC-modules in the resolution. The generator of $PC(-8)$ is X, and of $PC(-10)$ is Y. The Sq^2 action in homological degree 0 is a Koszul complex as before, and the Sq^2 action on the final term in the resolution of TU_2 has $Sq^2X = Y$, making the module $PC[Sq^2]$.

Thus $TO_3 = ZPC(-7)$. To make this explicit, recall

$$ZPC = PP \oplus I_2 \oplus PP(-8),$$

where the generators are Q, $p_{ij}Q$ and $p_{123}Q$. The module I_2 has a truncated Koszul resolution

$$0 \longleftarrow I_2 \longleftarrow 3PP(-6) \longleftarrow PP(-10) \longleftarrow 0.$$

The picture for this is

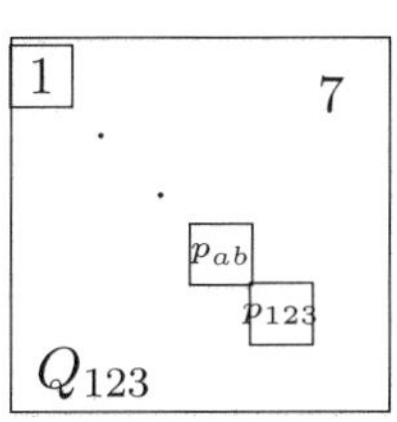

For TO_2, there is an exact sequence

$$0 \longleftarrow TO_2 \longleftarrow 3ZPC(-6) \longleftarrow PP(-10) \oplus 3PP(-12) \oplus 3PP(-14) \oplus PP(-16) \longleftarrow 0.$$

The generators in homological degree 0 are named p_SQ_{ij} as above, and the generators in homological degree 1 are $Y = m_\emptyset$, m_a, m_{ab}, m_{123}. In pictures, this is

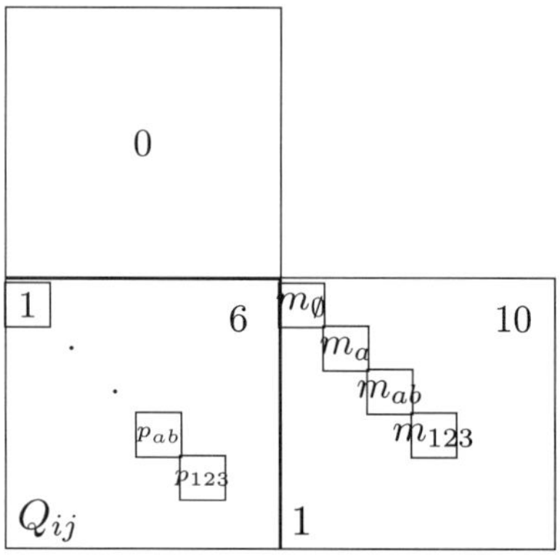

Turning to maps, it is convenient to separate into the parts in degrees congruent to 2 mod 4 and those congruent to 0 mod 4.

Starting in degrees congruent to 2 mod 4, we have a resolution

$$0 \longleftarrow TO_2^{\equiv 2(4)} \longleftarrow 3PP(-6) \oplus 3PP(-14) \longleftarrow PP(-10) \oplus 3PP(-14) \longleftarrow 0.$$

The bases are given in homological degree 0 by Q_{ij} in degree -6, and $p_{123}Q_{ij}$ in degree -14. In homological degree 1 we have $m_\emptyset = Y$ in degree -10, and m_{ij} in degree -14. A short calculation (or see Lemma 11.3.1) shows

$$z_1Q_{23} + z_2Q_{13} + z_3Q_{12} \longleftarrow m_\emptyset = Y$$

and

$$z_1z_2Q_{23} + z_1z_2Q_{13} + p_{123}Q_{12} \longleftarrow m_{12},$$

with similar formulae for m_{13} and m_{23}. Changing bases, we see that the generators in degree -14 are all redundant and we have a Koszul differential, so that

$$TO_2^{\equiv 2(4)} \cong I_2,$$

with top in degree -6.

In degrees congruent to 0 mod 4, we have a resolution

$$0 \longleftarrow TO_2^{\equiv 0(4)} \longleftarrow 3I_2(-6) \longleftarrow 3PP(-12) \oplus PP(-16) \longleftarrow 0$$

The generators of $3I_2(-6)$ are $p_{ab}Q_{ij}$ in degree -12, the generators of $3PP(-12)$ are $m_i = z_iX + y_iY$, and the generator of $PP(-16)$ is $m_{123} = p_{123}X + y_{123}Y$. A short calculation (or see Lemma 11.3.1) shows

$$p_{12}Q_{13} + p_{13}Q_{12} \longleftarrow m_1,$$

with m_2 and m_3 introducing other symmetrizing relations, and

$$z_1p_{23}Q_{23} + z_2p_{13}Q_{13} + z_3p_{12}Q_{12} \longleftarrow m_{123}.$$

12.4.D. Local cohomology I. We may begin with the routine piece.

LEMMA 12.4.1.

$$H_I^i(I_2) = H_I^i(TO_2^{\equiv 2(4)}) = \begin{cases} \mathrm{Start}(10)I_2^\vee & i = 3 \\ \mathbb{F}_2(2) & i = 2 \\ 0 & \textit{otherwise} \end{cases}$$

Proof: This is local duality at its most transparent. Note that

$$H_I^*(PP) = H_I^3(PP) = PP^\vee(12).$$

Apply local cohomology to the resolution of I_2:

$$0 \longleftarrow H_I^3(I_2) \longleftarrow 3PP^\vee(6) \longleftarrow PP^\vee(2) \longleftarrow H_I^2(I_2) \longleftarrow 0.$$

Taking vector space duals we see that the dual of $H_I^*(I_2)$ comes from the Koszul complex. □

Since $TO_2^{\equiv 2(4)} \cong I_2$ this is also the local cohomology of $TO_2^{\equiv 2(4)}$. Similarly, a shift gives the local cohomology of TO_3.

COROLLARY 12.4.2.

$$H_I^i(TO_3) = \begin{cases} PP^\vee(5) \oplus \mathrm{Start}(3)I_2^\vee \oplus PP^\vee(-3) & i=3 \\ \mathbb{F}_2(-5) & i=2 \\ 0 & \textit{otherwise} \end{cases}$$

12.4.E. Local cohomology II. Turning to $TO_2^{\equiv 0(4)}$, the situation is a little more complicated. Applying local cohomology to the resolution of $TO_2^{\equiv 0(4)}$ gives the exact sequence

$$0 \longleftarrow H_I^3(TO_2^{\equiv 0(4)}) \longleftarrow 3\mathrm{Start}(4)I_2^\vee \xleftarrow{\alpha} 3PP^\vee(0) \oplus PP^\vee(-4)$$
$$\longleftarrow H_I^2(TO_2^{\equiv 0(4)}) \longleftarrow 3\mathbb{F}_2(-4) \longleftarrow 0.$$

Thus

$$H_I^i(TO_2^{\equiv 0(4)}) = \begin{cases} \mathrm{cok}(\alpha) & i=3 \\ \ker(\alpha) \oplus 3\mathbb{F}_2(-4) & i=2 \\ 0 & \text{otherwise} \end{cases}$$

To complete the calculation we must identify the map α explicitly, and this requires a more careful examination of local duality. We give a more general discussion in Subsection 12.5.F, but the following special case is easily verified.

LEMMA 12.4.3. *If*

$$f : P(a-6) \longrightarrow I_2$$

corresponds to $\lambda_{12}p_{12} + \lambda_{13}p_{13} + \lambda_{23}p_{23}$ *(so that each* λ_{ij} *has degree* a*) then the induced map* $f_* : H_I^3(PP(a-6)) \longrightarrow H_I^3(I_2)$ *has dual*

$$H_I^3(PP(a-6))^\vee \xleftarrow{f_*^\vee} H_I^3(I_2)^\vee.$$

Explicitly, the dual

$$PP(-a-6) \xleftarrow{f_*^\vee} \mathrm{Finish}(-10)I_2$$

is given by

$$z_1\lambda_{13} + z_2\lambda_{23} \longleftarrow \check{p}_{12}$$
$$z_1\lambda_{12} + z_3\lambda_{23} \longleftarrow \check{p}_{13}$$
$$z_2\lambda_{12} + z_3\lambda_{13} \longleftarrow \check{p}_{23},$$

where the $\check{p}_{ij}$ *are the generators of the combinatorially dual copy* $H_I^2(I_2)$ *of* I_2.

Using the lemma it is easy to check that the dual of the map α is the map

$$\alpha^\vee : 3\mathrm{Start}(-4)I_2 \longrightarrow 3PP(0) \oplus PP(4)$$

given by the following matrix

$$\left(\begin{array}{c|cccc} & m_1 & m_2 & m_3 & m_{123} \\ & & & & \\ \check{p}_{12}Q_{12} & z_1 & z_2 & 0 & 0 \\ \check{p}_{13}Q_{12} & 0 & z_3 & 0 & z_1z_3 \\ \check{p}_{23}Q_{12} & z_3 & 0 & 0 & z_2z_3 \\ \check{p}_{12}Q_{13} & 0 & 0 & z_2 & z_1z_2 \\ \check{p}_{13}Q_{13} & z_1 & 0 & z_3 & 0 \\ \check{p}_{23}Q_{13} & z_2 & 0 & 0 & z_2z_3 \\ \check{p}_{12}Q_{23} & 0 & 0 & z_1 & z_1z_2 \\ \check{p}_{13}Q_{23} & 0 & z_1 & 0 & z_1z_3 \\ \check{p}_{23}Q_{23} & 0 & z_2 & z_3 & 0 \end{array}\right)$$

LEMMA 12.4.4. *The local cohomology of $TO_2^{\equiv 0(4)}$ is concentrated in degrees* 2 *and* 3. *More precisely,*

$$H^3_I(TO_2^{\equiv 0(4)}) \cong PP^\vee(4) \oplus PP^\vee(12)$$

The dual of $H^2_I(TO_2^{\equiv 0(4)})$ is generated as a PP-module in degrees 4 *and* 0, *and is elementary abelian of rank* 1, 6, 7, 7, 7, ... *in degrees* 4, 0, −4, −8, −12, −16, ...

Proof: By considering Hilbert series, we see that we need only prove that the kernel of $\alpha^\vee$ is the free module $PP(-4) \oplus PP(-12)$ as stated. In fact in degree -4 the elements

$$A = \check{p}_{12}Q_{12} + \check{p}_{13}Q_{13} + \check{p}_{23}Q_{23}$$

and in degree -12

$$B = (\check{p}_{13} + \check{p}_{23})z_{12}Q_{12} + (\check{p}_{12} + \check{p}_{23})z_{13}Q_{13} + (\check{p}_{12} + \check{p}_{13})z_{23}Q_{23}$$

are easily seen to be in the kernel. It is easy to check these are independent, since if $\alpha A+\beta B = 0$ we look at the coefficient of Q_{12} to see $\alpha\check{p}_{12}+\beta z_1z_2\check{p}_{13}+\beta z_1z_2\check{p}_{23} = 0$; but all relations are multiples of $z_3\check{p}_{12} + z_2\check{p}_{13} + z_1\check{p}_{23} = 0$. It follows that A and B generate a free submodule. It remains to check they span the kernel. The interested reader can repeat the verification (it can be done by hand on one sheet of A4, or verified by a computer algebra package). □

12.4.F. The local cohomology spectral sequence. We have now calculated the $E^{1\frac{1}{2}}$-page of the local cohomology spectral sequence. This is nearly the E^2-page, but the differential $d^1 : H^1_{JO}(\overline{QO}) \longrightarrow H^2_{JO}(TO)$ has not yet been taken into account. We display the spectral sequence in low degrees (≤ 20) in Figure 12.4.4.

First note that each column is bounded below. The H^0 column has lowest entry in degree -6, the H^1 column in degree -4, the H^2 column in degree -5 and the H^3 column in degree -3. This remains true even at E^2, since the $d^1 : H^1_{JO}(\overline{QO})_{-4} \longrightarrow H^2_{JO}(TO)_{-4}$ cannot be injective, and the codomain is not the lowest group in the H^2 column. Accordingly, the regularity of $ko^*_{V(3)}$ is -7.

FIGURE 12.4.4. The $E^{1\frac{1}{2}}$-page of the local cohomology spectral sequence for $ko_*(BV(3))$.

H^3_{JO}	H^2_{JO}	H^1_{JO}	H^0_{JO}	
2^{21}	2^7	$[2^{13}] \oplus [2^{12}]^3 \oplus [2^{11}]$	$\mathbb{Z}$	20
2^{35}	0	0	0	
2^{15}	0	0	2^8	
2^{31}	0	0	2^8	
2^{13}	2^7	$[2^{10}]^3 \oplus [2^9]^3 \oplus [2^8]$	$\mathbb{Z}$	16
2^{24}	0	0	0	
2^8	0	0	0	
2^{21}	0	0	0	
2^7	2^7	$[2^9]^3 \oplus [2^8]^3 \oplus [2^7]$	$\mathbb{Z}$	12
2^{15}	0	0	0	
2^3	0	0	2^8	
2^{13}	0	0	2^8	
2^3	2^7	$[2^6]^3 \oplus [2^5]^3 \oplus [2^4]$	$\mathbb{Z}$	8
2^8	0	0	0	
0	0	0	0	
2^7	0	0	0	
2^1	2^7	$[2^5]^3 \oplus [2^4]^3 \oplus [2^3]$	$\mathbb{Z}$	4
2^3	0	0	0	
0	2	0	2^8	
2^3	0	0	2^8	
0	2^6	$[2^2]^3 \oplus [2]^3 \oplus [1]$	$\mathbb{Z}$	0
0	0	0	0	
0	0	0	0	
2	0	0	0	
	2^4	$[2]^3 \oplus [1]^3 \oplus [4]$	0	-4
	2		0	
			2	
				-8

Now consider differentials. The principal fact is that the spectral sequence converges to $ko_*(BV(3))$, which is bounded below. The reader should note the slope -1 line below which the E^∞-page must be zero. Similarly, we know $ko_0(BG) \xrightarrow{\cong} KO_0(BG)$ is an isomorphism so that the only non-zero entry $E^\infty_{p,-p}$ has $p = 0$, where it is $E^2_{0,0}$. This forces certain differentials. Specifically,

- there must be an isomorphic d^2 leaving H^0_{-6}
- the d^1 and d^2 leaving H^1_{-4} must both be surjective and their common kernel is 0
- the d^1 and d^2 leaving H^1_0 must both be surjective and their common kernel is 0
- the d^2 and d^3 leaving H^0_1 must both be surjective and their common kernel is 2^4

The other elementary fact is that $ko_1BG = ko_1 \oplus H_1(BG;\mathbb{Z})$, so that in our case $ko_1(BV(3)) = 2^4$. This means that the only non-zero entry $E^\infty_{p+1,-p}$ has $p = 0$, where it is $E^\infty_{0,1}$. This forces one more differential

- the d^3 leaving H^1_2 must be surjective with kernel 2^7.

LEMMA 12.4.5. *There are no further differentials leaving the H^0-column.*

Proof: We know that ko_* is a summand, so it remains only to discuss the classes in degrees 1 and 2 mod 8. By sparsity in the H^1 and H^2 columns, the only possible differential is d^3. Since $ko_*(BG)[1/\beta] = KO_*(BG)$, we know that all of the 2^8 survives in sufficiently large degrees. Since the image in KO^G_* is a ko_*-module, it suffices to deal with degrees 9 and 10.

For this we use the fact that the differentials are maps of ko^*_V-modules. Every element of H^3_n for $n = 10$ or 11 is detected by some map $H^3_n \longrightarrow H^3_{n-4}$ coming from multiplication by an element λ of ko^4_V, so if $d^3(x)$ is non-zero in H^3_n and detected by λ there would have to be a non-zero $d^3(\lambda x)$. However λx lies in the zero group. □

We have dealt with the two entries of H^2 which are not in degrees divisible by 4.

LEMMA 12.4.6. *The differential $d^1 : H^1_{JO}(\overline{QO}) \longrightarrow H^2_{JO}(TO)$ is surjective in degrees divisible by 4, and therefore*

$$\ker(d^1 : H^1_{JO}(\overline{QO}) \longrightarrow H^2_{JO}(TO)) = 2H^1_{JO}(\overline{QO}).$$

Proof: The codomain is annihilated by 2, so d^1 factors through

$$\overline{d}^1 : H^1_{JO}(\overline{QO})/2 \longrightarrow H^2_{JO}(TO).$$

In degrees divisible by 4 these are isomorphic finite groups in each degree, so it is equivalent to show that $\overline{d}^1$ is an isomorphism. We consider the dual

$$(H^1_{JO}(\overline{QO})/2)^\vee \xleftarrow{(\overline{d}^1)^\vee} H^2_{JO}(TO)^\vee.$$

We already know it is surjective in degrees -4 and 0, so it suffices to show that $(H^1_{JO}(\overline{QO})/2)^\vee$ is generated in these degrees. Here generation refers to generation over ko^*_V, since we began with extension of modules over this ring. In degree -8 we have $ko^8_V = JU^4$, which is diagonal in the sense that it consists of all characters zero at the identity whose values are multiples of 16.

If we identify all local cohomology in degrees which are multiples of 4 with a quotient of JU^4 then generators are given by characters $(02^40000000), \ldots, (00000002^4)$, and elements of ko_V^8 act by 1/16 of their values. Thus we can act by any integer function zero at the identity, and evidently all generators can be obtained in this way. Similarly for local cohomology in degrees 2 mod 4, using JU^6 rather than JU^4. □

LEMMA 12.4.7. *The non-zero differentials $d^2 : H^1_{JO}(\overline{QO}) \longrightarrow H^3_{JO}(TO)$ have maximal rank, which is 1 from degree −4, 3 from degree 0 and 7 from higher degrees.*

Proof: The codomain is annihilated by 2, so this factors through

$$\overline{d}^2 : 2H^1_{JO}(\overline{QO})/2 \longrightarrow H^3_{JO}(TO).$$

The claim is that this is injective. We consider the dual

$$(2H^1_{JO}(\overline{QO})/2)^\vee \stackrel{(\overline{d}^2)^\vee}{\longleftarrow} H^3_{JO}(TO)^\vee,$$

and show that it is surjective.

We already know it is surjective in degrees −4 and 0, so it suffices to show $(2H^1_{JO}(\overline{QO})/2)^\vee$ is generated in these degrees. This is very similar to the verification in 12.4.6, and details are left to the reader. □

This gives the E^∞ term of the spectral sequence.

12.4.G. Conclusion: the *ko*-homology groups. The values for the homology groups $ko_*(BV(3))$ can be read off from the local cohomology spectral sequence and are given in Figure 12.4.5. The coefficients are described by the following generating functions in the four congruences classes mod 4. These are read off from the local cohomology calculations. As usual, t is of codegree −1 and $u = t^4$.

$$a(t) = \sum_s a_s t^s = \frac{u(1+u^2)}{(1-u)^3}$$

$$b(t) = \sum_s b_s t^s = \frac{t^{-3}(1+u^2)}{(1-u)^3}$$

$$c(t) = \sum_s c_s t^s = \frac{t^{10}(3-u)}{(1-u)^3}$$

$$d(t) = \sum_s d_s t^s = \frac{t^3(3-u)}{(1-u)^3}.$$

It is interesting to note that $b(t) = t^{-7}a(t)$ and $d(t) = t^{-7}c(t)$, which reflects isomorphisms at the level of PP-modules.

There are a number of potential additive extensions with kernel in the H^r column and the quotient in the H^0 or H^1 column (but not both). To see these are split, we note that the H^r column is detected by the map $ko_*(BV) \longrightarrow H_*(BV;\mathbb{F}_2)$.

FIGURE 12.4.5. The ko homology of $BV(3)$

n	$ko_n(BV(3))$
$8k+7 \geq 7$	$[2^{4k+4}]^3 \oplus [2^{4k+3}]^3 \oplus [2^{4k+2}] \oplus 2^{c_{8k+10}}$
$8k+6 \geq 6$	$2^{(b_{8k+9}-7)}$
$8k+5 \geq 5$	$2^{a_{8k+8}}$
$8k+4 \geq 4$	$\mathbb{Z} \oplus 2^{d_{8k+7}}$
$8k+3 \geq 3$	$[2^{4k+3}]^3 \oplus [2^{4k+2}]^3 \oplus [2^{4k+1}] \oplus 2^{c_{8k+6}}$
$8k+2 \geq 10$	$2^8 \oplus 2^{(b_{8k+5}-7)}$
$8k+1 \geq 9$	$2^8 \oplus 2^{a_{8k+4}}$
$8k+0 \geq 8$	$\mathbb{Z} \oplus 2^{d_{8k+3}}$
15	$[2^8]^3 \oplus [2^7]^3 \oplus [2^6] \oplus 2^{15}$
14	2^{24}
13	2^{13}
12	$\mathbb{Z} \oplus 2^{24}$
11	$[2^7]^3 \oplus [2^6]^3 \oplus [2^5] \oplus 2^8$
10	2^{22}
9	2^{15}
8	$\mathbb{Z} \oplus 2^{15}$
7	$[2^4]^3 \oplus [2^3]^3 \oplus [2^2] \oplus 2^3$
6	2^6
5	2^3
4	$\mathbb{Z} \oplus 2^8$
3	$[2^3]^3 \oplus [2^2]^3 \oplus [2]$
2	2^7
1	2^4
0	$\mathbb{Z}$

12.5. Calculation of $ko_*(BV)$ in rank 4

Before the reader commits effort to this section, we warn that the calculation requires substantial computer assistance, and the account of the calculation is about 32 pages long. On the other hand, after the work of previous chapters, the work consists of a little combinatorics of pairs of subsets of $\{1, 2, 3, 4\}$ together with systematic book-keeping. The material is included to show that despite the size of the calculation and the combinatorial intricacy, we have reduced it to a matter of patient persistence. On the other hand, the answer demonstrates that we have already lost the relative simplicity of the answer in rank 3.

12.5.A. Overview. We will apply the local cohomology spectral sequence to the known value of ko_V^*. To calculate the E^2-term we need to calculate local cohomology. This will occupy several subsections, and we will display part of the local cohomology spectral sequence in Subsection 12.5.N when this is done.

We have the short exact sequences

$$0 \longrightarrow T \longrightarrow ko_V^* \longrightarrow \overline{QO} \longrightarrow 0$$

where $\overline{QO}$ is the image in periodic complex K-theory, and

$$0 \longrightarrow \tau \longrightarrow T \longrightarrow TO \longrightarrow 0$$

where τ consists of the η-multiples.

First of all, the module τ is bounded below, so torsion, and

$$H_I^*(\tau) = H_I^0(\tau) = \tau.$$

Indeed, τ is 2^{16} in positive degrees congruent to 1 or 2 mod 8, 2^5 in degree -6, 2 in degree -7, and otherwise it is zero. In the following section we will calculate the local cohomology of $\overline{QO}$, and then turn to that of TO.

12.5.B. The local cohomology of $\overline{QO}$. From the Bockstein spectral sequence we find that $\overline{QO}$ is $RO(G) = RU(G)$ in degrees 0, 8, 16, ..., and $RSp(G) = 2RU(G)$ in degrees 4, 12, 20, ... In negative degrees it is $2JU$, JU^4, JU^6, JU^8, ... in degrees -4, -8, -12, -16, ... We identify elements of $RU(G)$ with their characters, displayed as a vector starting with the value at the identity. The augmentation ideal is the radical of the principal ideal generated by

$$(p^*)^2 \in JU^4, \text{ where } p^* = (0444444444444444),$$

so the local cohomology is given by the exact sequence

$$0 \longrightarrow H^0_{JO}(\overline{QO}) \longrightarrow \overline{QO} \longrightarrow \overline{QO}[1/p^*] \longrightarrow H^1_{JO}(\overline{QO}) \longrightarrow 0.$$

Evidently the kernel of inverting p^* consists of characters supported at the identity, so $H^0_{JO}(\overline{QO})$ is just $\mathbb{Z}$ in degrees 0, 4, 8, 12, 16, ...

By way of variation, and to save on linear algebra, we may calculate local cohomology of $\overline{QO}$ from the known local cohomology of QU [**28**, 4.4.7] as follows. First note that the map $\overline{QO} \longrightarrow QU$ is a monomorphism, with image in degrees which are multiples of 4. In calculating the local cohomology it suffices to work with the ring $\mathbb{Z}[(p^*)^2]$, and we have a short exact sequence of modules

$$0 \longrightarrow \overline{QO} \longrightarrow QU^{\equiv 0(4)} \longrightarrow C \longrightarrow 0.$$

This is especially convenient since C is zero in degrees below -4, so that $H^*_{(p^*)}(C) = H^0_{(p^*)}(C) = C$. Thus we have the exact sequence

$$0 \longrightarrow H^0_{JO}(\overline{QO}) \longrightarrow H^0_{JU}(QU)^{\equiv 0} \longrightarrow C \longrightarrow H^1_{JO}(\overline{QO}) \longrightarrow H^1_{JU}(QU)^{\equiv 0} \longrightarrow 0$$

It is easy to evaluate C: it is zero in degree divisible by 8, it is $RU/RSp \cong 2^{16}$ in degrees of the form $8k+4 \geq 4$, and it is $JU^2/2JU \cong 2^{11}$ in degree -4. It is easy to see that the map $\overline{QO} \longrightarrow QU^{\equiv 0(4)}$ in 0th local cohomology has cokernel $\mathbb{Z}/2$ in degrees $8k+4$. Since we know the rank is at most 15, this determines the answer in most degrees.

$$H^1_{JO}(\overline{QO})_i = \begin{cases} \langle 2^{17} \rangle & i = -4 \\ [2^{4k+3}]^4 \oplus [2^{4k+2}]^6 \oplus [2^{4k+1}]^4 \oplus [2^{4k}] & i = 8k \geq 0 \\ [2^{4k+6}]^4 \oplus [2^{4k+5}]^6 \oplus [2^{4k+4}]^4 \oplus [2^{4k+3}] & i = 8k+4 \geq 0 \\ 0 & \text{otherwise} \end{cases}$$

The notation $\langle 2^{17} \rangle$ for degree -4 refers to an abelian group of order 2^{17} of unspecified structure. The ambiguity is of secondary importance, since it must be cancelled out in the course of the local cohomology spectral sequence.

12.5.C. Resolutions of TO. To calculate the local cohomology of TO we use the resolutions we constructed in the Chapter 11. Recapitulating, in rank 4 we have a direct sum decomposition $TU = TU_2 \oplus TU_3 \oplus TU_4$, as PC-modules. Here TU_2 is generated by the elements q_{ij}, TU_3 by the elements q_{ijk} and $TU_4 = PC(-8)$ is generated by q_{1234}.

To identify TO we need to consider the action of Sq^2 and take cycles. In odd degrees, we have the term TU_3, which is also a summand as a $PC[Sq^2]$-module. We give the analysis in Subsections 12.5.D, 12.5.G and 12.5.H. In even degrees, we have a short exact sequence of Sq^2-modules

$$0 \longrightarrow TU_2 \longrightarrow TU_{24} \longrightarrow TU_4 \longrightarrow 0.$$

with $Sq^2 q_{1234} = \sum_{i<j} y_i y_j q_{\hat{i}\hat{j}}$.

As remarked above, $TU_4 = PC(-8)$, so that $TO_4 = ZPC(-8)$. To make this explicit, recall from Lemma 11.2.1

$$ZPC = PP \oplus I_2 \oplus I_3 \oplus I_4,$$

where PP is generated by Q, I_2 by the elements $p_{ij}Q$, I_3 by the elements $p_{ijk}Q$ and $I_4 = PP(-10)$ by $p_{1234}Q$.

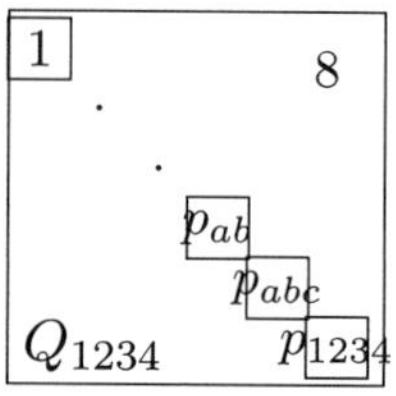

Returning to the exact sequence for TU_{24}, from our analysis of the spectral sequence of the filtration in Corollary 9.8.3 and Proposition 9.7.2, we find that $H(TU_4) = \mathbb{F}_2(-8)$ and $H(TU_2) = 6\mathbb{F}_2(-6) \oplus \mathbb{F}_2(-10)$ with the boundary map being non-trivial. Thus $H(TU_{24}) = 6\mathbb{F}_2(-6)$, and there is an exact sequence

$$0 \longrightarrow ZTU_2 \longrightarrow ZTU_{24} \longrightarrow ZTU_4 \longrightarrow \mathbb{F}_2(-8) \longrightarrow 0,$$

giving the short exact sequence

$$0 \longrightarrow TO_2 \longrightarrow TO_{24} \longrightarrow (I_1 \oplus I_2 \oplus I_3 \oplus I_4)(-8) \longrightarrow 0.$$

The modules I_1, I_2, I_3 and I_4 have truncated Koszul resolutions

$$0 \longleftarrow I_1 \longleftarrow 4PP(-4) \longleftarrow 6PP(-8) \longleftarrow 4PP(-12) \longleftarrow PP(-16) \longleftarrow 0,$$
$$0 \longleftarrow I_2 \longleftarrow 6PP(-6) \longleftarrow 4PP(-10) \longleftarrow PP(-14) \longleftarrow 0,$$

and

$$0 \longleftarrow I_3 \longleftarrow 4PP(-8) \longleftarrow PP(-12) \longleftarrow 0.$$

The analysis of TO_2 is a major undertaking, and we defer it to Subsection 12.5.I below.

12.5.D. Overview of TO_3. We start the detailed study with odd degrees. For TU_3 we have the resolution

$$0 \longleftarrow TU_3 \longleftarrow 4PC(-7) \longleftarrow PC(-9) \oplus PC(-11) \longleftarrow 0,$$

The generators in homological degree 0 are Q_{ijk}. The generator of $PC(-9)$ is X, and of $PC(-11)$ is Y. The Sq^2 action in homological degree 0 is the Koszul complex of PP, and the Sq^2 action on the fnal term in the resolution has $Sq^2X = Y$, making the module $PC[Sq^2]$.

Passing to Sq^2 cycles, since Sq^2 is exact on $PC[Sq^2]$, there is an exact sequence

$$0 \longleftarrow TO_3 \longleftarrow 4ZPC(-7) \longleftarrow ZPC[Sq^2](-9) \longleftarrow 0$$

We have already described ZPC in Subsection 12.5.C above, so the generators in homological degree 0 are the elements Q_{ijk} in degree -7, $p_{ab}Q_{ijk}$ in degree -13, $p_{abc}Q_{ijk}$ in degree -15 and the elements $p_{1234}Q_{ijk}$ in degree -15.

In homological degree 1 we have

$$ZPC[Sq^2](-9) = PP(-11) \oplus 4PP(-13) \oplus 6PP(-15) \oplus 4PP(-17) \oplus PP(-19),$$

so that the generators are $Y = m_\emptyset$ in degree -11, m_a in degree -13, m_{ab} in degree -15, m_{abc} in degree -17 and m_{1234} in degree -19.

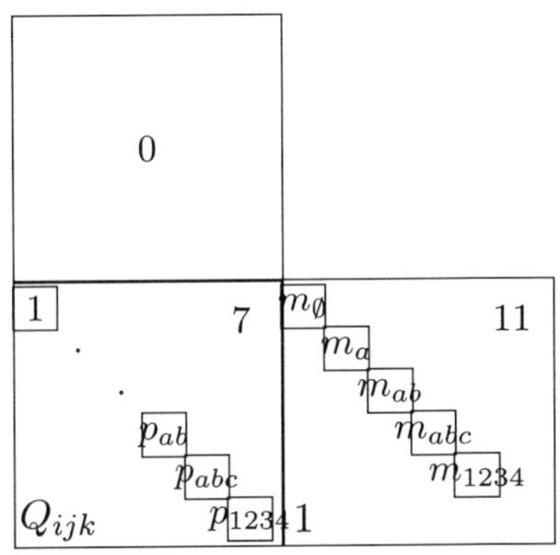

Turning to maps, it is convenient to separate into the parts in degrees congruent to 1 mod 4 and those congruent to 3 mod 4. We will deal with each of these in turn when we have recorded some facts about local cohomology.

12.5.E. Local cohomology I. We may begin with the routine piece. For definiteness, we recall that I_1, I_2, I_3 and I_4 have tops in degrees $-4, -6, -8$ and -10 respectively.

LEMMA 12.5.1. *The local cohomology of I_1 is as follows.*

$$H^i_I(I_1) = \begin{cases} \mathrm{Start}(16)I_4^\vee & i=4 \\ \mathbb{F}_2(0) & i=1 \\ 0 & \text{otherwise} \end{cases}$$

The local cohomology of I_2 is as follows.

$$H^i_I(I_2) = \begin{cases} \mathrm{Start}(14)I_3^\vee & i=4 \\ \mathbb{F}_2(2) & i=2 \\ 0 & \text{otherwise} \end{cases}$$

The local cohomology of I_3 is as follows.

$$H^i_I(I_3) = \begin{cases} \mathrm{Start}(12)I_2^\vee & i=4 \\ \mathbb{F}_2(4) & i=3 \\ 0 & \text{otherwise} \end{cases}$$

Proof: This is local duality at its most transparent. We will prove the statement for I_2; the other cases are exactly analogous.

Note that $H_I^*(PP) = H_I^4(PP) = PP^\vee(16)$, and apply local cohomology to the resolution of I_2:

$$0 \longleftarrow H_I^4(I_2) \longleftarrow 6PP^\vee(10) \longleftarrow 4PP^\vee(6) \longleftarrow PP^\vee(2) \longleftarrow H_I^2(I_2) \longleftarrow 0.$$

Taking vector space duals we see that the dual of $H_I^*(I_2)$ comes from the Koszul complex. □

12.5.F. Local duality. It rather clarifies the situation to give the general case. We thus work with the polynomial ring PP in r variables of degree -4, so that $H_I^r(PP) = PP^\vee(4r)$, and the trivial form of local duality states that if

$$f : F \longrightarrow F'$$

is a map of free modules given by a matrix A with respect to some choice of bases, then the map

$$(f_*)^\vee : H_I^r(F)^\vee \longleftarrow H_I^r(F')^\vee$$

is the PP-dual of f, and given with respect to the dual bases by the same matrix A.

Now consider the modules I_k generated by the classes p_S for $|S| = k$ in degree $-2k-2$. This is isomorphic to the image of

$$\binom{r}{k} PP \longrightarrow \binom{r}{k-1} PP$$

in the Koszul complex, where the map is given by

$$x_S \longmapsto \sum_s z_s x_{S \setminus s}.$$

LEMMA 12.5.2. *If*

$$f : P(a-2k-2) \longrightarrow I_k$$

corresponds to $\sum_{|S|=k} \lambda_S p_S$ *(so that each* λ_S *has degree* a*) then the induced map* $f_* : H_I^r(PP(a-2k-2)) \longrightarrow H_I^r(I_k)$ *has dual*

$$H_I^r(PP(a-2k-2))^\vee \overset{f_*^\vee}{\longleftarrow} H_I^r(I_k)^\vee.$$

Explicitly, the dual

$$PP(-a+2k+2-4r) \overset{f_*^\vee}{\longleftarrow} \mathrm{Finish}(2k-2-4r) I_{r-k+1}$$

is given by

$$\sum_{t \in T} z_t \lambda_{T^c t} \longleftarrow \check{p}_T,$$

where $T^c t$ *is the complement of* T *together with the element* t*, and where the elements* $\check{p}_T$ *for* $|T| = r-k+1$ *give a generating set for this combinatorially dual copy of* I_{r-k+1}.

Proof: We take the diagram

$$\begin{array}{ccc} P(a-2k-2) & \xrightarrow{f} & I_k \\ & \searrow^{\tilde{f}} & \downarrow \\ & & \binom{r}{k} PP(-2k-2). \end{array}$$

where $\tilde{f}(\iota) = \sum_S \lambda_S P_S$. and apply local cohomology

$$\begin{array}{ccc} H_I^r(P(a-2k-2))^\vee & \xleftarrow[(f_*)^\vee]{} & H_I^r(I_k)^\vee \\ & \nwarrow_{(\tilde{f}_*)^\vee} & \uparrow \\ & & \binom{r}{k} H_I^r(PP(-2k-2))^\vee \end{array}$$

Identifying the dual basis to $(P_S)_{|S|=k}$ with $(P_{S^c})_{|S^c|=r-k}$ (so that P_S corresponds to P_{S^c}), we find $(\tilde{f}_*)^\vee$ htakes P_{S^c} to $\lambda_S \iota$. On the other hand, the right hand vertical takes a basis element $\check{P}_T$ with $|T| = r-k+1$ to $\sum_{t\in T} z_t P_{T\setminus t}$. □

12.5.G. TO_3 in degrees congruent to 1 mod 4. In this subsection we describe the calculation of $H_I^*(TO_3^{\equiv 1(4)})$. Although we give more precise module theoretic information, it is useful to record our conclusion in terms of Hilbert series.

PROPOSITION 12.5.3. *The module $TO_3^{\equiv 1(4)}$ is of depth 3, and its top local cohomology module has dual with Hilbert series*

$$[H_I^4(TO_3^{\equiv 1(4)})^\vee] = \frac{t^5(4-u+10u^2-5u^3)}{(1-u)^4},$$

so that $H_I^4(TO_3^{\equiv 1(4)})$ has dimensions

$$4, 15, 46, 105, 200, 339, 530, 781, 1100, \ldots$$

in degrees $5, 9, 13, 17, 21, 25, \ldots$. *The third local cohomology group $H_I^3(TO_3^{\equiv 1(4)})$ has dimension* 5, 10, 15, 15, 15, 15, 15, ... *in degrees* -3, 1, 5, 9, 13, 17, ...

In degrees congruent to 1 mod 4, we have a resolution

$$0 \longleftarrow TO_3^{\equiv 1(4)} \longleftarrow 4[PP(-7) \oplus I_3(-7)] \longleftarrow$$
$$PP(-11) \oplus 6PP(-15) \oplus PP(-19) \longleftarrow 0.$$

The bases are given in homological degree 0 by Q_{ijk} in degree -7, and $p_{abc}Q_{ijk}$ in degree -15. In homological degree 1 we have $m_\emptyset = Y$ in degree -11, m_{ij} in degree -15 and m_{1234} in degree -19.

From Lemma 11.3.1, we see

$$z_1 Q_{234} + z_2 Q_{134} + z_3 Q_{124} + z_4 Q_{123} \longleftarrow m_\emptyset = Y$$

and

$$z_1 z_2 Q_{234} + z_1 z_2 Q_{134} + p_{123} Q_{124} + p_{124} Q_{123} \longleftarrow m_{12},$$

with similar formulae for m_{13}, m_{14}, m_{23}, m_{24} and m_{34}. Finally we see

$$z_1 p_{234} Q_{234} + z_2 p_{134} Q_{134} + z_3 p_{124} Q_{124} + z_4 p_{123} Q_{123} \longleftarrow m_{1234}.$$

Notice that the truncated Koszul complex

$$0 \longleftarrow \mathrm{Finish}(-7)I_3 \longleftarrow 4PP(-7) \longleftarrow PP(-11) \longleftarrow 0$$

is a subcomplex of the resolution of $TO_3^{\equiv 1(4)}$, so that there is a short exact sequence

$$0 \longrightarrow TO_3^{\equiv 1(4),Kos} \longrightarrow TO_3^{\equiv 1(4)} \longrightarrow TO_3^{\equiv 1(4),rem} \longrightarrow 0,$$

where

$$TO_3^{\equiv 1(4),Kos} \cong \mathrm{Finish}(-7)I_3.$$

and there is a short exact sequence

$$0 \longleftarrow TO_3^{\equiv 1(4),rem} \longleftarrow 4I_3(-7) \longleftarrow 6PP(-15) \oplus PP(-19) \longleftarrow 0.$$

The counterpart to $TO_3^{\equiv 1(4),rem}$ was zero in rank 3.

There seems to be no practical advantage to treating the two parts separately, so we apply local cohomology to the complete resolution, and obtain the exact sequence

$$\begin{aligned}0 \longleftarrow H_I^4(TO_3^{\equiv 1(4)}) \longleftarrow 4[PP^\vee(9) \oplus \mathrm{Start}(5)I_2^\vee] \xleftarrow{\beta} \\ PP^\vee(5) \oplus 6PP^\vee(1) \oplus PP^\vee(-3) \longleftarrow H_I^3(TO_3^{\equiv 1(4)}) \longleftarrow 4\mathbb{F}_2(-3) \longleftarrow 0.\end{aligned}$$

Thus

$$H_I^i(TO_3^{\equiv 1(4)}) = \begin{cases} \mathrm{cok}(\beta) & i = 4 \\ \ker(\beta) \oplus 4\mathbb{F}_2(-3) & i = 3 \\ 0 & \text{otherwise} \end{cases}$$

To complete the calculation we must identify the map β explicitly, and this requires a more careful examination of local duality, as described in Subsection 12.5.F. From this we infer that a map $f : PP(-a-10) \longrightarrow I_3$ given by $f(\iota) = \sum_{ij} \lambda_{ijk} p_{ijk}$ induces the map

$$PP(-a-10) \xleftarrow{f_*^\vee} \mathrm{Finish}(-5)I_2$$

given by

$$z_1\lambda_{134} + z_2\lambda_{234} \longleftarrow \check{p}_{12}$$

Using the lemma it is easy to check that the dual of the map β is the map

$$\beta^\vee : 4PP(-9) \oplus 4\mathrm{Finish}(-5)I_2 \longrightarrow PP(-5) \oplus 6PP(-1) \oplus PP(3)$$

given by the following 8×28 matrix

	$m_\emptyset$	m_{12}	m_{13}	m_{14}	m_{23}	m_{24}	m_{34}	m_{1234}
Q_{123}	z_4			z_1z_4		z_2z_4	z_3z_4	
Q_{124}	z_3		z_1z_3		z_2z_3		z_3z_4	
Q_{134}	z_2	z_1z_2			z_2z_3	z_2z_4		
Q_{234}	z_1	z_1z_2	z_1z_3	z_1z_4				
$\check{p}_{12}Q_{123}$			z_1		z_2			
$\check{p}_{13}Q_{123}$		z_1			z_3			
$\check{p}_{14}Q_{123}$					z_4			z_1z_4
$\check{p}_{23}Q_{123}$		z_2	z_3					
$\check{p}_{24}Q_{123}$			z_4					z_2z_4
$\check{p}_{34}Q_{123}$		z_4						z_3z_4
$\check{p}_{12}Q_{124}$				z_1		z_2		
$\check{p}_{13}Q_{124}$						z_3		z_1z_3
$\check{p}_{14}Q_{124}$		z_1				z_4		
$\check{p}_{23}Q_{124}$				z_3				z_2z_3
$\check{p}_{24}Q_{124}$		z_2		z_4				
$\check{p}_{34}Q_{124}$		z_3						z_3z_4
$\check{p}_{12}Q_{134}$							z_2	z_1z_2
$\check{p}_{13}Q_{134}$				z_1			z_3	
$\check{p}_{14}Q_{134}$			z_1				z_4	
$\check{p}_{23}Q_{134}$				z_2				z_2z_3
$\check{p}_{24}Q_{134}$			z_2					z_2z_4
$\check{p}_{34}Q_{134}$			z_3	z_4				
$\check{p}_{12}Q_{234}$							z_1	z_1z_2
$\check{p}_{13}Q_{234}$						z_1		z_1z_3
$\check{p}_{14}Q_{234}$					z_1			z_1z_4
$\check{p}_{23}Q_{234}$						z_2	z_3	
$\check{p}_{24}Q_{234}$					z_2		z_4	
$\check{p}_{34}Q_{234}$					z_3	z_4		

The bottom right 7×24 block bears a resemblance to the matrix of α in rank 3, as described in Subsection 12.4.E.

By construction, the cokernel of $\beta^\vee$ (contributing to $H_I^3(TO_3^{\equiv 1(4)})^\vee)$ is generated in degrees 3, -1 and -5. Using Macaulay, one finds that and it has dimensions

$$1, 10, 15, 15, 15, 15, 15, 15, \ldots$$

in degrees

$$3, -1, -5, -9, -13, -17, -21, \ldots$$

It is not hard to verify the first three terms by hand. A calculation with Hilbert series shows the kernel of $\beta^\vee$ (viz $H_I^4(TO_3^{\equiv 1(4)})^\vee$) has Hilbert series

$$\frac{t^5(4 - u + 10u^2 - 5u^3)}{(1-u)^4},$$

where t is of degree -1 and $u = t^4$. This gives dimensions

$$4, 15, 46, 105, 200, 339, 530, 781, 1100, \ldots$$

in degrees

$$-5, -9, -13, -17, -21, \ldots.$$

12.5.H. TO_3 in degrees congruent to 3 mod 4. In this subsection we describe the calculation of $H_I^*(TO_3^{\equiv 3(4)})$. Although we give more precise module theoretic information, it is useful to record our conclusion in terms of Hilbert series.

PROPOSITION 12.5.4. *The module $TO_3^{\equiv 3(4)}$ is of depth 2, and its top local cohomology module has dual with Hilbert series*

$$[H_I^4(TO_3^{\equiv 3(4)})^\vee] = \frac{t^{11}4(5-4u+u^2)}{(1-u)^4},$$

so that $H_I^4(TO_3^{\equiv 3(4)})$ has dimensions

$$20, 64, 140, 256, 420, 640, \ldots$$

in degrees 11, 15, 19, 23, 27, 31, ... *The third local cohomology group is a four dimensional vector space*

$$H_I^3(TO_3^{\equiv 3(4)}) \cong 4\mathbb{F}_2(3).$$

The second local cohomology group is a four dimensional vector space

$$H_I^2(TO_3^{\equiv 3(4)}) \cong 4\mathbb{F}_2(-5).$$

In degrees congruent to 3 mod 4, we have a resolution

$$0 \longleftarrow TO_3^{\equiv 3(4)} \longleftarrow 4[I_2(-7) \oplus PP(-17)] \longleftarrow 4PP(-13) \oplus 4PP(-17) \longleftarrow 0.$$

The generators in homological degree 0 are the elements $p_{ab}Q_{ijk}$ in degree -13 for the copies of $I_2(-7)$, and the elements $p_{1234}Q_{ijk}$ in degree -17 for the copies of $PP(-17)$. In homological degree 1, the generators of $4PP(-13)$ are $m_i = z_iX + y_iY$, and the generators of $4PP(-17)$ are $m_{ijk} = p_{ijk}X + y_{ijk}Y$.

A short calculation (or see Lemma 11.3.1) shows

$$p_{12}Q_{134} + p_{13}Q_{124} + p_{14}Q_{123} \longleftarrow m_1,$$

with m_2, m_3 and m_4 introducing similar relations. Similarly,

$$z_1p_{23}Q_{234} + z_2p_{13}Q_{134} + z_3p_{12}Q_{124} + p_{1234}Q_{123} \longleftarrow m_{123},$$

and similarly m_{124}, m_{134} and m_{234}. Thus the generators m_{ijk} allow us to remove four copies of $PP(-17)$ in degrees 0 and 1 to obtain the exact sequence

$$0 \longleftarrow TO_3^{\equiv 3(4)} \longleftarrow 4I_2(-7) \longleftarrow 4PP(-13) \longleftarrow 0.$$

The generators of $4PP(-13)$ are m_1, m_2, m_3 and m_4 with differential given by the same formula as above.

Now consider the local cohomology of $TO_3^{\equiv 3(4)}$.

Applying local cohomology to the resolution of $TO_3^{\equiv 3(4)}$ gives the exact sequence

$$0 \longleftarrow H_I^4(TO_3^{\equiv 3(4)}) \longleftarrow 4\mathrm{Start}(7)I_3^\vee \stackrel{\gamma}{\longleftarrow} 4PP^\vee(3) \longleftarrow H_I^3(TO_3^{\equiv 3(4)}) \longleftarrow 0$$

Thus

$$H_I^i(TO_3^{\equiv 3(4)}) = \begin{cases} \mathrm{cok}(\gamma) & i = 4 \\ \ker(\gamma) & i = 3 \\ 4\mathbb{F}_2(-5) & i = 2 \\ 0 & \text{otherwise} \end{cases}$$

To complete the calculation we must identify the map γ explicitly, and this requires the examination of local duality from Subsection 12.5.F. From this we infer that a map $f : PP(-a-10) \longrightarrow I_2$ given by $f(\iota) = \sum_{ij} \lambda_{ij} p_{ij}$ induces the map

$$PP(-a-10) \overset{f_*^\vee}{\longleftarrow} \mathrm{Finish}(-14)I_3$$

given by

$$z_1\lambda_{14} + z_2\lambda_{24} + z_3\lambda_{34} \longleftarrow \check{p}_{123}$$

$$z_1\lambda_{13} + z_2\lambda_{23} + z_4\lambda_{34} \longleftarrow \check{p}_{124}$$

$$z_1\lambda_{12} + z_3\lambda_{23} + z_4\lambda_{24} \longleftarrow \check{p}_{134}$$

$$z_2\lambda_{12} + z_3\lambda_{13} + z_4\lambda_{14} \longleftarrow \check{p}_{234}$$

Applying the local duality statement, we find that the dual of the map γ is the map

$$\gamma^\vee : 4\mathrm{Finish}(-7)I_3 \longrightarrow 4PP(-3)$$

given by the following 16×4 matrix

	m_1	m_2	m_3	m_4
$\check{p}_{123}Q_{123}$	z_1	z_2	z_3	
$\check{p}_{124}Q_{123}$			z_4	
$\check{p}_{134}Q_{123}$		z_4		
$\check{p}_{234}Q_{123}$	z_4			
$\check{p}_{123}Q_{124}$				z_3
$\check{p}_{124}Q_{124}$	z_1	z_2		z_4
$\check{p}_{134}Q_{124}$		z_3		
$\check{p}_{234}Q_{124}$	z_3			
$\check{p}_{123}Q_{134}$				z_2
$\check{p}_{124}Q_{134}$			z_2	
$\check{p}_{134}Q_{134}$	z_1		z_3	z_4
$\check{p}_{234}Q_{134}$	z_2			
$\check{p}_{123}Q_{234}$				z_1
$\check{p}_{124}Q_{234}$			z_1	
$\check{p}_{134}Q_{234}$		z_1		
$\check{p}_{234}Q_{234}$		z_2	z_3	z_4

Using the rows with a single entry we see the cokernel is a quotient of $\mathbb{F}_2[z_1] \oplus \mathbb{F}_2[z_2] \oplus \mathbb{F}_2[z_3] \oplus \mathbb{F}_2[z_4]$. Adding three rows with a z_1 in the m_1 column then shows that z_1 is in the image, and similarly for the other three columns. The cokernel (viz $H_I^3(TO_3^{\equiv 3(4)})^\vee$) is therefore $4\mathbb{F}_2(-3)$. This in turn gives the kernel (viz $H_I^4(TO_3^{\equiv 3(4)})^\vee$) Hilbert series

$$\frac{t^{11}4(5-4u+u^2)}{(1-u)^4},$$

where u is of degree -4. This starts

$$20, 64, 140, 256, 420, 640, \ldots$$

in degrees

$$-11, -15, -19, -23, -27, -31, \ldots.$$

Some generators in degree -11 are the 12 of the form $z_3\check{p}_{234}Q_{123} + z_4\check{p}_{234}Q_{124}$ and the 12 of the form $z_1\check{p}_{123}Q_{123} + z_1\check{p}_{124}Q_{124}$. One may verify by hand that they span a 20-dimensional space in degree -11. Using Macaulay we verify that the kernel has a resolution of the form

$$0 \longleftarrow ker(\gamma^\vee) \longleftarrow 20PP(-11) \longleftarrow 16PP(-15) \longleftarrow 4PP(-19) \longleftarrow 0,$$

which gives the simplest possible explanation of the Hilbert series.

12.5.I. TO_2 in rank 4. Finally, we return to even degrees, with the hardest part of the local cohomology calculation. The reader should be warned that this calculation is very intricate and computer assistance is essential.

For TU_2 we have an exact sequence

$$0 \longleftarrow TU_2 \longleftarrow 6PC(-6) \longleftarrow 4[PC(-8) \oplus PC(-10)] \longleftarrow \\ PC(-10) \oplus PC(-12) \oplus PC(-14) \longleftarrow 0.$$

The Sq^2 action on PC makes it a Koszul complex for PP, and the Sq^2 actions on $PC(-9) \oplus PC(-11)$ and on $4[PC(-8) \oplus PC(-10)]$ make them free modules over $PC[Sq^2]$. The Sq^2-action on $PC(-10) \oplus PC(-12) \oplus PC(-14)$ makes it a sum $PC[Sq^2](-10) \oplus PC(-14)$. For brevity, we rewrite this as

$$0 \longleftarrow TU_2 \longleftarrow F_0 \longleftarrow F_1 \longleftarrow F_2 \longleftarrow 0,$$

so that

$$F_0 = 6PC(-6) \text{ with generators } Q_{ij},$$

$$F_1 = 4PC[Sq^2](-8) \text{ with } PC \text{ generators } X_{ijk}, Y_{ijk},$$

and

$$F_2 = PC[Sq^2](-10) \oplus PC(-14) \text{ with } PC \text{ generators } X, Y, Z.$$

There is then an exact sequence

$$0 \longleftarrow TO_2 \longleftarrow ZF_0 \oplus PP(-10) \longleftarrow ZF_1 \oplus PP(-12) \oplus 4PP(-14) \longleftarrow \\ ZF_2/(I_2' \oplus I_3') \oplus 6PP(-18) \longleftarrow 0.$$

In terms of pictures we have

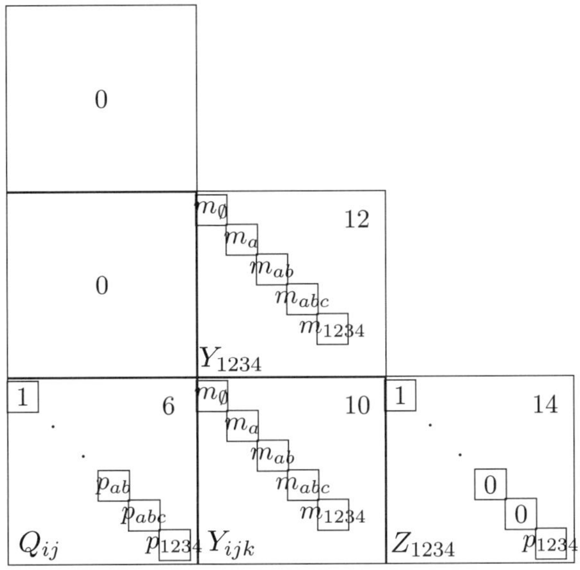

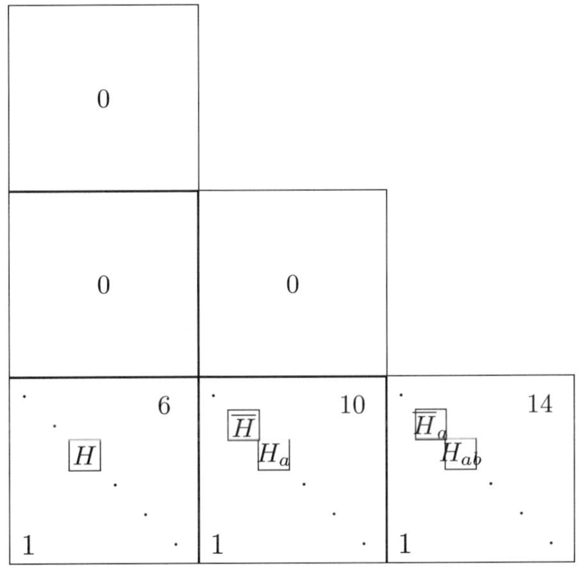

Splitting congruence classes mod 4 we see this is a sum of two resolutions of PP-modules.

12.5.J. TO_2 in degrees congruent to 0 mod 4. In this subsection we describe the calculation of $H_I^*(TO_2^{\equiv 0(4)})$. Although we give more precise module theoretic information, it is useful to record our conclusion in terms of Hilbert series.

PROPOSITION 12.5.5. *The module $TO_2^{\equiv 0(4)}$ is of depth 2, and its top local cohomology module has dual with Hilbert series*

$$[H_I^4(TO_2^{\equiv 0(4)})^\vee] = \frac{u^3(6-3u+u^2)}{(1-u)^4},$$

so that $H^4_I(TO_2^{\equiv 0(4)})$ *has dimensions*

$$6, 21, 49, 94, 160, 251, 371, \ldots$$

in degrees 12, 16, 20, 24, 28, ... *The third local cohomology group is a one dimensional vector space*

$$H^3_I(TO_2^{\equiv 0(4)}) \cong \mathbb{F}_2(4).$$

The second local cohomology group has $H^2_I(TO_2^{\equiv 0(4)})$ *has dimension* 1, 11, 14, 15, 15, 15, 15, ... *in degrees* −8, −4, 0, 4, 8, 12, ...

The rest of the subsection will describe the calculation. We start from the exact sequence

$$0 \longleftarrow TO_2^{\equiv 0(4)} \longleftarrow 6[I_2 \oplus I_4](-6) \longleftarrow 4[4PP(-12) \oplus 4PP(-16)] \oplus PP(-12) \longleftarrow$$
$$[PP(-12) \oplus 6PP(-16) \oplus PP(-20)] \oplus I_4(-14) \oplus 4PP(-16) \longleftarrow 0$$

We may tabulate the generators (where we write $m_B(R)$ for the generator $m_{B,R}$ of Section 11.3).

	Module	Generators
0	$6[I_2 \oplus I_4](-6)$	$p_{ab}Q_{ij}, p_{1234}Q_{ij}$
1	$4[4PP(-12) \oplus 4PP(-16)]$	$m_a(ijk), m_{abc}(ijk)$
1	$PP(-12)$	$\overline{H}$
2	$[PP(-12) \oplus 6PP(-16) \oplus PP(-20)]$	$Y = m_\emptyset, m_{ab}, m_{1234}$
2	$I_4(-14)$	$p_{1234}Z$
2	$4PP(-16)$	$\overline{H}_a$

The differential from degree 1 to degree 0 is given by

$$p_{12}Q_{13} + p_{13}Q_{12} \longleftarrow m_1(123)$$
$$p_{12}Q_{34} + p_{13}Q_{24} + p_{14}Q_{23} \longleftarrow m_1(234)$$
$$z_1p_{23}Q_{23} + z_2p_{13}Q_{13} + z_3p_{12}Q_{12} \longleftarrow m_{123}(123)$$
$$z_2p_{13}Q_{34} + z_3p_{12}Q_{24} + p_{1234}Q_{23} \longleftarrow m_{123}(234)$$
$$p_{12}Q_{34} + p_{13}Q_{24} + p_{14}Q_{23} + p_{23}Q_{14} + p_{24}Q_{13} + p_{34}Q_{12} \longleftarrow \overline{H}$$

The differential from degree 2 to degree 1 is given as follows. In verifying these, it is useful to note the general formula $d(m_B) = \sum_s m_{Bs}(\hat{s})$, and then decompose m_{Bs} for $s \in B$.

$$m_1(234) + m_2(134) + m_3(124) + m_4(123) \longleftarrow m_\emptyset$$
$$z_1m_2(234) + z_2m_1(134) + m_{123}(124) + m_{124}(123) \longleftarrow m_{12}$$
$$z_1m_{234}(234) + z_2m_{134}(134) + z_3m_{124}(124) + z_4m_{123}(123) \longleftarrow m_{1234}$$
$$\sum_{s,t} z_sz_tm_{\hat{s}}(\hat{t}) \longleftarrow p_{1234}Z$$
$$p_{12}Y_{134} + p_{13}Y_{124} + p_{14}Y_{123} + z_1\overline{H} \longleftarrow \overline{H}_1$$

In terms of the $m_a(R)$ basis, the image of $\overline{H}_1$ is

$$(z_1m_2(134) + z_2m_1(134)) + (z_1m_3(124) + z_3m_1(124)) +$$
$$(z_1m_4(123) + z_4m_1(123)) + z_1\overline{H}.$$

Looking at the differentials, we see that we may remove an exact complex

$$0 \longleftarrow 6I_4(-6) \longleftarrow 12PP(-16) \longleftarrow 6PP(-16) \longleftarrow 0.$$

More precisely, we first factor out the subcomplex consisting of the elements $m_{\hat{i}}(\hat{j})$ with $i < j$ and their differentials. Once this is done, we factor out the subcomplex consisting of the elements m_{ab} and their differentials. After this, we obtain

$$0 \longleftarrow TO_2^{\equiv 0(4)} \longleftarrow 6I_2(-6) \xleftarrow{\theta} G_1 \xleftarrow{\phi} G_2 \longleftarrow 0$$

where

$$G_1 = 16PP(-12) \oplus 4PP(-16) \oplus PP(-12)$$

and

$$G_2 = [PP(-12) \oplus PP(-20)] \oplus I_4(-14) \oplus 4PP(-16).$$

The remaining copies of $PC(-16)$ in homological degree 1 have a basis consisting of the four elements $m_{\hat{i}}(\hat{i})$. Thus θ will be given by a 21×36 matrix of polynomials and ϕ by the 7×21 matrix.

$\phi =$

	$m_\emptyset$	$\overline{H}_1$	$\overline{H}_2$	$\overline{H}_3$	$\overline{H}_4$	m_{1234}	$p_{1234}Z$
$m_1(\hat{1})$	1		z_2	z_3	z_4		
$m_1(\hat{2})$		z_2					$z_2z_3z_4$
$m_1(\hat{3})$		z_3					$z_2z_3z_4$
$m_1(\hat{4})$		z_4					$z_2z_3z_4$
$m_2(\hat{1})$			z_1				$z_1z_3z_4$
$m_2(\hat{2})$	1	z_1		z_3	z_4		
$m_2(\hat{3})$			z_3				$z_1z_3z_4$
$m_2(\hat{4})$			z_4				$z_1z_3z_4$
$m_3(\hat{1})$				z_1			$z_1z_2z_4$
$m_3(\hat{2})$				z_2			$z_1z_2z_4$
$m_3(\hat{3})$	1	z_1	z_2		z_4		
$m_3(\hat{4})$				z_4			$z_1z_2z_4$
$m_4(\hat{1})$					z_1		$z_1z_2z_3$
$m_4(\hat{2})$					z_2		$z_1z_2z_3$
$m_4(\hat{3})$					z_3		$z_1z_2z_3$
$m_4(\hat{4})$	1	z_1	z_2	z_3			
$\overline{H}$		z_1	z_2	z_3	z_4		
$m_{\hat{1}}(\hat{1})$						z_1	z_1^2
$m_{\hat{2}}(\hat{2})$						z_2	z_2^2
$m_{\hat{3}}(\hat{3})$						z_3	z_3^2
$m_{\hat{4}}(\hat{4})$						z_4	z_4^2

In discussing θ it is convenient to factorize it as

$$G_1 \xrightarrow{\theta'} A \xrightarrow{\theta''} 6I_2(-6),$$

where $A = \mathrm{im}(G_1 \longrightarrow 6I_2(-6))$. We then obtain an exact sequence

$$0 \longleftarrow H_I^4(A) \longleftarrow H_I^4(G_1) \xleftarrow{\phi_*} H_I^4(G_2) \longleftarrow H_I^3(A) \longleftarrow 0.$$

The $\mathbb{F}_2$-dual of ϕ_* is a map

$$(\phi_*)^\vee : 4PP(0) \oplus 17PP(-4) \longrightarrow PP(8) \oplus PP(4) \oplus 4PP(0) \oplus PP(-4)$$

given by the transpose of the above matrix for ϕ. It is not hard to see from the above formulae for differentials that the map from $4PP(0)$ (corresponding to the elements $m_{\hat{i}}(\hat{i})$) maps to z_i times the generator of $PP(4)$ (corresponding to m_{1234}) plus z_i^2 times the generator of $PP(8)$ (corresponding to $p_{1234}Z$). This map is mono in degrees 0 and -4. Accordingly, the map $(\phi_*)^\vee$ is monomorphic in degrees 0 and above. Thus the kernel is concentrated in degrees ≤ -4 and the cokernel is of dimensions 1, 5 and 14 in degrees 8, 4 and 0. In fact we may display the whole matrix, and use Macaulay to calculate that its cokernel (viz $H^3_I(A)^\vee$) has Hilbert series

$$1, 5, 14, 15, 15, 15, 15, 15, 15, 15, \ldots$$

in degrees 8, 4, 0, -4, ... Its kernel (viz $H^4_I(A)^\vee$) has Hilbert series

$$\frac{u(1 + 20u - 6u^2 - u^3)}{(1-u)^4}$$

giving dimensions

$$1, 24, 84, 195, 371, 626, 974, 1429, \ldots$$

in degrees

$$-4, -8, -12, -16, -20, \ldots$$

From the exact sequence

$$0 \longrightarrow A \xrightarrow{\theta''} 6I_2(-6) \longrightarrow TO_2^{\equiv 0(4)} \longrightarrow 0$$

we obtain the exact sequences

$$(7) \quad 0 \longleftarrow H^4_I(TO_2^{\equiv 0(4)}) \longleftarrow 6H^4_I(I_2)(-6) \xleftarrow{\theta''_*} H^4_I(A) \longleftarrow H^3_I(TO_2^{\equiv 0(4)}) \longleftarrow 0$$

and

$$0 \longleftarrow H^3_I(A) \longleftarrow H^2_I(TO_2^{\equiv 0(4)}) \longleftarrow 6\mathbb{F}_2(-4) \longleftarrow 0.$$

Since $H^4_I(I_2)(-6) = \mathrm{Start}(8)(I_3)^\vee$, $H^4_I(TO_2^{\equiv 0(4)})$ is zero below degree 8. Since the group $H^4_I(A)$ is zero below degree 4, the same is true for $H^3_I(TO_2^{\equiv 0(4)})$. Finally, from what we know about $H^3_I(A)$, we conclude that $H^2_I(TO_2^{\equiv 0(4)})$ is zero below degree -8 and of dimension 1, 11 and 14 in degrees -8, -4 and 0.

Next we identify the map θ''_* in Equation 7 above. We factorized $\theta : G_1 \longrightarrow 6I_2(-6)$ as a composite

$$G_1 \xrightarrow{\theta'} A \xrightarrow{\theta''} 6I_2(-6).$$

Since θ' is an epimorphism, and H^4_I is right exact, we find $\mathrm{im}(\theta''_*) = \mathrm{im}(\theta_*)$ in $H^4_I(6I_2(-6))$. Now, using our usual generators,

$$\theta : 17PP(-12) \oplus 4PP(-16) \longrightarrow 6I_2(-6)$$

is given by the matrix

	m_1				m_2				m_3				m_4					$m_{\hat{1}}$	$m_{\hat{2}}$	$m_{\hat{3}}$	$m_{\hat{4}}$
	$\hat{1}$	$\hat{2}$	$\hat{3}$	$\hat{4}$	$\hat{1}$	$\hat{2}$	$\hat{3}$	$\hat{4}$	$\hat{1}$	$\hat{2}$	$\hat{3}$	$\hat{4}$	$\hat{1}$	$\hat{2}$	$\hat{3}$	$\hat{4}$	$\overline{H}$	$\hat{1}$	$\hat{2}$	$\hat{3}$	$\hat{4}$
$Q_{12}p_{12}$																				z_4	z_3
$Q_{12}p_{13}$				1																	
$Q_{12}p_{14}$			1																		
$Q_{12}p_{23}$								1													
$Q_{12}p_{24}$							1														
$Q_{12}p_{34}$											1					1	1				
$Q_{13}p_{12}$				1																	
$Q_{13}p_{13}$																			z_4		z_2
$Q_{13}p_{14}$		1																			
$Q_{13}p_{23}$												1									
$Q_{13}p_{24}$						1										1	1				
$Q_{13}p_{34}$										1											
$Q_{14}p_{12}$			1																		
$Q_{14}p_{13}$		1																			
$Q_{14}p_{14}$																			z_3	z_2	
$Q_{14}p_{23}$						1					1						1				
$Q_{14}p_{24}$															1						
$Q_{14}p_{34}$														1							
$Q_{23}p_{12}$								1													
$Q_{23}p_{13}$												1									
$Q_{23}p_{14}$	1															1	1				
$Q_{23}p_{23}$																		z_4			z_1
$Q_{23}p_{24}$					1																
$Q_{23}p_{34}$									1												
$Q_{24}p_{12}$							1														
$Q_{24}p_{13}$	1										1						1				
$Q_{24}p_{14}$															1						
$Q_{24}p_{23}$					1																
$Q_{24}p_{24}$																		z_3		z_1	
$Q_{24}p_{34}$													1								
$Q_{34}p_{12}$	1					1											1				
$Q_{34}p_{13}$										1											
$Q_{34}p_{14}$														1							
$Q_{34}p_{23}$									1												
$Q_{34}p_{24}$													1								
$Q_{34}p_{34}$																		z_2	z_1		

The induced map

$$(\theta_*)^\vee : \mathrm{Finish}(-8)6I_3 \longrightarrow 17PP(-4) \oplus 4PP(0)$$

in H_I^4 follows from local duality (Lemma 12.5.2), and is given by following matrix

	m_1				m_2				m_3				m_4					$m_{\hat{1}}$	$m_{\hat{2}}$	$m_{\hat{3}}$	$m_{\hat{4}}$
	$\hat{1}$	$\hat{2}$	$\hat{3}$	$\hat{4}$	$\hat{1}$	$\hat{2}$	$\hat{3}$	$\hat{4}$	$\hat{1}$	$\hat{2}$	$\hat{3}$	$\hat{4}$	$\hat{1}$	$\hat{2}$	$\hat{3}$	$\hat{4}$	$\overline{H}$	$\hat{1}$	$\hat{2}$	$\hat{3}$	$\hat{4}$
$Q_{12}\check{p}_{123}$			z_1				z_2				z_3					z_3	z_3				
$Q_{12}\check{p}_{124}$				z_1				z_2			z_4					z_4	z_4				
$Q_{12}\check{p}_{134}$							z_4	z_3												z_1z_4	z_1z_3
$Q_{12}\check{p}_{234}$			z_4	z_3																z_2z_4	z_2z_3
$Q_{13}\check{p}_{123}$		z_1				z_2				z_3						z_2	z_2				
$Q_{13}\check{p}_{124}$										z_4		z_2							z_1z_4		z_1z_2
$Q_{13}\check{p}_{134}$				z_1		z_4						z_4				z_4	z_4				
$Q_{13}\check{p}_{234}$		z_4		z_2															z_3z_4		z_2z_3
$Q_{14}\check{p}_{123}$														z_3	z_2				z_1z_3	z_1z_2	
$Q_{14}\check{p}_{124}$		z_1				z_2					z_2			z_4			z_2				
$Q_{14}\check{p}_{134}$			z_1			z_3					z_3				z_4		z_3				
$Q_{14}\check{p}_{234}$		z_3	z_2																z_3z_4	z_2z_4	
$Q_{23}\check{p}_{123}$	z_1				z_2				z_3							z_1	z_1				
$Q_{23}\check{p}_{124}$									z_4			z_1						z_2z_4			z_1z_2
$Q_{23}\check{p}_{134}$					z_4			z_1										z_3z_4			z_1z_3
$Q_{23}\check{p}_{234}$	z_4							z_2				z_3				z_4	z_4				
$Q_{24}\check{p}_{123}$													z_3		z_1			z_2z_3		z_1z_2	
$Q_{24}\check{p}_{124}$	z_1				z_2						z_1		z_4				z_1				
$Q_{24}\check{p}_{134}$					z_3		z_1											z_3z_4		z_1z_4	
$Q_{24}\check{p}_{234}$	z_3						z_2				z_3				z_4		z_3				
$Q_{34}\check{p}_{123}$													z_2	z_1				z_2z_3	z_1z_3		
$Q_{34}\check{p}_{124}$									z_2	z_1								z_2z_4	z_1z_4		
$Q_{34}\check{p}_{134}$	z_1					z_1			z_3				z_4				z_1				
$Q_{34}\check{p}_{234}$	z_2					z_2				z_3				z_4			z_2				

The cokernel of $(\theta_*)^\vee$ is the same if we compose with the epimorphism

$$24PP(-8) \longrightarrow \text{Finish}(-8)6I_3,$$

which we can calculate with Macaulay. It has Hilbert series

$$[\text{cok}(\theta_*^\vee)] = \frac{4 + 17u - 24u^2 + 12u^3 - 3u^4 + u^5}{(1-u)^4}.$$

Now the exact sequence

$$0 \longrightarrow H_I^4(TO_2^{\equiv 0(4)})^\vee \longrightarrow \text{Finish}(-8)6I_3 \longrightarrow$$
$$17PP(-4) \oplus 4PP(0) \longrightarrow \text{cok}(\theta_*^\vee) \longrightarrow 0$$

gives

$$[H_I^4(TO_2^{\equiv 0(4)})^\vee] = \frac{u^3(6 - 3u + u^2)}{(1-u)^4},$$

and then the exact sequence (7) gives the striking conclusion

$$[H_I^3(TO_2^{\equiv 0(4)})^\vee] = \frac{u(1 - 4u + 6u^2 - 4u^3 + u^4)}{(1-u)^4} = u.$$

Indeed, one might hope for a more direct proof!

12.5.K. Local cohomology of TO_{24} in degrees 0 mod 4. In this subsection we see how the local cohomology of $TO_2^{\equiv 0(4)}$ is modified to obtain that of $TO_{24}^{\equiv 0(4)}$. Although we give more precise module theoretic information, it is useful to record our conclusion in terms of Hilbert series.

PROPOSITION 12.5.6. *The module $TO_{24}^{\equiv 0(4)}$ is of depth 2, and its top local cohomology module has dual with Hilbert series*

$$[H_I^4(TO_{24}^{\equiv 0(4)})^\vee] = \frac{u(6-3u+7u^2-3u^3+u^4)}{(1-u)^4},$$

so that $H_I^4(TO_{24}^{\equiv 0(4)})$ has dimensions

$$6, 21, 55, 115, 209, 345, 531, 775, \ldots$$

in degrees 4, 8, 12, 16, 20, 24, 28, ... *The third local cohomology group is a two dimensional vector space*

$$H_I^3(TO_{24}^{\equiv 0(4)}) \cong \mathbb{F}_2(4) \oplus \mathbb{F}_2(-4)$$

The second local cohomology group has $H_I^2(TO_{24}^{\equiv 0(4)})$ has dimension 11, 14, 15, 15, 15, 15, ... *in degrees* −4, 0, 4, 8, 12, 16, ...

The exact sequence

$$0 \longleftarrow (I_1 \oplus I_3)(-8) \longleftarrow TO_{24}^{\equiv 0(4)} \longleftarrow TO_2^{\equiv 0(4)} \longleftarrow 0$$

gives exact sequences

$$0 \longleftarrow \mathrm{Start}(8)I_4^\vee \oplus \mathrm{Start}(4)I_2^\vee \longleftarrow H_I^4(TO_{24}^{\equiv 0(4)}) \longleftarrow H_I^4(TO_2^{\equiv 0(4)}) \longleftarrow$$
$$\mathbb{F}_2(-4) \longleftarrow H_I^3(TO_{24}^{\equiv 0(4)}) \longleftarrow H_I^3(TO_2^{\equiv 0(4)}) \longleftarrow 0$$

and

$$0 \longleftarrow H_I^2(TO_{24}^{\equiv 0(4)}) \longleftarrow H_I^2(TO_2^{\equiv 0(4)}) \longleftarrow \mathbb{F}_2(-8) \longleftarrow H_I^1(TO_{24}^{\equiv 0(4)}) \longleftarrow 0$$

The first of these splits into two short exact sequences since $H_I^4(TO_2^{\equiv 0(4)})$ is zero below degree 8. Thus we find

$$H_I^4(TO_{24}^{\equiv 0(4)}) \cong H_I^4(TO_2^{\equiv 0(4)}) \oplus \mathrm{Start}(8)I_4^\vee \oplus \mathrm{Start}(4)I_2^\vee$$

and

$$H_I^3(TO_{24}^{\equiv 0(4)}) \cong H_I^3(TO_2^{\equiv 0(4)}) \oplus \mathbb{F}_2(-4).$$

In the second exact sequence, we do not need to calculate the map from $\mathbb{F}_2(-8)$. If zero it would contribute two copies of $\mathbb{F}_2$ to the local cohomology spectral sequence in $(H_I^2)_{-8}$ and $(H_I^1)_{-8}$. Connectivity of $ko_*(BV(4))$ would require that they must be cancelled, and there is no group available to do so. It follows that the map is an isomorphism, and the Proposition 12.5.6 now follows from Proposition 12.5.5.

12.5.L. TO_2 in degrees congruent to 2 mod 4. In this subsection we describe the calculation of $H_I^*(TO_2^{\equiv 2(4)})$. Although we give more precise module theoretic information, it is useful to record our conclusion in terms of Hilbert series.

PROPOSITION 12.5.7. *The module $TO_2^{\equiv 2(4)}$ is of depth 2, and its top local cohomology module has dual with Hilbert series*

$$[H_I^4(TO_2^{\equiv 2(4)})^\vee] = \frac{t^6(1+4u^2-u^3)}{(1-u)^4},$$

so that $H_I^4(TO_2^{\equiv 2(4)})$ has dimensions

$$1, 4, 14, 35, 71, 126, 200, 309, \ldots$$

in degrees 6, 10, 14, 18, 22, ... *In fact*

$$H_I^4(TO_2^{\equiv 2(4)})^\vee = PP(-6) \oplus \text{Finish}(-14)I_3.$$

The second and third local cohomology group are finite dimensional; in fact

$$H_I^3(TO_2^{\equiv 2(4)}) = 0$$

and

$$H_I^2(TO_2^{\equiv 2(4)}) = \mathbb{F}_2(2).$$

The rest of the subsection will describe the calculation. We start from the exact sequence

$$\begin{aligned} 0 \longleftarrow TO_2^{\equiv 2(4)} \longleftarrow 6[PP \oplus I_3](-6) \oplus PP(-10) \longleftarrow \\ 4[4PP(-10) \oplus 6PP(-14) \oplus PP(-18)] \oplus 4PP(-14) \longleftarrow \\ [4PP(-14) \oplus 4PP(-18)] \oplus PP(-14) \oplus 6PP(-18) \longleftarrow 0 \end{aligned}$$

We may tabulate the generators (where we write $m_B(R)$ for the generator $m_{B,R}$ of Section 11.3).

	Module	Generators
0	$6[PP \oplus I_3](-6)$	$Q_{ij}, p_{abc}Q_{ij}$
0	$PP(-10)$	H
1	$4[PP(-10) \oplus 6PP(-14) \oplus PP(-18)]$	$Y(ijk) = m_\emptyset(ijk), m_{ab}(ijk), m_{1234}(ijk)$
1	$4PP(-14)$	H_a
2	$[4PP(-14) \oplus 4PP(-18)]$	m_a, m_{abc}
2	$PP(-14)$	Z
2	$6PP(-18)$	H_{ab}

The differential from degree 1 to degree 0 is given by

$$z_1Q_{23} + z_2Q_{13} + z_3Q_{12} \longleftarrow m_\emptyset(123)$$
$$z_1z_2(Q_{23} + Q_{13}) + p_{123}Q_{12} \longleftarrow m_{12}(123)$$
$$z_1z_2Q_{34} + p_{123}Q_{24} + p_{124}Q_{23} \longleftarrow m_{12}(234)$$
$$z_1p_{234}Q_{23} + z_2p_{134}Q_{13} + z_3p_{124}Q_{12} \longleftarrow m_{1234}(123)$$
$$z_1H + z_1z_2Q_{34} + z_1z_3Q_{24} + z_1z_4Q_{23} + p_{123}Q_{14} + p_{124}Q_{13} + p_{134}Q_{12} \longleftarrow H_1$$

The differential from degree 2 to degree 1 is as follows. In verifying these, it is useful to note the general formula $d(m_B) = \sum_s m_{Bs}(\hat{s})$, and then decompose m_{Bs} for $s \in B$.

$$z_1m_\emptyset(234) + m_{12}(134) + m_{13}(124) + m_{14}(123) \longleftarrow m_1$$
$$z_1m_{23}(234) + z_2m_{13}(134) + z_3m_{12}(124) + m_{1234}(123) \longleftarrow m_{123}$$
$$z_1Y(234) + z_2Y(134) + z_3Y(124) + z_4Y(123) \longleftarrow Z$$
$$z_1z_2Y(234) + z_1z_2Y(134) + p_{123}Y(124) + p_{124}Y(123) + z_1H_2 + z_2H_1 \longleftarrow H_{12}$$

The image of H_{12} in terms of the $m_{ab}(ijk)$-basis is

$$\begin{aligned} z_1z_2Y(234) + z_1z_2Y(134) + z_1m_{23}(124) + z_2m_{13}(124) + z_3m_{12}(124) + \\ z_1m_{24}(123) + z_2m_{14}(123) + z_4m_{12}(123) + p_{123}Y(124) + p_{124}Y(123) + z_1H_2 + z_2H_1. \end{aligned}$$

The differential on the m_{ijk} allows us to delete $4PP(-18)$ in homological degrees 1 and 2, leaving

$$0 \longleftarrow TO_2^{\equiv 2(4)} \longleftarrow 6I_3(-6) \oplus \overline{G}_0 \xleftarrow{\overline{\theta}} \overline{G}_1 \xleftarrow{\overline{\phi}} \overline{G}_2 \longleftarrow 0$$

where

$$\overline{G}_0 = 6PP(-6) \oplus PP(-10)$$

$$\overline{G}_1 = 4[PP(-10) \oplus 6PP(-14)] \oplus 4PP(-14)$$

and

$$\overline{G}_2 = [4PP(-14)] \oplus PP(-14) \oplus 6PP(-18).$$

Thus $\overline{\theta}$ will be given by a 31×32 matrix of polynomials and $\overline{\phi}$ by the 32×11 matrix

$\overline{\phi} =$

	m_1	m_2	m_3	m_4	Z	H_{12}	H_{13}	H_{14}	H_{23}	H_{24}	H_{34}
$Y(\hat{1})$	z_1				z_1	z_1z_2	z_1z_3	z_1z_4			
$Y(\hat{2})$		z_2			z_2	z_1z_2			z_2z_3	z_2z_4	
$Y(\hat{3})$			z_3		z_3		z_1z_3		z_2z_3		z_3z_4
$Y(\hat{4})$				z_4	z_4			z_1z_4		z_2z_4	z_3z_4
$m_{12}(\hat{1})$		1							z_3	z_4	
$m_{13}(\hat{1})$			1						z_2		z_4
$m_{14}(\hat{1})$				1						z_2	z_3
$m_{23}(\hat{1})$									z_1		
$m_{24}(\hat{1})$										z_1	
$m_{34}(\hat{1})$											z_1
$m_{12}(\hat{2})$	1						z_3	z_4			
$m_{13}(\hat{2})$							z_2				
$m_{14}(\hat{2})$								z_2			
$m_{23}(\hat{2})$			1				z_1				z_4
$m_{24}(\hat{2})$				1				z_1			z_3
$m_{34}(\hat{2})$											z_2
$m_{12}(\hat{3})$						z_3					
$m_{13}(\hat{3})$	1					z_2		z_4			
$m_{14}(\hat{3})$								z_3			
$m_{23}(\hat{3})$		1				z_1				z_4	
$m_{24}(\hat{3})$										z_3	
$m_{34}(\hat{3})$				1				z_1		z_2	
$m_{12}(\hat{4})$						z_4					
$m_{13}(\hat{4})$							z_4				
$m_{14}(\hat{4})$	1					z_2	z_3				
$m_{23}(\hat{4})$									z_4		
$m_{24}(\hat{4})$		1				z_1			z_3		
$m_{34}(\hat{4})$			1				z_1		z_2		
H_1						z_2	z_3	z_4			
H_2						z_1			z_3	z_4	
H_3							z_1		z_2		z_4
H_4								z_1		z_2	z_3

In discussing $\overline{\theta}$ it is convenient to factorize it as

$$\overline{G}_1 \xrightarrow{\overline{\theta}'} \overline{A} \xrightarrow{\overline{\theta}''} 6I_3(-6) \oplus \overline{G}_0,$$

where $\overline{A} = \mathrm{im}(\overline{G}_1 \longrightarrow 6I_3(-6) \oplus \overline{G}_0)$. We then obtain two long exact local cohomology sequences. The first reads

$$0 \longleftarrow H_I^4(\overline{A}) \longleftarrow H_I^4(\overline{G}_1) \xleftarrow{\overline{\phi}_*} H_I^4(\overline{G}_2) \longleftarrow H_I^3(\overline{A}) \longleftarrow 0.$$

The $\mathbb{F}_2$-dual of $\overline{\phi}_*$ is a map

$$(\overline{\phi}_*)^\vee : 4PP(-6) \oplus 28PP(-2) \longrightarrow 5PP(-2) \oplus 6PP(2)$$

given by the transpose of the above matrix for $\overline{\phi}$. The only obvious thing is that the kernel is zero above degree -2 and the cokernel is of dimension 6 in degree 2 and non-zero in degree -2. It may be verified that in degree -2 it of maximal rank (viz 28), so that the kernel is 0 in that degree and the cokernel is one dimensional. With the aid of Macaulay one may check that the kernel (viz $H_I^4(\overline{A})^\vee$) has Hilbert series

$$\frac{t^6(36 - 18u + 2u^2 + u^3)}{(1-u)^4}$$

so that it has dimensions

$$36, 126, 290, 549, 924, 1436, 2106, \ldots$$

in degrees

$$-6, -10, -14, -18, \ldots$$

It has a resolution making it appear like 7 copies of I_2, 6 of them finishing in degree -6 and 1 of them finishing in degree -10.

The cokernel (viz $H_I^3(\overline{A})^\vee$) is of dimension 6 in degree 2, dimension 1 in degree -2 and zero elsewhere.

The second long exact local cohomology sequence reads

$$(8) \quad 0 \longleftarrow H_I^4(TO_2^{\equiv 2(4)}) \longleftarrow 6H_I^4(I_3)(-6) \oplus H_I^4(\overline{G}_0) \xleftarrow{\overline{\theta}''_*} H_I^4(\overline{A}) \longleftarrow$$
$$H_I^3(TO_2^{\equiv 2(4)}) \longleftarrow 6\mathbb{F}_2(-2) \xleftarrow{\overline{\theta}''_*} H_I^3(\overline{A}) \longleftarrow$$
$$H_I^2(TO_2^{\equiv 2(4)}) \longleftarrow 0$$

We do not need to calculate the map $6\mathbb{F}_2(-2) \xleftarrow{\overline{\theta}''_*} H_I^3(\overline{A})$ because any deviation from being an isomorphism in degree -2 will give two isomorphic groups local groups. In the local cohomology spectral sequence they would appear in $(H_I^3)_{-2}$ and $(H_I^2)_{-2}$. By connectivity they would have to be cancelled, and the rest of the calculation will show there are no other groups available to do this. It follows that the map is an isomorphism in degree -2. Hence we find $H_I^2(TO_2^{\equiv 2(4)})$ is zero except in degree 2, where it is of dimension 1 and $H_I^3(TO_2^{\equiv 2(4)})$ is zero below degree 6. The module $H_I^4(TO_2^{\equiv 2(4)})$ is also zero below degree 6 and non-zero in degree 6.

Next we turn to the map $\overline{\theta}''_*$ on H_I^4. It is convenient to factorize $\overline{\theta} : \overline{G}_1 \longrightarrow 6I_3(-6) \oplus \overline{G}_0$ as a composite

$$\overline{G}_1 \xrightarrow{\overline{\theta}'} \overline{A} \xrightarrow{\overline{\theta}''} 6I_3(-6) \oplus \overline{G}_0.$$

Since $\overline{\theta}'$ is an epimorphism, and H_I^4 is right exact, we find $\mathrm{im}(\overline{\theta}''_*) = \mathrm{im}(\overline{\theta}_*)$ in $H_I^4(6I_3(-6) \oplus \overline{G}_0)$. Now, using our usual generators,

$$\overline{\theta} : 4PP(-10) \oplus 28PP(-14) \longrightarrow 6I_3(-6) \oplus 6PP(-6) \oplus PP(-10)$$

is given by the following matrix (given in two halves because of its size)

	Y				$m_{ab}(\hat{1})$						$m_{ab}(\hat{2})$					
	$\hat{1}$	$\hat{2}$	$\hat{3}$	$\hat{4}$	12	13	14	23	24	34	12	13	14	23	24	34
Q_{12}			z_4	z_3												
Q_{13}		z_4		z_2									z_1z_4		z_2z_4	z_3z_4
Q_{14}		z_3	z_2									z_1z_3		z_2z_3		z_3z_4
Q_{23}	z_4			z_1			z_1z_4		z_2z_4	z_3z_4						
Q_{24}	z_3		z_1			z_1z_3		z_2z_3		z_3z_4						
Q_{34}	z_2	z_1			z_1z_2			z_2z_3	z_2z_4		z_1z_2	z_1z_3	z_1z_4			
H																
$Q_{12}p_{\hat{1}}$																
$Q_{12}p_{\hat{2}}$																
$Q_{12}p_{\hat{3}}$																
$Q_{12}p_{\hat{4}}$																
$Q_{13}p_{\hat{1}}$														1		
$Q_{13}p_{\hat{2}}$												1				
$Q_{13}p_{\hat{3}}$											1					
$Q_{13}p_{\hat{4}}$																
$Q_{14}p_{\hat{1}}$															1	
$Q_{14}p_{\hat{2}}$													1			
$Q_{14}p_{\hat{3}}$																
$Q_{14}p_{\hat{4}}$											1					
$Q_{23}p_{\hat{1}}$								1								
$Q_{23}p_{\hat{2}}$						1										
$Q_{23}p_{\hat{3}}$					1											
$Q_{23}p_{\hat{4}}$																
$Q_{24}p_{\hat{1}}$									1							
$Q_{24}p_{\hat{2}}$							1									
$Q_{24}p_{\hat{3}}$																
$Q_{24}p_{\hat{4}}$					1											
$Q_{34}p_{\hat{1}}$										1						
$Q_{34}p_{\hat{2}}$																1
$Q_{34}p_{\hat{3}}$							1								1	
$Q_{34}p_{\hat{4}}$						1								1		

	$m_{ab}(\hat{3})$						$m_{ab}(\hat{4})$						H			
	12	13	14	23	24	34	12	13	14	23	24	34	1	2	3	4
Q_{12}			z_1z_4		z_2z_4	z_3z_4		z_1z_3		z_2z_3		z_3z_4			z_3z_4	z_3z_4
Q_{13}							z_1z_2			z_2z_3	z_2z_4			z_2z_4		z_2z_4
Q_{14}	z_1z_2			z_2z_3	z_2z_4									z_2z_3	z_2z_3	
Q_{23}							z_1z_2	z_1z_3	z_1z_4				z_1z_4			z_1z_4
Q_{24}	z_1z_2	z_1z_3	z_1z_4										z_1z_3		z_1z_3	
Q_{34}													z_1z_2	z_1z_2		
H													z_1	z_2	z_3	z_4
$Q_{12}p_{\hat{1}}$				1							1			1		
$Q_{12}p_{\hat{2}}$		1							1				1			
$Q_{12}p_{\hat{3}}$	1															
$Q_{12}p_{\hat{4}}$							1									
$Q_{13}p_{\hat{1}}$												1			1	
$Q_{13}p_{\hat{2}}$																
$Q_{13}p_{\hat{3}}$									1				1			
$Q_{13}p_{\hat{4}}$								1								
$Q_{14}p_{\hat{1}}$						1										1
$Q_{14}p_{\hat{2}}$																
$Q_{14}p_{\hat{3}}$			1													
$Q_{14}p_{\hat{4}}$		1											1			
$Q_{23}p_{\hat{1}}$																
$Q_{23}p_{\hat{2}}$												1			1	
$Q_{23}p_{\hat{3}}$											1			1		
$Q_{23}p_{\hat{4}}$										1						
$Q_{24}p_{\hat{1}}$																
$Q_{24}p_{\hat{2}}$						1										1
$Q_{24}p_{\hat{3}}$					1											
$Q_{24}p_{\hat{4}}$				1										1		
$Q_{34}p_{\hat{1}}$																
$Q_{34}p_{\hat{2}}$																
$Q_{34}p_{\hat{3}}$																1
$Q_{34}p_{\hat{4}}$															1	

The induced map

$$(\overline{\theta}_*)^\vee : \text{Finish}(-6)6I_2 \oplus 6PP(-10) \oplus PP(-6) \longrightarrow 4PP(-6) \oplus 28PP(-2)$$

in top local cohomology follows from Local Duality (Lemma 12.5.2), and is given by the following matrix (displayed in two halves because of its size)

	Y				$m_{ab}(\hat{1})$						$m_{ab}(\hat{2})$					
	$\hat{1}$	$\hat{2}$	$\hat{3}$	$\hat{4}$	12	13	14	23	24	34	12	13	14	23	24	34
Q_{12}			z_4	z_3												
Q_{13}		z_4		z_2									z_1z_4		z_2z_4	z_3z_4
Q_{14}		z_3	z_2									z_1z_3		z_2z_3		z_3z_4
Q_{23}	z_4			z_1			z_1z_4		z_2z_4	z_3z_4						
Q_{24}	z_3		z_1			z_1z_3		z_2z_3		z_3z_4						
Q_{34}	z_2	z_1			z_1z_2			z_2z_3	z_2z_4		z_1z_2	z_1z_3	z_1z_4			
H																
$Q_{12}\check{p}_{12}$																
$Q_{12}\check{p}_{13}$																
$Q_{12}\check{p}_{14}$																
$Q_{12}\check{p}_{23}$																
$Q_{12}\check{p}_{24}$																
$Q_{12}\check{p}_{34}$																
$Q_{13}\check{p}_{12}$												z_1		z_2		
$Q_{13}\check{p}_{13}$											z_1			z_3		
$Q_{13}\check{p}_{14}$														z_4		
$Q_{13}\check{p}_{23}$											z_2	z_3				
$Q_{13}\check{p}_{24}$												z_4				
$Q_{13}\check{p}_{34}$											z_4					
$Q_{14}\check{p}_{12}$													z_1		z_2	
$Q_{14}\check{p}_{13}$															z_3	
$Q_{14}\check{p}_{14}$											z_1				z_4	
$Q_{14}\check{p}_{23}$													z_3			
$Q_{14}\check{p}_{24}$											z_2		z_4			
$Q_{14}\check{p}_{34}$											z_3					
$Q_{23}\check{p}_{12}$						z_1		z_2								
$Q_{23}\check{p}_{13}$					z_1			z_3								
$Q_{23}\check{p}_{14}$								z_4								
$Q_{23}\check{p}_{23}$					z_2	z_3										
$Q_{23}\check{p}_{24}$						z_4										
$Q_{23}\check{p}_{34}$					z_4											
$Q_{24}\check{p}_{12}$							z_1		z_2							
$Q_{24}\check{p}_{13}$									z_3							
$Q_{24}\check{p}_{14}$					z_1				z_4							
$Q_{24}\check{p}_{23}$							z_3									
$Q_{24}\check{p}_{24}$					z_2		z_4									
$Q_{24}\check{p}_{34}$					z_3											
$Q_{34}\check{p}_{12}$										z_2						z_1
$Q_{34}\check{p}_{13}$							z_1			z_3					z_1	
$Q_{34}\check{p}_{14}$						z_1				z_4				z_1		
$Q_{34}\check{p}_{23}$							z_2								z_2	z_3
$Q_{34}\check{p}_{24}$						z_2								z_2		z_4
$Q_{34}\check{p}_{34}$						z_3	z_4							z_3	z_4	

	$m_{ab}(\hat{3})$						$m_{ab}(\hat{4})$						H			
	12	13	14	23	24	34	12	13	14	23	24	34	1	2	3	4
Q_{12}			z_1z_4		z_2z_4	z_3z_4		z_1z_3		z_2z_3		z_3z_4			z_3z_4	z_3z_4
Q_{13}							z_1z_2			z_2z_3	z_2z_4			z_2z_4		z_2z_4
Q_{14}	z_1z_2			z_2z_3	z_2z_4									z_2z_3	z_2z_3	
Q_{23}							z_1z_2	z_1z_3	z_1z_4				z_1z_4			z_1z_4
Q_{24}	z_1z_2	z_1z_3	z_1z_4										z_1z_3		z_1z_3	
Q_{34}													z_1z_2	z_1z_2		
H													z_1	z_2	z_3	z_4
$Q_{12}\check{p}_{12}$		z_1		z_2					z_1		z_2		z_1	z_2		
$Q_{12}\check{p}_{13}$	z_1			z_3							z_3			z_3		
$Q_{12}\check{p}_{14}$				z_4			z_1				z_4			z_4		
$Q_{12}\check{p}_{23}$	z_2	z_3							z_3				z_3			
$Q_{12}\check{p}_{24}$		z_4					z_2		z_4				z_4			
$Q_{12}\check{p}_{34}$	z_4						z_3									
$Q_{13}\check{p}_{12}$												z_2			z_2	
$Q_{13}\check{p}_{13}$									z_1			z_3	z_1		z_3	
$Q_{13}\check{p}_{14}$								z_1				z_4			z_4	
$Q_{13}\check{p}_{23}$									z_2				z_2			
$Q_{13}\check{p}_{24}$								z_2								
$Q_{13}\check{p}_{34}$								z_3	z_4				z_4			
$Q_{14}\check{p}_{12}$						z_2										z_2
$Q_{14}\check{p}_{13}$			z_1			z_3										z_3
$Q_{14}\check{p}_{14}$		z_1				z_4							z_1			z_4
$Q_{14}\check{p}_{23}$			z_2													
$Q_{14}\check{p}_{24}$		z_2											z_2			
$Q_{14}\check{p}_{34}$		z_3	z_4										z_3			
$Q_{23}\check{p}_{12}$												z_1			z_1	
$Q_{23}\check{p}_{13}$											z_1			z_1		
$Q_{23}\check{p}_{14}$										z_1						
$Q_{23}\check{p}_{23}$											z_2	z_3		z_2	z_3	
$Q_{23}\check{p}_{24}$										z_2		z_4			z_4	
$Q_{23}\check{p}_{34}$										z_3	z_4			z_4		
$Q_{24}\check{p}_{12}$						z_1										z_1
$Q_{24}\check{p}_{13}$					z_1											
$Q_{24}\check{p}_{14}$				z_1										z_1		
$Q_{24}\check{p}_{23}$					z_2	z_3										z_3
$Q_{24}\check{p}_{24}$				z_2		z_4								z_2		z_4
$Q_{24}\check{p}_{34}$				z_3	z_4									z_3		
$Q_{34}\check{p}_{12}$																
$Q_{34}\check{p}_{13}$																z_1
$Q_{34}\check{p}_{14}$															z_1	
$Q_{34}\check{p}_{23}$																z_2
$Q_{34}\check{p}_{24}$															z_2	
$Q_{34}\check{p}_{34}$															z_3	z_4

The cokernel of $(\overline{\theta}_*)^\vee$ is the same if we compose with the epimorphism

$$36PP(-6) \longrightarrow \text{Finish}(-6)6I_3,$$

which we can calculate with Macaulay. It has Hilbert series

$$[\text{cok}(\overline{\theta}_*^\vee)] = \frac{t^2(28 - 32u + 18u^2 - 2u^3 - u^4)}{(1-u)^4}.$$

Now the exact sequence

$$0 \longrightarrow H_I^4(TO_2^{\equiv 2(4)})^\vee \longrightarrow \mathrm{Finish}(-6)6I_2 \oplus 6PP(-10) \oplus PP(-6)$$
$$\longrightarrow 4PP(-6) \oplus 28PP(-2) \longrightarrow \mathrm{cok}(\overline{\theta}_*^\vee) \longrightarrow 0$$

gives

$$[H_I^4(TO_2^{\equiv 2(4)})^\vee] = \frac{t^6(1+4u^2-u^3)}{(1-u)^4}.$$

This suggests an obvious conjecture for the structure of the local cohomology module, which we now verify.

LEMMA 12.5.8.

$$H_I^4(TO_2^{\equiv 2(4)})^\vee = PP(-6) \oplus \mathrm{Finish}(-14)I_3.$$

Proof: It is easy to see that there is a summand $PP(-6)$. Indeed, we inspect the matrix to see that the term $PP(-6)$, which is the row corresponding to H, can be obtained as the sum of the rows $p_{ij}Q_{ij}$.

For the term I_3 we need to find 4 generators in degree -14 satisfying the required relation. In fact the element

$$\begin{aligned} g_1 = \; & z_2Q_{12} + z_3Q_{13} + z_4Q_{14} + \\ & z_2(z_4\check{p}_{13} + z_3\check{p}_{14})Q_{12} + z_3(z_4\check{p}_{12} + z_2\check{p}_{14})Q_{13} + z_4(z_3\check{p}_{12} + z_2\check{p}_{13})Q_{14} + \\ & z_2z_3(\check{p}_{24} + \check{p}_{34})Q_{23} + z_2z_4(\check{p}_{23} + \check{p}_{34})Q_{24} + z_3z_4(\check{p}_{23} + \check{p}_{24}Q_{34}) \end{aligned}$$

and the corresponding elements g_2, g_3, g_4 satisfy the relation $z_1g_1 + z_2g_2 + z_3g_3 + z_4g_4 = 0$. □

Now we observe

$$[H_I^4(TO_2^{\equiv 2(4)})^\vee + [H_I^4(\overline{A})^\vee] = [H_I^4(6I_3(-6) \oplus \overline{G}_0)^\vee],$$

so the exact sequence (8) gives the startling conclusion that the kernel of $\overline{\theta}_*''$ in H_I^4 is zero. Since $\overline{\theta}_*''$ is epimorphic in H_I^3, we find

$$H_I^3(TO_2^{\equiv 2(4)}) = 0$$

Once again, one might hope for a more direct proof!

12.5.M. Local cohomology of TO_{24} in degrees 2 mod 4. In this subsection we show how the local cohomology of $TO_2^{\equiv 2(4)}$ is modified to obtain that of $TO_{24}^{\equiv 2(4)}$. Although we give more precise module theoretic information, it is useful to record our conclusion in terms of Hilbert series.

PROPOSITION 12.5.9. *The module $TO_{24}^{\equiv 2(4)}$ is of depth 2 and the structure of the local cohomology modules is as follows*

$$H_I^4(TO_2^{\equiv 2(4)})^\vee = PP^\vee(6) \oplus \mathrm{Start}(14)I_3^\vee \oplus \mathrm{Start}(6)I_3^\vee \oplus PP^\vee(-2).$$

$$H_I^3(TO_{24}^{\equiv 2(4)}) = 0$$

and

$$H_I^2(TO_{24}^{\equiv 2(4)}) \cong \mathbb{F}_2(-6) \oplus \mathbb{F}_2(2).$$

The top local cohomology module has dual with Hilbert series

$$[H^4_I(TO_{24}^{\equiv 2(4)})^\vee] = \frac{t^{-2}(1+5u^2-u^3+4u^4-u^5)}{(1-u)^4},$$

so that $H^4_I(TO_{24}^{\equiv 2(4)})$ *has dimensions*

$$1, 4, 15, 39, 85, 161, 275, 435, \ldots$$

in degrees -2, 2, 6, 10, 14, 18, 22, ...

The exact sequence

$$0 \longleftarrow (I_2 \oplus I_4)(-8) \longleftarrow TO_{24}^{\equiv 2(4)} \longleftarrow TO_2^{\equiv 2(4)} \longleftarrow 0$$

gives exact sequences

$$0 \longleftarrow \mathrm{Start}(6)I_3^\vee \oplus PP^\vee(-2) \longleftarrow H^4_I(TO_{24}^{\equiv 2(4)}) \longleftarrow H^4_I(TO_2^{\equiv 2(4)}) \longleftarrow 0$$

and

$$0 \longleftarrow H^3_I(TO_{24}^{\equiv 2(4)}) \longleftarrow H^3_I(TO_2^{\equiv 2(4)}) \longleftarrow \mathbb{F}_2(-6)$$
$$\longleftarrow H^2_I(TO_{24}^{\equiv 2(4)}) \longleftarrow H^2_I(TO_2^{\equiv 2(4)}) \longleftarrow 0$$

The map from $\mathbb{F}_2(-6)$ must be zero since $H^3_I(TO_2^{\equiv 2(4)})$ is zero below degree -2. Thus we find

$$H^4_I(TO_{24}^{\equiv 2(4)}) \cong H^4_I(TO_2^{\equiv 2(4)}) \oplus \mathrm{Start}(6)I_3^\vee \oplus PP^\vee(-2),$$
$$H^3_I(TO_{24}^{\equiv 2(4)}) \cong H^3_I(TO_2^{\equiv 2(4)})$$

and

$$H^2_I(TO_{24}^{\equiv 2(4)}) \cong H^2_I(TO_2^{\equiv 2(4)}) \oplus \mathbb{F}_2(-6).$$

Combining this with Proposition 12.5.7 gives the stated results.

12.5.N. The local cohomology spectral sequence. We have now calculated the $E^{1\frac{1}{2}}$-page of the local cohomology spectral sequence for $ko_*(BV(4))$. This is nearly the E^2-page, but the differential $d^1 : H^1_{JO}(\overline{QO}) \longrightarrow H^2_{JO}(TO)$ has not yet been taken into account. We display the spectral sequence in low degrees (≤ 20) in Figure 12.5.6. The entries are filled in by combining Propositions 12.5.3, 12.5.4, 12.5.6 and 12.5.9.

Once again we find the striking way in which algebraically unrelated exceptional terms appear in appropriate degrees to cancel as required by the connectivity of $ko_*(BV(4))$. In the diagram we have marked certain differentials that we discuss below. There are precisely three non-trivial d^4 differentials; for typographical reasons they are indicated with an asterisk at their starting and finishing points.

First note that each column is bounded below. The H^0 column has lowest entry in degree -7, the H^1 column in degree -4, the H^2 column in degree -6, the H^3 column in degree -4 and the H^4 column in degree -2. This remains true even on the E^2-page, since the $d^1 : H^1_{JO}(\overline{QO})_{-4} \longrightarrow H^2_{JO}(TO)_{-4}$ cannot be injective, and the codomain is not the lowest group in the H^2 column. Accordingly, the regularity of $ko^*_{V(3)}$ is -8.

Now consider differentials. The principal fact is that the spectral sequence converges to $ko_*(BV(4))$, which is bounded below. Since we need one more piece of information than in lower ranks, we record the input in a lemma.

FIGURE 12.5.6. The $E^{1\frac{1}{2}}$-page of the local cohomology spectral sequence for $ko_*(BV(4))$.

H^4_{JO}	H^3_{JO}	H^2_{JO}	H^1_{JO}	H^0_{JO}	
2^{209}	0	2^{15}	$[2^{14}]^4 \oplus [2^{13}]^6 \oplus [2^{12}]^4 \oplus [2^{11}]$	$\mathbb{Z}$	20
2^{140}	0	0	0	0	
2^{161}	0	0	0	2^{16}	
2^{105}	2^{15}	0	0	2^{16}	
2^{115}	0	2^{15}	$[2^{11}]^4 \oplus [2^{10}]^6 \oplus [2^9]^4 \oplus [2^8]$	$\mathbb{Z}$	16
2^{64}	0	0	0	0	
2^{85}	0	0	0	0	
2^{46}	2^{15}	0	0	0	
$2^{55}*$	0	2^{15}	$[2^{10}]^4 \oplus [2^9]^6 \oplus [2^8]^4 \oplus [2^7]$	$\mathbb{Z}$	12
2^{20}	0	0	0	0	
2^{39}	0	0	0	2^{16}	
2^{15}	2^{15}	0	0	$2^{16}*$	
2^{21}	0	2^{15}	$[2^7]^4 \oplus [2^6]^6 \oplus [2^5]^4 \oplus [2^4]$	$\mathbb{Z}$	8
0	0	0	0	0	
2^{15}	0	0	0	0	
2^4*	2^{15}	0	0	0	
2^6*	2	2^{15}	$[2^6]^4 \oplus [2^5]^6 \oplus [2^4]^4 \oplus [2^3]$	$\mathbb{Z}$	4
0	2^4	0	0	0	
2^4	0	2	0	$2^{16}*$	
0	2^{10}	0	0	$2^{16}*$	
0	0	2^{14}	$[2^3]^4 \oplus [2^2]^6 \oplus [2]^4 \oplus [2^0]$	$\mathbb{Z}$	0
0	0	0	0	0	
2	0	0	0	0	
	2^5	0	0	0	
	2	2^{11}	$\langle 2^{17} \rangle$	0	−4
		2^4		0	
		2		2^5	
				2	
					−8

LEMMA 12.5.10. *(i)* $ko_i(BV(4)) = 0$ *for* $i < 0$.
(ii) $ko_0(BV(4)) = \mathbb{Z}$.
(iii) $ko_1(BV(4)) = 2^5$.
(iv) $ko_2(BV(4)) = 2^a$ *for some* $a \leq 11$.

Proof: Parts (i) and (ii) follow from the Hurewicz theorem. Parts (iii) and (iv) follow from the E^2-term of the Atiyah-Hirzebruch spectral sequence.

Alternatively, all four parts follow from the bottom four degrees of the η-Bockstein spectral sequence deducing $ko_*(BV(4))$ from $ku_*(BV(4))$ (as displayed in Figure 10.3.7). □

The reader should note the slope -1 line below which E^∞ must be zero. Similarly, we know $ko_0(BG) \xrightarrow{\cong} KO_0(BG)$ is an isomorphism so that the only non-zero entry $E^\infty_{p,-p}$ has $p = 0$, where it is $E^2_{0,0}$.

There is a unique pattern of differentials in total degrees ≤ 2 forced by the groups stated in the lemma. This also shows that $a = 11$ in Part (iv). Specifically, with the relevant part of Lemma 12.5.10 cited in each case, we have the following.

- there must be an isomorphic d^2 leaving H^0_{-7} (Part (i)).
- the d^2 and d^3 leaving H^0_{-6} must both be surjective and their common kernel is 0 (Part (i)).
- the d^1, d^2 and d^3 leaving H^1_{-4} must both all be surjective and their common kernel is 0 (Part (i)).
- the d^1, d^2 and d^3 leaving H^1_0 must both all be surjective and their common kernel is 0 (Part (i)).
- the d^2, d^3 and d^4 leaving H^0_1 must all be surjective and their common kernel is 2^5 (Part (ii) or Part (iii)).
- the d^3 and d^4 leaving H^0_2 must both be surjective and their common kernel is 2^{11} (Part (iii) or Part (iv)).
- the d^1, d^2 and d^3 leaving H^1_4 must all be surjective (Part (iv)).

We now turn to the general behaviour of the differentials in high degrees.

LEMMA 12.5.11. *The differential*

$$d^1 : (H^1_I)_{4k} \longrightarrow (H^2_I)_{4k}$$

is an epimorphism for all k.

Proof: First note that the differential factors through $H^1_I/2H^1_I$, to give a map

$$\overline{d}^1 : (H^1_I/2H^1_I)_{4k} \longrightarrow (H^2_I)_{4k}.$$

Observe that (except possibly in degree 4), the domain and codomain are $\mathbb{F}_2$-vector spaces of the same dimension.

It is convenient to consider the dual map

$$(\overline{d}^1)^\vee : (H^2_I)^\vee \longrightarrow (H^1_I/2H^1_I)^\vee,$$

since it relates finitely generated PP-modules. Since we have already noted that connectivity forces $(\overline{d}^1)^\vee$ to be an isomorphism in the top two degrees, it suffices to show that the PP-module $(H^1_I/2H^1_I)^\vee$ is generated in codegrees 0 and -4.

In fact the part in negative degrees is entirely generated in degree 0: from Subsection 12.5.B it is enough to show the corresponding result for the part of $H^1_I(QU)$

in degrees divisible by 4. This can be seen in [**28**, Section 4.4], which fortuitously contains a depiction of the rank 4 case as Figure 4.3. □

LEMMA 12.5.12. *The differential*

$$d^2 : (2H_I^1)_{4k} \longrightarrow (H_I^3)_{4k+1}$$

is an epimorphism for all k.

Proof: From the previous lemma, we see that (except possibly in degree -4) the domain of the differential is $2H_I^1$, and we again note that the differential factors through $2H_I^1/4H_I^1$ to give a map

$$\overline{d}^2 : (2H_I^1/4H_I^1)_{4k} \longrightarrow (H_I^3)_{4k+1}.$$

Observe that the domain and codomain are vector spaces of the same dimension for all k.

It is convenient to consider the dual map

$$(\overline{d}^2)^\vee : (H_I^3)^\vee \longrightarrow (2H_I^1/4H_I^1)^\vee,$$

since it relates two finitely generated PP-modules. Since we have already noted that $(\overline{d}^2)^\vee$ is an isomorphism for $k = 1$ (from the known value of $ko_3(BV(4))$, and since all degrees ≤ -5 of $(H_I^3)^\vee$ are generated by degree -5, it suffices to show that $((2H_I^1/4H_I^1)^\vee)^{\leq -4}$ generated in codegree -4.

As in the proof of Lemma 12.5.11, from Subsection 12.5.B it is enough to show the corresponding result for the part of $H_I^1(QU)$ in degrees divisible by 4. This can be seen in [**28**, Section 4.4], which fortuitously contains a depiction of the rank 4 case as Figure 4.3. □

LEMMA 12.5.13. *The differential*

$$d^3 : (4H_I^1)_{4k} \longrightarrow (H_I^4)_{4k+2}$$

is of rank 15 for all $k \geq 1$.

Proof: From the previous lemma, we see that the domain of the differential is $4H_I^1$, and we again note that the differential factors through $4H_I^1/8H_I^1$ to give a map

$$\overline{d}^3 : (4H_I^1/8H_I^1)_{4k} \longrightarrow (H_I^4)_{4k+2}.$$

Observe that the domain and codomain are vector spaces of dimension 15 for all $k \geq 1$.

It is convenient to consider the dual map

$$(\overline{d}^3)^\vee : (H_I^4)^\vee \longrightarrow (4H_I^1/8H_I^1)^\vee,$$

since it relates two finitely generated PP-modules. Since we have already noted that $(\overline{d}^3)^\vee$ is an isomorphism for $k = 1$, it suffices to show that the PP-module $((4H_I^1/8H_I^1)^\vee)^{\leq -4}$ is generated in codegree -4.

As in the proof of Lemma 12.5.11, from Subsection 12.5.B it is enough to show the corresponding result for the part of $H_I^1(QU)$ in degrees divisible by 4. This can be seen in [**28**, Section 4.4], which fortuitously contains a depiction of the rank 4 case as Figure 4.3. □

LEMMA 12.5.14. *The only differentials leaving the H^0-column come from H^0_{-7}, H^0_{-6}, H^0_1, H^0_2 and H^0_9.*

Proof: We know that ko_* is a summand, so it remains only to discuss the classes in degrees 1 and 2 mod 8. We need only discuss degrees ≥ 10. By sparsity in the H^1 and H^2 columns, the only possible differential is

$$d^4 : H^0_{8k+i} \longrightarrow H^4_{8k+i+3}$$

for $i = 1$ or 2. Since $ko_*(BG)[1/\beta] = KO_*(BG)$, we know that all of the 2^{16} survives in sufficiently large degrees. Since the image in KO^G_* is a ko_*-module, it suffices to deal with degrees 10 and 17.

For this we use the fact that the differentials are maps of ko^*_V-modules. Every element of H^4_n for $n = 13$ or 20 is detected by some map $H^4_n \longrightarrow H^4_{n-4}$ coming from multiplication by an element λ of ko^4_V. For $n = 13$ we refer to Subsection 12.5.G to see there is an exact sequence

$$0 \longrightarrow H^4_I(TO_3^{\equiv 1(4)})^\vee \longrightarrow M \longrightarrow N$$

of PP-modules, where $M = 4[PP(-9) \oplus \mathrm{Finish}(-5)I_2]$, and N is a free module. Now suppose $x \in (H^4_I(TO_3^{\equiv 1(4)})^\vee_{-13}$. Since M is generated in degrees -5 and -9, we find a linear form λ and $y \in M_{-9}$ so that $\lambda y = x$. Since N is free, λ acts as a monomorphism on N, and hence $y \in H^4_I(TO_3^{\equiv 1(4)})^\vee$.

A precisely similar argument applies for $n = 20$. We refer to Subsection 12.5.K to see that $H^4_I(TO_{24}^{\equiv 0(4)})$ is the sum of $H^4_I(TO_2^{\equiv 0(4)})$ and two other factors whose duals are generated in degrees -4 and -8 respectively. We then refer to Subsection 12.5.J to see that there is an exact sequence

$$0 \longrightarrow H^4_I(TO_2^{\equiv 0(4)})^\vee \longrightarrow M \longrightarrow N$$

of PP-modules, where $M = 6\mathrm{Finish}(-8)I_3$ and N is a free module. The argument is finished as before.

Finally, if $d^4(x)$ is non-zero in H^4_n and detected by λ there would have to be a non-zero $d^4(\lambda x)$. However λx lies in the zero group. □

REMARK 12.5.15. Note that since $H^4_I(TO_2^{\equiv 0(4)})$ starts in degree 12, we cannot make the corresponding argument for $n = 12$. However, it is of dimension 6 in that degree, so the differential must be of rank $e \leq 6$.

12.5.O. Conclusion: the ko-homology groups. First, it is helpful to display the spectral sequence with the effect of the known differentials taken into account.

The values for the homology groups $ko_*(BV(4))$ can be read off from the local cohomology spectral sequence and are given in Figure 12.5.8. The number e is the rank of the d^4 from H^0_9, and $0 \leq e \leq 6$. It follows from the Bockstein spectral sequence (i.e., from Chapter 10) that $e = 0$. The coefficients are described by the following generating functions in the four congruence classes mod 4. These can be read off from the local cohomology calculations. As usual, t is of codegree -1 and $u = t^4$.

$$a(t) = \sum_s a_s t^s = \frac{u(6 - 3u + 7u^2 - 3u^3 + u^4)}{(1-u)^4}$$

FIGURE 12.5.7. The E^∞-page of the local cohomology spectral sequence for $ko_*(BV(4))$.

2^{209}			$[2^{11}]^4 \oplus [2^{10}]^6 \oplus [2^9]^4 \oplus [2^8]$	$\mathbb{Z}$	20
2^{140}			0	0	
2^{146}			0	2^{16}	
2^{105}			0	2^{16}	
2^{115}			$[2^8]^4 \oplus [2^7]^6 \oplus [2^6]^4 \oplus [2^5]$	$\mathbb{Z}$	16
2^{64}			0	0	
2^{70}			0	0	
2^{46}			0	0	
2^{55-e}			$[2^7]^4 \oplus [2^6]^6 \oplus [2^5]^4 \oplus [2^4]$	$\mathbb{Z}$	12
2^{20}			0	0	
2^{24}			0	2^{16}	
2^{15}			0	2^{16-e}	
2^{21}			$[2^4]^4 \oplus [2^3]^6 \oplus [2^2]^4 \oplus [2^1]$	$\mathbb{Z}$	8
			0	0	
			0	0	
			0	0	
			$[2^3]^4 \oplus [2^2]^6 \oplus [2^1]^4 \oplus [2^0]$	$\mathbb{Z}$	4
				0	
				2^{11}	
				2^5	
				$\mathbb{Z}$	0
					−4
					−8
H^4_{JO}	H^3_{JO}	H^2_{JO}	H^1_{JO}	H^0_{JO}	

FIGURE 12.5.8. The ko homology of $BV(4)$

n	$ko_n(BV(4))$
$8k+7 \geq 7$	$[2^{4k+4}]^4 \oplus [2^{4k+3}]^6 \oplus [2^{4k+2}]^4 \oplus [2^{4k+1}] \oplus 2^{d_{8k+11}}$
$8k+6 \geq 6$	$2^{(c_{8k+10}-15)}$
$8k+5 \geq 5$	$2^{b_{8k+9}}$
$8k+4 \geq 4$	$\mathbb{Z} \oplus 2^{a_{8k+8}}$
$8k+3 \geq 3$	$[2^{4k+4}]^4 \oplus [2^{4k+3}]^6 \oplus [2^{4k+2}]^4 \oplus [2^{4k+1}] \oplus 2^{d_{8k+7}}$
$8k+2 \geq 10$	$2^{16} \oplus 2^{(c_{8k+6}-15)}$
$8k+1 \geq 17$	$2^{16} \oplus 2^{b_{8k+5}}$
$8k+0 \geq 16$	$\mathbb{Z} \oplus 2^{a_{8k+4}}$
15	$[2^8]^4 \oplus [2^7]^6 \oplus [2^6]^4 \oplus [2^5] \oplus 2^{140}$
14	2^{146}
13	2^{105}
12	$\mathbb{Z} \oplus 2^{115}$
11	$[2^7]^4 \oplus [2^6]^6 \oplus [2^5]^4 \oplus [2^4] \oplus 2^{64}$
10	2^{86}
9	$2^{62-e}(0 \leq e \leq 6)$
8	$\mathbb{Z} \oplus 2^{55-e}(0 \leq e \leq 6)$
7	$[2^4]^4 \oplus [2^3]^6 \oplus [2^2]^4 \oplus [2] \oplus 2^{20}$
6	2^{24}
5	2^{15}
4	$\mathbb{Z} \oplus 2^{21}$
3	$[2^3]^4 \oplus [2^2]^6 \oplus [2]^4$
2	2^{11}
1	2^5
0	$\mathbb{Z}$

$$b(t) = \sum_s b_s t^s = \frac{t^5(3+u+10u^2-7u^3+u^4)}{(1-u)^4}$$

$$c(t) = \sum_s c_s t^s = \frac{t^{-2}(1+5u^2-u^3+4u^4-u^5)}{(1-u)^4}$$

$$d(t) = \sum_s d_s t^s = \frac{t^{11}4(5-4u+u^2)}{(1-u)^4}.$$

There are a number of potential additive extensions with kernel in the H^r column and the quotient in the H^0 or H^1 column (but not both). To see these are split, we note that the H^r column is detected by the map $ko_*(BV) \longrightarrow H_*(BV;\mathbb{F}_2)$.

12.6. Discussion of the results

After the above calculations, it seems worth making the following remarks. The local cohomology spectral sequence is concentrated on the H^0, H^1, ..., H^r columns. Here are some statements that would describe the behaviour of the local cohomology spectral sequence and give more information about $ko_*(BV(r))$; although we have proved some of them, we collect them under the heading of a conjecture, and explain their status in due course.

CONJECTURE 12.6.1. (i) The module τ is bounded below and therefore torsion (proved in Chapter 9).
(ii) The module TO is of depth 2.
(iii) The module $\overline{QO}$ is of dimension 1 (proved in Chapter 10).
(iv) The local cohomology groups $H^1_I(\overline{QO})$ are of rank $\leq 2^r - 1$. (proved in Chapter 10).
(v) The part of the spectral sequence on columns H^2, H^3, ..., H^r is elementary abelian (implied by (iii)).
(vi) The regularity of ko^*_V is $-r-4$, and witnessed by H^2_{-r-2}.
(vii) All non-trivial differentials come from the H^0 or H^1 columns.
(viii) There are a finite number of differentials leaving the H^0 column.
(ix) All differentials from H^1 are of maximal rank.
(x) For $2 \leq s \leq r-1$, the local cohomology group H^s behaves as follows. In large degrees it is concentrated in degrees $s-2$ mod 4, and is of dimension $2^r - 1$ in each such degree.
(xi) The E^∞-term is concentrated on the H^0, H^1 and H^r columns.
(xii) The part of the E^∞-term on the H^0 column is concentrated in degrees 0, 1, 2 and 4 mod 8. It is $\mathbb{Z}$ in degrees $4k \geq 0$, and elementary abelian of dimension 2^r in sufficiently large degrees of the form $8k+1$, $8k+2$.
(xiii) The part of the E^∞ term on the H^1 column is concentrated in degrees 0 mod 4, and

$$E^\infty_{*,-1} = 2^{r-1} H^1_I(\overline{QO}).$$

(xiv) There is a filtration

$$0 \subseteq F_{-2} \subseteq F_{-1} \subseteq F_0 = ko_*(BV(r))$$

with

- F_0/F_{-1} is concentrated in degrees 0, 1, 2 and 4 mod 8. It is $\mathbb{Z}$ in degrees $4k \geq 0$, and elementary abelian of dimension 2^r in sufficiently large degrees of the form $8k+1$, $8k+2$.
- $F_{-1}/F_{-2} = \Sigma^{-1} 2^{r-1} H^1_I(\overline{QO})$
- $F_{-2} = \Sigma^{-r} H^r_I(TO)'$, where the prime indicates a quotient of the local cohomology group. In large degrees, it agrees with the local cohomology group except in degrees $r-2$ mod 4, where we take the quotient of local cohomology group by an elementary abelian group of fixed dimension $2^r - 1$.

The conjectures are of various degrees of difficulty. Indeed, (i), (iii), (iv) and (v) are already known. There are various relations between the other conjectures.

Ideally, one would prove the results without giving complete calculations of local cohomology. Indeed, we may define

$$QOh = \mathrm{im}(ko_*(BV(r)) \longrightarrow KO_*(BV(r)),$$

and then define STh by the exact sequence

$$0 \longrightarrow STh \longrightarrow ko_*(BV(r)) \longrightarrow QOh \longrightarrow 0.$$

If the conjectures are true,

$$QOh = ko_*(BV(4))/F_{-2} \text{ and } STh = F_{-2}.$$

One can obtain estimates for QOh by using the diagram

$$\begin{array}{ccccccc} 0 & \longrightarrow & K & \longrightarrow & ko_*(BV(r)) & \longrightarrow & KO_*(BV(r)) \\ & & \downarrow & & \downarrow & & \downarrow \\ 0 & \longrightarrow & \mathrm{Start}(2)TU^\vee & \longrightarrow & ku_*(BV(r)) & \longrightarrow & KU_*(BV(r)), \end{array}$$

the known values of $ku_*(BV(r))$ and the Bockstein spectral sequence. Indeed, the homology Bockstein spectral sequence of Chapter 10 shows that, with an error of bounded rank from representation theory, the reduced homology of $BV(r)$ is $\mathrm{Start}(2)(TU/Sq^2)^\vee$. It therefore has Hilbert serives

$$t^{-2}[TO^\vee] + (1 - t^{-2})[TU^\vee].$$

On the other hand, the local cohomology spectral sequence shows that, with an error of bounded rank from representation theory, and from any differentials or surviving groups on the columns between 2 and $r - 1$, the reduced homology of $BV(r)$ is $\Sigma^{-r} H_I^r(TO)$.

REMARK 12.6.2. For more general groups we know (i) is false (for example it fails for the dihedral groups). We expect (ii) and (iii) to be true as stated, and (iv) with the bound being one less than the number of conjugacy classes. Again Part (v) follows. We also expect (viii) to be true in general.

The best we could hope for the higher columns would be that columns $-2, \ldots,$ $-(d-1)$ are zero at E^∞, where d is the depth of $H^*(BG; \mathbb{F}_2)$, and these columns might be detected in $H_*(BG; \mathbb{F}_2)$. However, we really do not have enough evidence to speculate.

APPENDIX A

Ext Charts

A.1. Introduction

Since the only prime of interest in the chapter is $p = 2$, we shall write H^* for mod 2 cohomology. The mod 2 cohomology of ko is $H^*ko = \mathcal{A}//\mathcal{A}(1) = \mathcal{A}\otimes_{\mathcal{A}(1)}\mathbb{F}_2$, and therefore, the Adams spectral sequences for ko homology and cohomology are

$$\mathrm{Ext}_{\mathcal{A}}^{s,t}(H^*ko \otimes H^*X, \mathbb{F}_2) \Longrightarrow ko_{t-s}X\widehat{_{(2)}}$$

and

$$\mathrm{Ext}_{\mathcal{A}}^{s,t}(H^*ko, H^*X) \Longrightarrow ko^{-(t-s)}X\widehat{_{(2)}}.$$

The untwisting isomorphism $(\mathcal{A} \otimes_{\mathcal{A}(1)} \mathbb{F}_2) \otimes H^*X \cong \mathcal{A} \otimes_{\mathcal{A}(1)} H^*X$, and the fact that $\mathcal{A}$ is free as an $\mathcal{A}(1)$-module, allow us to rewrite these as

$$\mathrm{Ext}_{\mathcal{A}(1)}^{s,t}(\mathbb{F}_2, H^*X) \Longrightarrow ko^{-(t-s)}X\widehat{_{(2)}} \qquad \text{and} \qquad \mathrm{Ext}_{\mathcal{A}(1)}^{s,t}(H^*X, \mathbb{F}_2) \Longrightarrow ko_{t-s}X\widehat{_{(2)}}.$$

Similarly, the fact that $H^*ku = \mathcal{A}//E(1)$ gives

$$\mathrm{Ext}_{E(1)}^{s,t}(\mathbb{F}_2, H^*X) \Longrightarrow ku^{-(t-s)}X\widehat{_{(2)}} \qquad \text{and} \qquad \mathrm{Ext}_{E(1)}^{s,t}(H^*X, \mathbb{F}_2) \Longrightarrow ku_{t-s}X\widehat{_{(2)}}.$$

We shall generally omit the completion $(-)\widehat{_{(2)}}$ to simplify the notation, expecting the reader to recall that this is what one should expect from a functor defined using mod 2 cohomology.

As is customary, we will display the Adams spectral sequence with the topological degree $t - s$ displayed horizontally and the homological degree s vertically. We shall display codegrees as negative degrees, so that, for example, ko^nX will be displayed in column $-n$ while ko_nX will be displayed in column n. This has the merit of consistency with respect to the action of the coefficients:

$$\begin{array}{ccc}
\mathrm{Ext}_{\mathcal{A}(1)}^{s_1,t_1}(\mathbb{F}_2, \mathbb{F}_2) \otimes \mathrm{Ext}_{\mathcal{A}(1)}^{s_2,t_2}(\mathbb{F}_2, H^*X) & \Longrightarrow & ko^{-(t_1-s_1)} \otimes ko^{-(t_2-s_2)}X \\
\downarrow & & \downarrow \\
\mathrm{Ext}_{\mathcal{A}(1)}^{s_1+s_2,t_1+t_2}(\mathbb{F}_2, H^*X) & \Longrightarrow & ko^{-((t_1-s_1)+(t_2-s_2))}X
\end{array}$$

and

$$\begin{array}{ccc}
\mathrm{Ext}_{\mathcal{A}(1)}^{s_1,t_1}(\mathbb{F}_2, \mathbb{F}_2) \otimes \mathrm{Ext}_{\mathcal{A}(1)}^{s_2,t_2}(H^*X, \mathbb{F}_2) & \Longrightarrow & ko_{(t_1-s_1)} \otimes ko_{(t_2-s_2)}X \\
\downarrow & & \downarrow \\
\mathrm{Ext}_{\mathcal{A}(1)}^{s_1+s_2,t_1+t_2}(H^*X, \mathbb{F}_2) & \Longrightarrow & ko_{((t_1-s_1)+(t_2-s_2))}X
\end{array}$$

We shall generally write projective resolutions of $\mathcal{A}(1)$-modules, obtaining injective resolutions by duality. Recall that $\mathcal{A}(1)^\vee = \Sigma^{-6}\mathcal{A}(1)$.

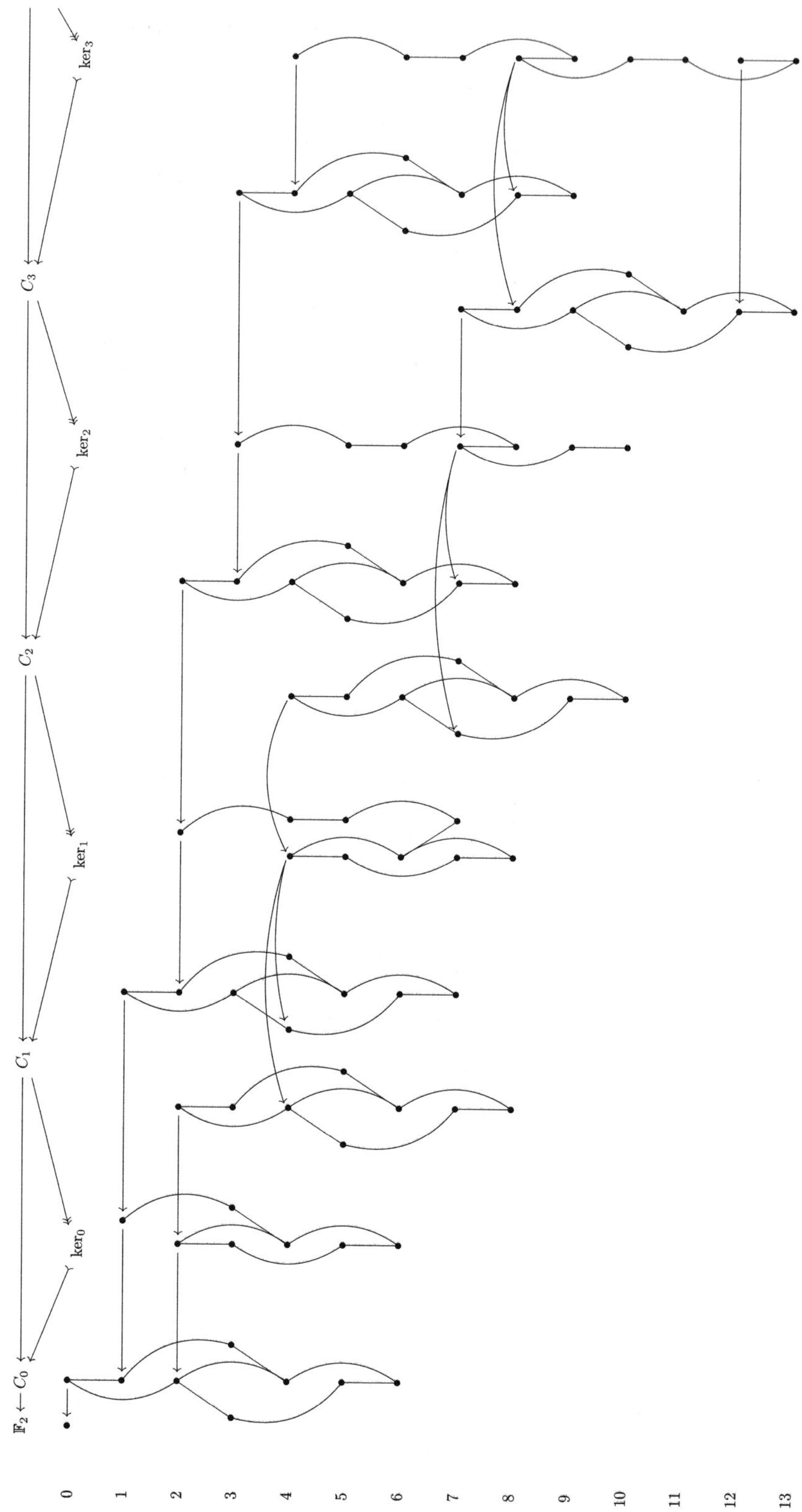

FIGURE A.1.1. The minimal $\mathcal{A}(1)$ resolution of $\mathbb{F}_2$

FIGURE A.1.2. The minimal $\mathcal{A}(1)$ resolution of $\mathbb{F}_2$

$$\Sigma^8\mathcal{A} \oplus \Sigma^{12}\mathcal{A} \oplus \Sigma^{16}\mathcal{A} \oplus \Sigma^{20}\mathcal{A} \oplus \Sigma^{24}\mathcal{A}$$

$$\Big\downarrow \begin{bmatrix} Sq^1 & Sq^2Sq^1Sq^2 & 0 & 0 & 0 \\ 0 & Sq^1 & Sq^2Sq^1Sq^2 & 0 & 0 \\ 0 & 0 & Sq^1 & Sq^2Sq^1Sq^2 & 0 \\ 0 & 0 & 0 & Sq^1 & Sq^2Sq^1Sq^2 \end{bmatrix}$$

$$\Sigma^7\mathcal{A} \oplus \Sigma^{11}\mathcal{A} \oplus \Sigma^{15}\mathcal{A} \oplus \Sigma^{19}\mathcal{A}$$

$$\Big\downarrow \begin{bmatrix} Sq^1 & Sq^2Sq^1Sq^2 & 0 & 0 \\ 0 & Sq^1 & Sq^2Sq^1Sq^2 & 0 \\ 0 & 0 & Sq^1 & Sq^2Sq^1Sq^2 \\ 0 & 0 & 0 & Sq^1Sq^2 \end{bmatrix}$$

$$\Sigma^6\mathcal{A} \oplus \Sigma^{10}\mathcal{A} \oplus \Sigma^{14}\mathcal{A} \oplus \Sigma^{16}\mathcal{A}$$

$$\Big\downarrow \begin{bmatrix} Sq^1 & Sq^2Sq^1Sq^2 & 0 & 0 \\ 0 & Sq^1 & Sq^2Sq^1Sq^2 & 0 \\ 0 & 0 & Sq^1 & Sq^1Sq^2 \\ 0 & 0 & 0 & Sq^2 \end{bmatrix}$$

$$\Sigma^5\mathcal{A} \oplus \Sigma^9\mathcal{A} \oplus \Sigma^{13}\mathcal{A} \oplus \Sigma^{14}\mathcal{A}$$

$$\Big\downarrow \begin{bmatrix} Sq^1 & Sq^2Sq^1Sq^2 & 0 & 0 \\ 0 & Sq^1 & Sq^2Sq^1Sq^2 & 0 \\ 0 & 0 & Sq^1 & Sq^2 \end{bmatrix}$$

$$\Sigma^4\mathcal{A} \oplus \Sigma^8\mathcal{A} \oplus \Sigma^{12}\mathcal{A}$$

$$\Big\downarrow \begin{bmatrix} Sq^1 & Sq^2Sq^1Sq^2 & 0 \\ 0 & Sq^1 & Sq^2Sq^1Sq^2 \end{bmatrix}$$

$$\Sigma^3\mathcal{A} \oplus \Sigma^7\mathcal{A}$$

$$\Big\downarrow \begin{bmatrix} Sq^1 & Sq^2Sq^1Sq^2 \\ 0 & Sq^1Sq^2 \end{bmatrix}$$

$$\Sigma^2\mathcal{A} \oplus \Sigma^4\mathcal{A}$$

$$\Big\downarrow \begin{bmatrix} Sq^1 & Sq^1Sq^2 \\ 0 & Sq^2 \end{bmatrix}$$

$$\Sigma^1\mathcal{A} \oplus \Sigma^2\mathcal{A}$$

$$\Big\downarrow \begin{bmatrix} Sq^1 & Sq^2 \end{bmatrix}$$

$$\mathcal{A}$$

$$\Big\downarrow \epsilon$$

$$\mathbb{F}_2$$

The coefficients are

$$\mathrm{Ext}_{\mathcal{A}(1)}(\mathbb{F}_2,\mathbb{F}_2) = \frac{\mathbb{F}_2[h_0,h_1,a,b]}{(h_0h_1, h_1^3, h_1a, a^2-h_0^2b)} \Longrightarrow ko_* = \frac{\mathbb{Z}[\eta,\alpha,\beta]}{(2\eta,\eta^3,\eta\alpha,\alpha^2-4\beta)}$$

and

$$\mathrm{Ext}_{E(1)}(\mathbb{F}_2,\mathbb{F}_2) = \mathbb{F}_2[h_0,v] \Longrightarrow ku_* = \mathbb{Z}[v],$$

where the degrees (s,t) of the generators are $|h_0| = (1,1)$, $|h_1| = (1,2)$, $|a| = (3,7)$, $|b| = (4,12)$ and $|v| = (1,3)$. (See the next section for the calculation.) In the charts, vertical lines indicate multiplication by h_0, representing 2, lines of slope 1 indicate multiplication by h_1, detecting η, and, in $\mathrm{Ext}_{E(1)}$ charts, lines of slope 1/2 indicate multiplication by v, representing the complex Bott map $v \in ku_2$. Multiplication by v is explicitly shown only at the bottom of an h_0-tower.

We also compute the Postnikov tower of ko in this Appendix, because the calculation is best carried out in the category of $\mathcal{A}(1)$-modules. This is used in the analysis of the Atiyah-Hirzebruch spectral sequence in Section 2.9.

FIGURE A.1.3. $\mathrm{Ext}_{\mathcal{A}(1)}(\mathbb{F}_2,\mathbb{F}_2)$

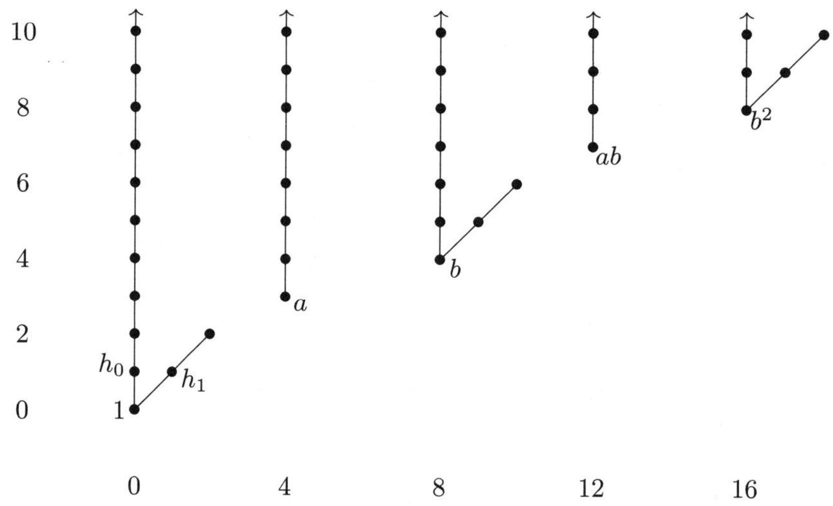

A.2. The trivial module

The coefficient ring $\mathrm{Ext}_{\mathcal{A}(1)}(\mathbb{F}_2,\mathbb{F}_2)$, over which all our other calculations are modules, may be deduced from the resolution

$$\mathbb{F}_2 \longleftarrow C_0 \longleftarrow C_1 \longleftarrow C_2 \longleftarrow C_3 \longleftarrow C_4 \longleftarrow \cdots$$

in Figures A.1.1 and A.1.2. The graphical representation, Figure A.1.1, shows C_0 through C_3, the kernels of the maps involved, and the epi-mono factorization of the differentials through the kernels. Note that the kernel of $C_2 \longleftarrow C_3$ contains an $\mathbb{F}_2$ in degree 12 linked to two $\mathcal{A}(1)//\mathcal{A}(0)$s. The resolution continues with the period 1 resolutions of the $\mathcal{A}(1)//\mathcal{A}(0)$s and the 12th suspension of the resolution $C_0 \longleftarrow C_1 \longleftarrow \cdots$. The free generator hitting the class in degree 12 in $\ker_3$ is dual to the element $b \in \mathrm{Ext}^{4,12}$ converging to the Bott element $\beta \in ko_8$.

The text version, Figure A.1.2, shows the resolution through C_8, by which point the periodicity is clearly apparent. The reader who undertakes these types of calculations will probably find that the graphical representation is the easiest to construct, and the actual formulas for the differentials are then easy to extract.

A.3. The integral Eilenberg-MacLane spectrum

The cohomology of $H\mathbb{Z}$ is $H^*H\mathbb{Z} = \mathcal{A}//E[Q_0] = \mathcal{A} \otimes_{\mathcal{A}(1)} \mathcal{A}(1)//E[Q_0]$, so the Adams spectral sequence collapses further to

$$\mathrm{Ext}_{E[Q_0]}(\mathbb{F}_2, \mathbb{F}_2) \Longrightarrow H\mathbb{Z}_* = \mathbb{Z}$$

(completed at 2, of course), consisting of a single h_0-tower in column 0.

A.4. The augmentation ideal

The augmentation ideal is the kernel $I = \ker(\mathcal{A}(1) \longrightarrow \mathbb{F}_2)$. (This is $\ker_0$ in Figure A.1.1.) Since $\mathrm{Ext}_{\mathcal{A}(1)}(\mathcal{A}(1), \mathbb{F}_2) = \mathbb{F}_2$ (in bidegree $(0,0)$), the long exact sequence simplifies to isomorphisms

$$\mathrm{Ext}^{s,t}_{\mathcal{A}(1)}(I, \mathbb{F}_2) \cong \mathrm{Ext}^{s+1,t}_{\mathcal{A}(1)}(\mathbb{F}_2, \mathbb{F}_2)$$

as displayed in Figure A.4.4. This is simply Ext for the trivial module with the bottom row removed. Replacing I by $\ker_i$ in the minimal resolution of $\mathbb{F}_2$ similarly gives $\mathrm{Ext}(\ker_i, \mathbb{F}_2)$ equal to $\mathrm{Ext}(\mathbb{F}_2, \mathbb{F}_2)$ with the bottom $i+1$ rows removed.

FIGURE A.4.4. $\mathrm{Ext}_{\mathcal{A}(1)}(I, \mathbb{F}_2)$

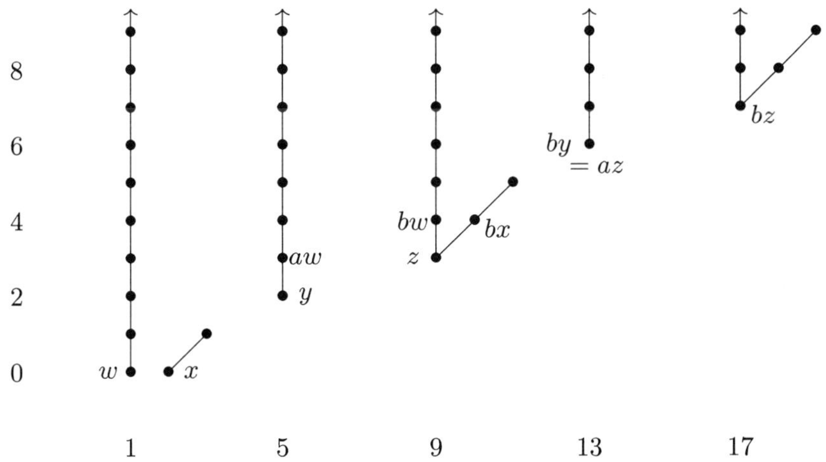

Dually, $\mathrm{Ext}_{\mathcal{A}(1)}(\mathbb{F}_2, \mathcal{A}(1)) = \Sigma^{0,-6}\mathbb{F}_2$, which gives an isomorphism between $\mathrm{Ext}^{0,t}(\mathbb{F}_2, I)$ and $\mathrm{Ext}^{0,t}(\mathbb{F}_2, \mathcal{A}(1))$, together with isomorphisms $\mathrm{Ext}^{s,t}(\mathbb{F}_2, \mathbb{F}_2) \cong \mathrm{Ext}^{s+1,t}(\mathbb{F}_2, I)$. The result is displayed in Figure A.4.5. The class in $\mathrm{Ext}^{0,-6}(\mathbb{F}_2, I)$ is annihilated by $\mathrm{Ext}(\mathbb{F}_2, \mathbb{F}_2)$.

FIGURE A.4.5. $\mathrm{Ext}_{\mathcal{A}(1)}(\mathbb{F}_2, I)$

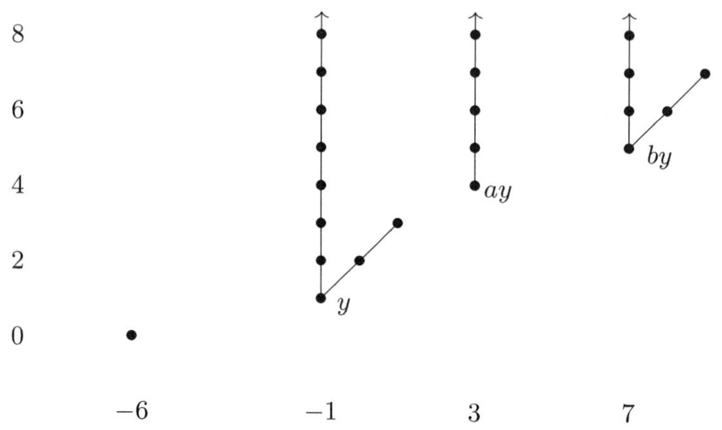

A.5. The Postnikov tower for ko

Algebraically, i.e., after applying $H\mathbb{F}_2^*$, the Postnikov tower for ko is a quotient of the Adams resolution. It is much simpler, since we are killing all of each $\mathbb{Z}$ at once, rather than filtering it by powers of (2), and Bott periodicity stands out quite prominently.

The calculation of the Postnikov tower is easy because the spectra and maps in it are all defined over $\mathcal{A}(1)$. Since $\mathcal{A} \otimes_{\mathcal{A}(1)} -$ is exact, we can compute the cohomology of the spectra and maps in the Postnikov tower by computing in the category of $\mathcal{A}(1)$-modules and tensoring up. In fact, we shall use only the fact that $H\mathbb{F}_2^* ko = \mathcal{A}//\mathcal{A}(1)$ and shall see that this forces the homotopy groups to be $\mathbb{Z}, \mathbb{Z}/2, \mathbb{Z}/2, 0, \mathbb{Z}, 0, 0, 0$ and then to repeat with period 8. If we were able to prove that ko is characterized by its mod 2 cohomology, this would give a proof of Bott periodicity, since it would show that $ko\langle 8\rangle = \Sigma^8 ko$

We shall also obtain the primary differentials in the Atiyah-Hirzebruch spectral sequence as a consequence.

THEOREM A.5.1. *The first four layers of the Postnikov tower of ko are as follows, after which they repeat this pattern suspended by 8 each time. The mod 2 cohomology of each sequence is given by applying $\mathcal{A} \otimes_{\mathcal{A}(1)} -$ to the short exact sequence of modules specified. In each case there is a unique map of $\mathcal{A}(1)$-modules which makes the sequence exact.*

- $ko \longrightarrow H\mathbb{Z} \longrightarrow \Sigma ko\langle 1\rangle$:
$$\mathbb{F}_2 \longleftarrow \mathcal{A}(1)/(Sq^1) \longleftarrow \Sigma^2\mathcal{A}(1)/(Sq^2)$$
- $ko\langle 1\rangle \longrightarrow \Sigma H\mathbb{F}_2 \longrightarrow \Sigma ko\langle 2\rangle$:
$$\Sigma\mathcal{A}(1)/(Sq^2) \longleftarrow \Sigma\mathcal{A}(1) \longleftarrow \Sigma^3\mathcal{A}(1)/(Sq^1Sq^2)$$
- $ko\langle 2\rangle \longrightarrow \Sigma^2 H\mathbb{F}_2 \longrightarrow \Sigma ko\langle 4\rangle$:
$$\Sigma^2\mathcal{A}(1)/(Sq^1Sq^2) \longleftarrow \Sigma^2\mathcal{A}(1) \longleftarrow \Sigma^5\mathcal{A}(1)/(Sq^1, Sq^2Sq^1Sq^2)$$
- $ko\langle 4\rangle \longrightarrow \Sigma^4 H\mathbb{Z} \longrightarrow \Sigma ko\langle 8\rangle = \Sigma^9 ko$:
$$\Sigma^4\mathcal{A}(1)/(Sq^1, Sq^2Sq^1Sq^2) \longleftarrow \Sigma^4\mathcal{A}(1)/(Sq^1) \longleftarrow \Sigma^9\mathbb{F}_2$$

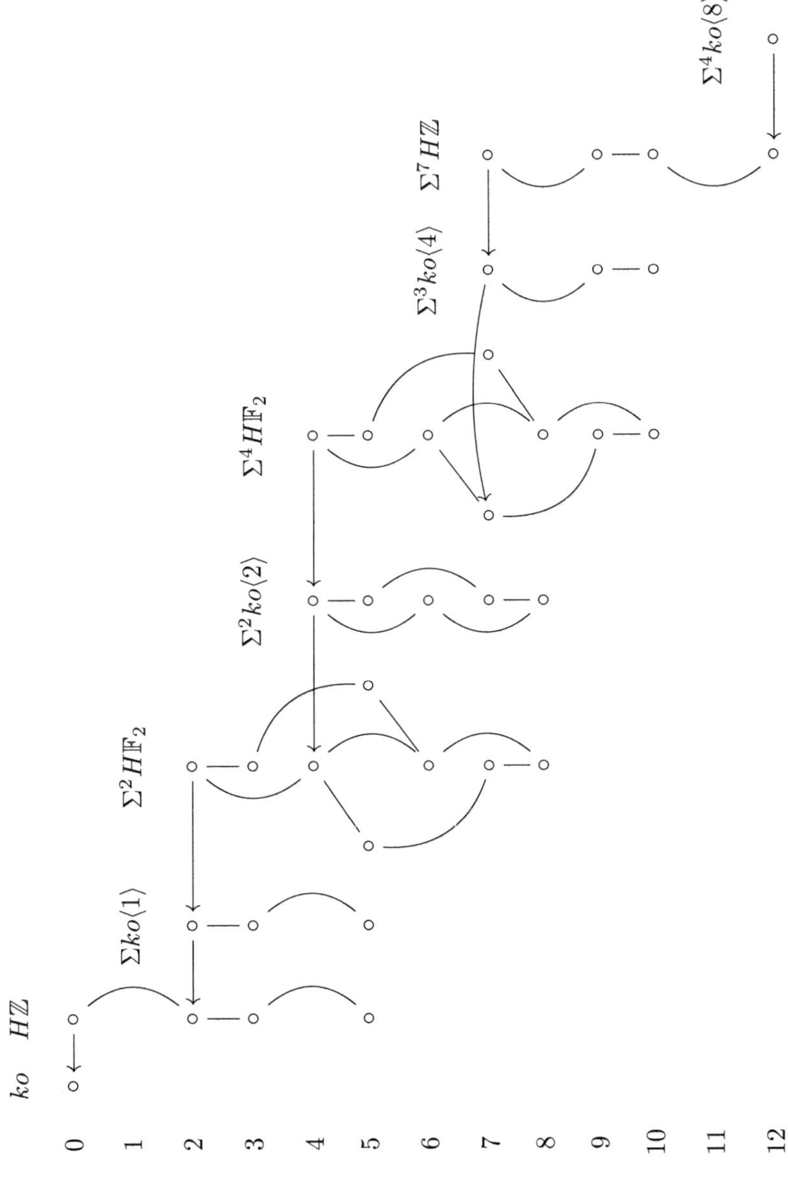

FIGURE A.5.6. The mod 2 cohomology of the Postnikov tower for ko is obtained by applying $\mathcal{A} \otimes_{\mathcal{A}(1)} (-)$ to these modules.

Proof: This is immediate from Figure A.5.6, which shows the $\mathcal{A}(1)$-modules and maps involved. □

COROLLARY A.5.2. *The primary differentials in the Atiyah-Hirzebruch spectral sequences $H^p(X, ko_q) \Longrightarrow ko^{p-q}(X)$ and $H^p(X, KO_q) \Longrightarrow KO^{p-q}(X)$ are as follows:*

- $H^p(X, ko_{8i}) \xrightarrow{d_2} H^{p+2}(X, ko_{8i+1})$ *is* $H^p(X, \mathbb{Z}) \xrightarrow{Sq^2} H^{p+2}(X, \mathbb{Z}/2)$,
- $H^p(X, ko_{8i+1}) \xrightarrow{d_2} H^{p+2}(X, ko_{8i+2})$ *is* $H^p(X, \mathbb{Z}/2) \xrightarrow{Sq^2} H^{p+2}(X, \mathbb{Z}/2)$,
- $H^p(X, ko_{8i+2}) \xrightarrow{d_3} H^{p+3}(X, ko_{8i+4})$ *is* $H^p(X, \mathbb{Z}/2) \xrightarrow{Sq^3} H^{p+3}(X, \mathbb{Z})$, *and*
- $H^p(X, ko_{8i+4}) \xrightarrow{d_5} H^{p+5}(X, ko_{8i+8})$ *is* $H^p(X, \mathbb{Z}) \xrightarrow{Sq^5} H^{p+5}(X, \mathbb{Z})$.

Proof: The maps between adjacent layers of the Postnikov tower are
- $Sq^2 : H\mathbb{Z} \longrightarrow \Sigma^2 H\mathbb{Z}/2$
- $Sq^2 : \Sigma H\mathbb{Z}/2 \longrightarrow \Sigma^3 H\mathbb{Z}/2$
- $Sq^1 Sq^2 = Sq^3 : \Sigma^2 H\mathbb{Z}/2 \longrightarrow \Sigma^5 H\mathbb{Z}$
- $Sq^2 Sq^1 Sq^2 = Sq^5 : \Sigma^4 H\mathbb{Z} \longrightarrow \Sigma^9 H\mathbb{Z}$

Naturally, these are integral lifts or extensions of the mod 2 classes we have specified. Writing them in this fashion emphasizes their behaviour when X is a space. For example, $d_5 : H^p(X, ko_{8i+4}) \longrightarrow H^{p+5}(X, ko_{8i+8})$ is 0 when $p < 5$ if X is a space. □

A.6. The inverted question mark

FIGURE A.6.7. The inverted question mark M

The 'inverted question mark' M which occurs in Lemma 8.2.4 is the three cell complex with cells in degrees 3, 5 and 6 and non-trivial $Sq^1 Sq^2$. It sits in the exact sequence

$$0 \longrightarrow \Sigma^8 \mathbb{F}_2 \longrightarrow \Sigma^3 \mathcal{A}(1)//\mathcal{A}(0) \longrightarrow M \longrightarrow 0$$

which makes it easy to calculate $\mathrm{Ext}_{\mathcal{A}(1)}(\mathbb{F}_2, M)$. The single h_0-tower in degree -8 in $\mathrm{Ext}_{\mathcal{A}(1)}(\mathbb{F}_2, \Sigma^3 \mathcal{A}(1)//\mathcal{A}(0))$ exactly cancels the bottom h_0-tower in the suspended coefficent ring $\mathrm{Ext}_{\mathcal{A}(1)}(\mathbb{F}_2, \Sigma^8 \mathbb{F}_2)$, so that the boundary map

$$\mathrm{Ext}^{s,t}_{\mathcal{A}(1)}(\mathbb{F}_2, M) \longrightarrow \mathrm{Ext}^{s+1,t}_{\mathcal{A}(1)}(\mathbb{F}_2, \mathbb{F}_2)$$

is an isomorphism in degrees larger than -7.

The degrees in this version of the Ext chart are relevant to the lowest degree copy of M occuring in the dihedral groups (Lemma 8.2.4). The crucial thing to note is that β acts monomorphically.

By the Postnikov tower computation, $ko \wedge M^\vee$ is a suspension of $ko\langle 1\rangle$. The isomorphism $\mathrm{Ext}_{\mathcal{A}(1)}(M^\vee, \mathbb{F}_2) \cong \mathrm{Ext}_{\mathcal{A}(1)}(\mathbb{F}_2, M)$ clearly shows that it has the correct homotopy.

FIGURE A.6.8. $\mathrm{Ext}_{\mathcal{A}(1)}(\mathbb{F}_2, M)$

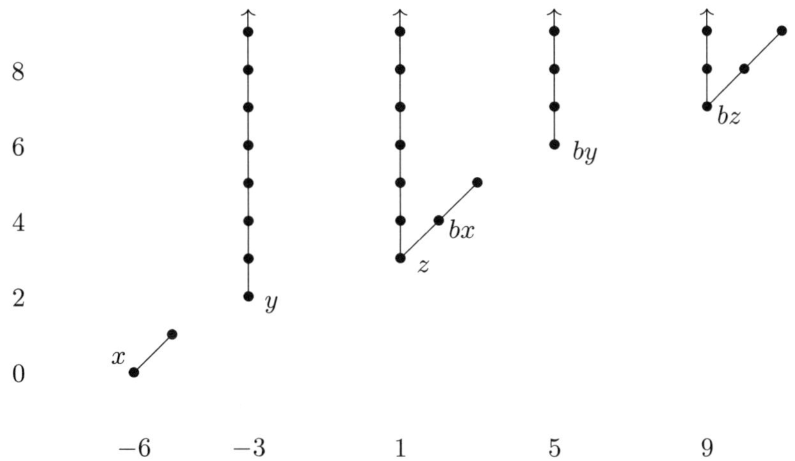

A.7. The question mark

FIGURE A.7.9. The question mark Q

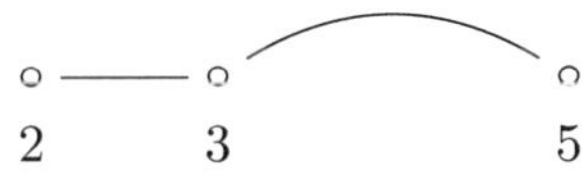

The question mark module Q is $M^\vee$ up to suspensions. We choose to suspend it 8 times to get an exact sequence dual to the one above

$$0 \longrightarrow Q \longrightarrow \mathcal{A}(1)//\mathcal{A}(0) \longrightarrow \mathbb{F}_2 \longrightarrow 0$$

which we use to calculate $\mathrm{Ext}_{\mathcal{A}(1)}(\mathbb{F}_2, Q)$. Here, the map in Ext induced by $\mathcal{A}(1)//\mathcal{A}(0) \longrightarrow \mathbb{F}_2$ must be zero for degree reasons and the result is therefore evident.

In this and the preceding cases, we have made implicit assertions about the action of the coefficients. These are relatively easy to verify using Yoneda products. In the two question marks, the situation of the modules in the Postnikov tower for ko gives the $\mathrm{Ext}_{\mathcal{A}(1)}(\mathbb{F}_2, \mathbb{F}_2)$ action. As above, we note that this chart shows the homotopy of $ko\langle 4\rangle$ after appropriate suspension.

FIGURE A.7.10. $\mathrm{Ext}_{\mathcal{A}(1)}(\mathbb{F}_2, Q)$

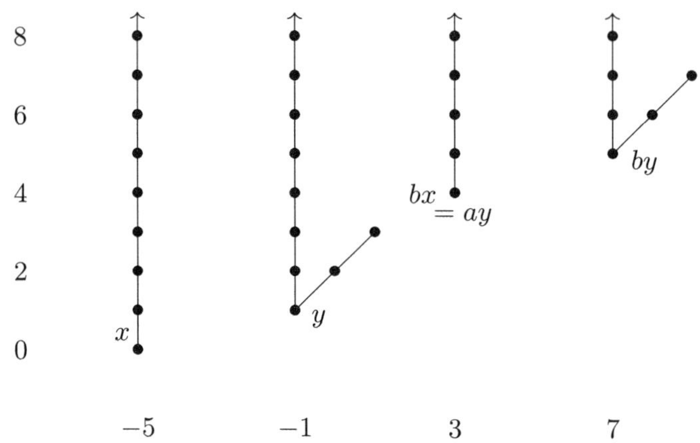

FIGURE A.8.11. The Joker J

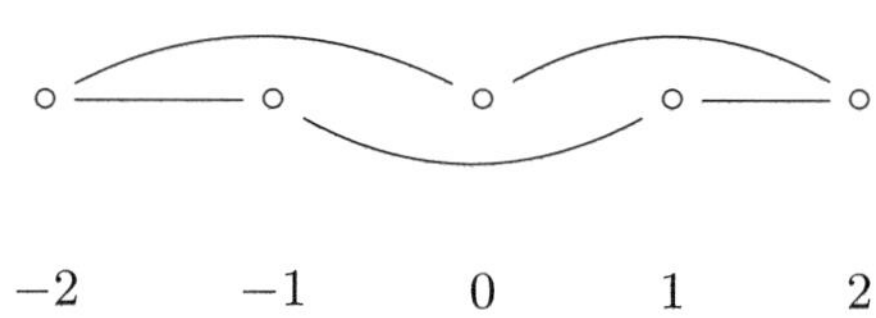

A.8. Joker

One presumes that the 'joker' J is so named because of its shape (Figure A.8.11) when drawn in the conventional way and because it generates the torsion in the Picard group $\mathrm{Pic}(\mathcal{A}(1))$ ([**6**]). It can be written as the quotient $J = \Sigma^{-2}\mathcal{A}(1)/(Sq^1Sq^2)$ or as $\Sigma^{-4}(Sq^2)$, a desuspension of the submodule of $\mathcal{A}(1)$ generated by Sq^2. The computation of a free $\mathcal{A}(1)$-resolution of J is of the same order of difficulty as that of $\mathbb{F}_2$, and bears a great deal of resemblance to it. The minimal resolution of J through homological degree 8 is shown in Figure A.8.13. This is sufficiently far that it is easy to extrapolate to the whole resolution.

The degrees in this version of the Ext chart are relevant to $O(2)$ and the dihedral groups. The notation is that used in Subsection 5.6.B. The crucial thing to note is that β acts non-trivially on the element s in degree -6.

A.9. Projective space

Let $P = (x) \subset \mathbb{F}_2[x] = H^*BC_2$ be the reduced mod 2 cohomology of BC_2. There is a short exact sequence

$$0 \longrightarrow P \longrightarrow R \longrightarrow \Sigma^{-1}\mathbb{F}_2 \longrightarrow 0$$

where R is the subset of the Laurent series $\mathbb{F}_2[x, x^{-1}]$ spanned by P and x^{-1}. The module R is filtered by submodules with associated graded $\oplus_{i\geq 0}\Sigma^{4i-1}\mathcal{A}(1)//\mathcal{A}(0)$, as shown in Figure A.9.14.

FIGURE A.8.12. $\mathrm{Ext}_{\mathcal{A}(1)}(\mathbb{F}_2, \Sigma^4 J)$

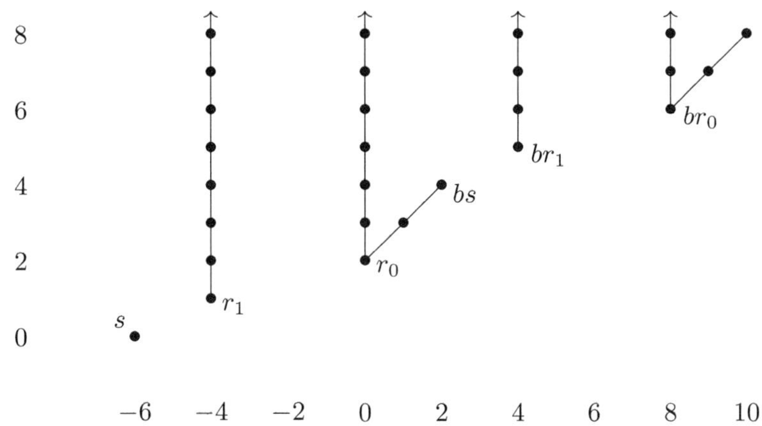

The filtration of R gives spectral sequences for calculating $\mathrm{Ext}_{\mathcal{A}(1)}(\mathbb{F}_2, R)$ and $\mathrm{Ext}_{\mathcal{A}(1)}(R, \mathbb{F}_2)$ which must collapse because they are concentrated in total degrees congruent to 0 mod 4 and -1 mod 4, respectively. Each term $\Sigma^{4i-1}\mathcal{A}(1)//\mathcal{A}(0)$ contributes an h_0-tower $\mathbb{F}_2[h_0]$ in degree $-(4i+4)$ in $\mathrm{Ext}_{\mathcal{A}(1)}(\mathbb{F}_2, R)$ and in degree $4i-1$ in $\mathrm{Ext}_{\mathcal{A}(1)}(R, \mathbb{F}_2)$, as in Figure A.9.15. It is a simple matter to use the Yoneda product to check that multiplication by α or β is always nonzero in $\mathrm{Ext}_{\mathcal{A}(1)}(R, \mathbb{F}_2)$, and is nonzero below degree -4 in $\mathrm{Ext}_{\mathcal{A}(1)}(\mathbb{F}_2, R)$.

The long exact sequences

$$0 \longrightarrow \mathrm{Ext}^0_{\mathcal{A}(1)}(\mathbb{F}_2, P) \longrightarrow \mathrm{Ext}^0_{\mathcal{A}(1)}(\mathbb{F}_2, R) \longrightarrow \mathrm{Ext}^0_{\mathcal{A}(1)}(\mathbb{F}_2, \Sigma^{-1}\mathbb{F}_2) \longrightarrow \mathrm{Ext}^1_{\mathcal{A}(1)}(\mathbb{F}_2, P) \longrightarrow \cdots$$

and

$$0 \longrightarrow \mathrm{Ext}^0_{\mathcal{A}(1)}(\Sigma^{-1}\mathbb{F}_2, \mathbb{F}_2) \longrightarrow \mathrm{Ext}^0_{\mathcal{A}(1)}(R, \mathbb{F}_2) \longrightarrow \mathrm{Ext}^0_{\mathcal{A}(1)}(P, \mathbb{F}_2) \longrightarrow \mathrm{Ext}^1_{\mathcal{A}(1)}(\Sigma^{-1}\mathbb{F}_2, \mathbb{F}_2) \longrightarrow \cdots$$

then allow a simple calculation of $\mathrm{Ext}_{\mathcal{A}(1)}(\mathbb{F}_2, P)$ and $\mathrm{Ext}_{\mathcal{A}(1)}(P, \mathbb{F}_2)$. A Yoneda product calculation shows that in $\mathrm{Ext}_{\mathcal{A}(1)}(\mathbb{F}_2, P)$, β is a regular element, and that α annihilates only the h_1-multiples. The result is shown in Figure A.9.16, which uses the notation of Theorem 5.6.6. It is an interesting fact that the relations in $ko^*_{O(1)}$ are exactly what one would predict from the relations in the associated graded, $\mathrm{Ext}_{\mathcal{A}(1)}(\mathbb{F}_2, \mathbb{F}_2[x]) \Longrightarrow ko^*BO(1)$. For example, $p_0p_1 = 2p_1$ and $\beta p_1 = \alpha p_0$. (Note that the Adams spectral sequence must collapse here for dimensional reasons.)

The long exact sequence involving $\mathrm{Ext}_{\mathcal{A}(1)}(P, \mathbb{F}_2)$ is not split exact, but is completely determined by the obvious fact that the nonzero class in $\mathrm{Ext}^{0,-1}_{\mathcal{A}(1)}(\Sigma^{-1}\mathbb{F}_2, \mathbb{F}_2)$ maps non-trivially. The value of $\mathrm{Ext}_{\mathcal{A}(1)}(P, \mathbb{F}_2)$ is then determined except for one set of unknown h_1 multiples. The first is from $\mathrm{Ext}^{1,3}$ to $\mathrm{Ext}^{2,5}$ and is non-zero by computing a small part of an $\mathcal{A}(1)$-free resolution of P. The rest then follow by the fact that β is a regular element in $\mathrm{Ext}_{\mathcal{A}(1)}(P, \mathbb{F}_2)$. The result is shown in Figure A.9.17, and immediately gives $ko_*(BO(1))$. Our calculations of the complex connective K-theory of elementary abelian groups in [**28**, Section 4.2] exploited

FIGURE A.8.13. The minimal $\mathcal{A}(1)$ resolution of the 'Joker' J

$$\Sigma^8\mathcal{A}\oplus\Sigma^{12}\mathcal{A}\oplus\Sigma^{16}\mathcal{A}\oplus\Sigma^{20}\mathcal{A}\oplus\Sigma^{22}\mathcal{A}$$

$$\Big\downarrow \begin{bmatrix} Sq^1 & Sq^2Sq^1Sq^2 & 0 & 0 & 0 \\ 0 & Sq^1 & Sq^2Sq^1Sq^2 & 0 & 0 \\ 0 & 0 & Sq^1 & Sq^2Sq^1Sq^2 & 0 \\ 0 & 0 & 0 & Sq^1 & Sq^1Sq^2 \\ 0 & 0 & 0 & 0 & Sq^2 \end{bmatrix}$$

$$\Sigma^7\mathcal{A}\oplus\Sigma^{11}\mathcal{A}\oplus\Sigma^{15}\mathcal{A}\oplus\Sigma^{19}\mathcal{A}\oplus\Sigma^{20}\mathcal{A}$$

$$\Big\downarrow \begin{bmatrix} Sq^1 & Sq^2Sq^1Sq^2 & 0 & 0 & 0 \\ 0 & Sq^1 & Sq^2Sq^1Sq^2 & 0 & 0 \\ 0 & 0 & Sq^1 & Sq^2Sq^1Sq^2 & 0 \\ 0 & 0 & 0 & Sq^1 & Sq^2 \end{bmatrix}$$

$$\Sigma^6\mathcal{A}\oplus\Sigma^{10}\mathcal{A}\oplus\Sigma^{14}\mathcal{A}\oplus\Sigma^{18}\mathcal{A}$$

$$\Big\downarrow \begin{bmatrix} Sq^1 & Sq^2Sq^1Sq^2 & 0 & 0 \\ 0 & Sq^1 & Sq^2Sq^1Sq^2 & 0 \\ 0 & 0 & Sq^1 & Sq^2Sq^1Sq^2 \end{bmatrix}$$

$$\Sigma^5\mathcal{A}\oplus\Sigma^{9}\mathcal{A}\oplus\Sigma^{13}\mathcal{A}$$

$$\Big\downarrow \begin{bmatrix} Sq^1 & Sq^2Sq^1Sq^2 & 0 \\ 0 & Sq^1 & Sq^2Sq^1Sq^2 \\ 0 & 0 & Sq^1Sq^2 \end{bmatrix}$$

$$\Sigma^4\mathcal{A}\oplus\Sigma^{8}\mathcal{A}\oplus\Sigma^{10}\mathcal{A}$$

$$\Big\downarrow \begin{bmatrix} Sq^1 & Sq^2Sq^1Sq^2 & 0 \\ 0 & Sq^1 & Sq^1Sq^2 \\ 0 & 0 & Sq^2 \end{bmatrix}$$

$$\Sigma^3\mathcal{A}\oplus\Sigma^{7}\mathcal{A}\oplus\Sigma^{8}\mathcal{A}$$

$$\Big\downarrow \begin{bmatrix} Sq^1 & Sq^2Sq^1Sq^2 & 0 \\ 0 & Sq^1 & Sq^2 \end{bmatrix}$$

$$\Sigma^2\mathcal{A}\oplus\Sigma^{6}\mathcal{A}$$

$$\Big\downarrow \begin{bmatrix} Sq^1 & Sq^2Sq^1Sq^2 \end{bmatrix}$$

$$\Sigma\mathcal{A}$$

$$\Big\downarrow \begin{bmatrix} Sq^1Sq^2 \end{bmatrix}$$

$$\Sigma^{-2}\mathcal{A}$$

$$\Big\downarrow \epsilon$$

$$J$$

FIGURE A.9.14. The $\mathcal{A}(1)$-module $R \subset \mathbb{F}_2[x, x^{-1}]$

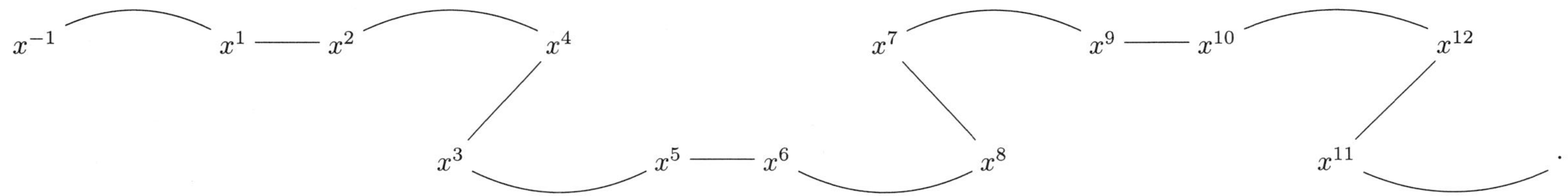

FIGURE A.9.15. $\mathrm{Ext}_{\mathcal{A}(1)}(\mathbb{F}_2, R)$ and $\mathrm{Ext}_{\mathcal{A}(1)}(R, \mathbb{F}_2)$, respectively

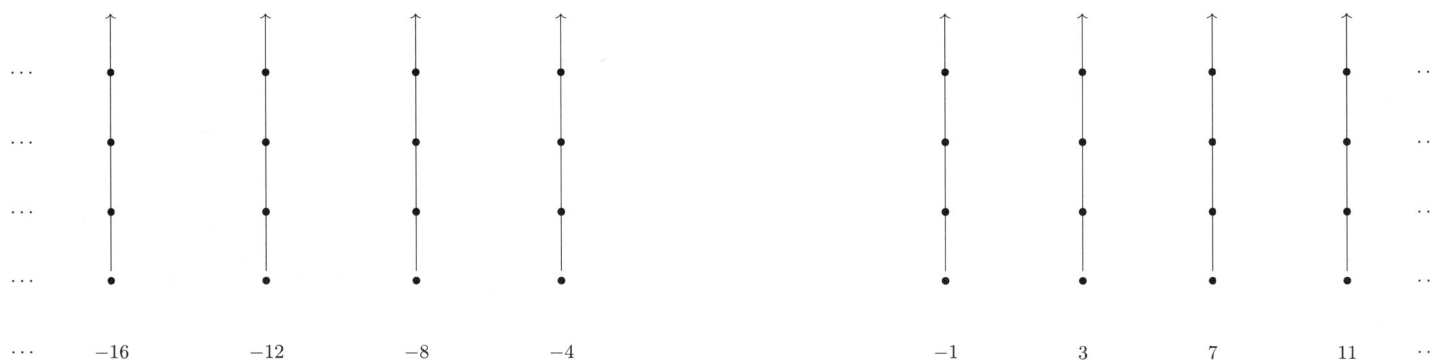

FIGURE A.9.16. $\mathrm{Ext}_{\mathcal{A}(1)}(\mathbb{F}_2, P) \Longrightarrow \widetilde{ko}^* BO(1)$

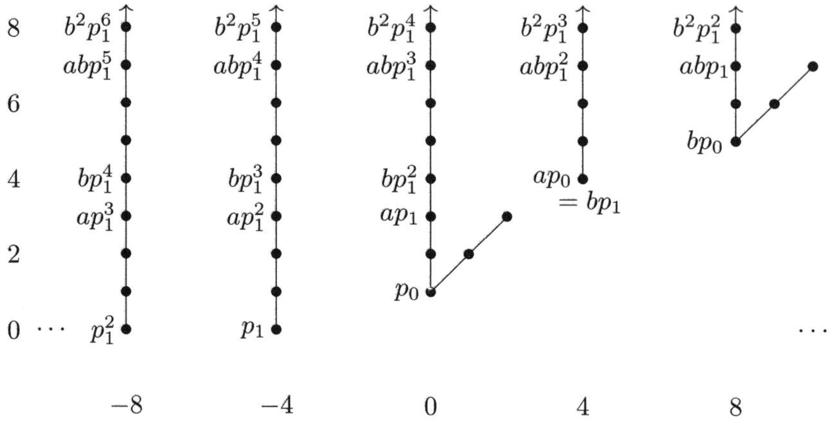

FIGURE A.9.17. $\mathrm{Ext}_{\mathcal{A}(1)}(P, \mathbb{F}_2) \Longrightarrow \widetilde{ko}_* BO(1)$

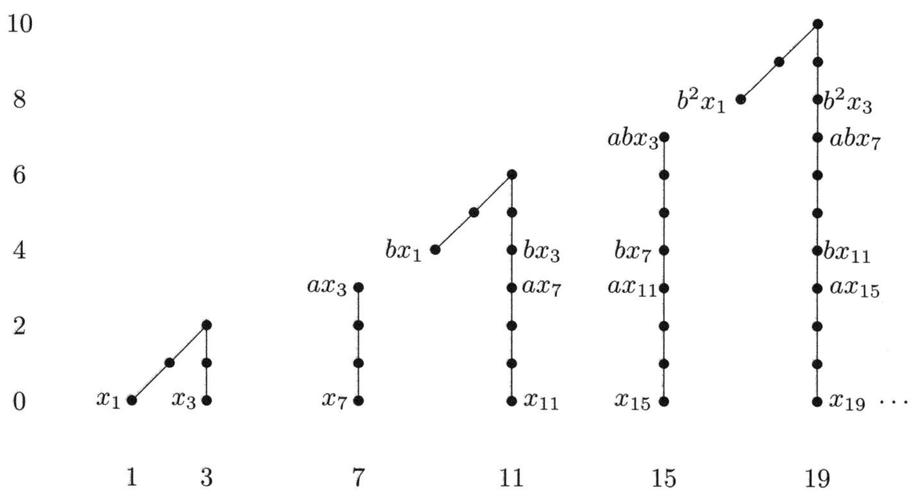

(among other things) Erich Ossa's observation ([**87**]) that P is idempotent as an $E(1)$-module: $P \otimes P$ is isomorphic to the direct sum of $\Sigma^2 P$ and a free $E(1)$-module. Iterating this shows that QU for an elementary abelian group is simply a sum of copies of the rank 1 case suitably suspended and indexed by the representation ring. It also implies that the Adams spectral sequence will collapse. See Johnson and Wilson ([**65**]) and Bruner ([**23**]) for further discussion of the complex case.

The real case is more subtle, as P is no longer idempotent as an $\mathcal{A}(1)$-module. Instead, the stable type of the tensor powers of P cycles with period 4. The first correct published account of this seems to be the thesis of Cherng-Yih Yu ([**99**]). Related matters are also discussed in [**23**]. The simplest way to prove this periodicity is to show that the $(n+1)^{\mathrm{st}}$ tensor power of P is stably equivalent to the nth Adams cover of P, and to observe that this is periodic with period 4. The first of these facts is immediate from the fact that $P \otimes R$ is $\mathcal{A}(1)$ free, while the second can be done quite economically, and is worked out carefully in [**24**]. We do not pursue this any further here, because we have other ways to get at ko^*BV and ko_*BV. We should point out that the analysis of the stable type of H^*BV as an $\mathcal{A}(1)$-module is the easy part. The parameterization of the free summands in H^*BV is a daunting task, and this is reflected in the complexity of the analysis in Chapters 9 through 12.

A.10. $\eta - c - R$ sequence in Ext

There is an exact analogue of the $\eta - c - R$ sequence in the E_2 term of the Adams spectral sequence. It arises from the fact that the $\eta - c - R$ sequence $\Sigma ko \xrightarrow{\eta} ko \xrightarrow{c} ku \xrightarrow{R} \Sigma^2 ko$ can be obtained by smashing ko with the cofiber sequence $S^1 \xrightarrow{\eta} S \longrightarrow C\eta \longrightarrow S^2$. (See Case 3 in Subsection 4.1.C.) In the Adams spectral sequence

$$\mathrm{Ext}_{\mathcal{A}}(H^*ko, H^*X) \cong \mathrm{Ext}_{\mathcal{A}(1)}(\mathbb{F}_2, H^*X) \Longrightarrow ko^*X,$$

this corresponds to applying $\mathrm{Ext}_{\mathcal{A}(1)}(-, H^*X)$ to the short exact sequence

$$0 \longleftarrow \mathbb{F}_2 \longleftarrow H^*C\eta \longleftarrow \Sigma^2\mathbb{F}_2 \longleftarrow 0.$$

Using the isomorphism $H^*C\eta \cong \mathcal{A}(1)//E(1)$, the 'middle' term in the resulting long exact sequence can be rewritten

$$\mathrm{Ext}_{\mathcal{A}(1)}(H^*C\eta, H^*X) \cong \mathrm{Ext}_{\mathcal{A}(1)}(\mathcal{A}(1)//E(1), H^*X) \cong \mathrm{Ext}_{E(1)}(\mathbb{F}_2, H^*X)$$

giving us a long exact sequence

$$\cdots \xrightarrow{h_1} \mathrm{Ext}^{s,t}_{\mathcal{A}(1)}(\mathbb{F}_2, H^*X) \xrightarrow{c} \mathrm{Ext}^{s,t}_{E(1)}(\mathbb{F}_2, H^*X) \xrightarrow{R}$$
$$\mathrm{Ext}^{s,t-2}_{\mathcal{A}(1)}(\mathbb{F}_2, H^*X) \xrightarrow{h_1} \mathrm{Ext}^{s+1,t}_{\mathcal{A}(1)}(\mathbb{F}_2, H^*X) \xrightarrow{c} \cdots$$

converging to

$$\cdots \xrightarrow{\eta} ko^nX \xrightarrow{c} ku^nX \xrightarrow{R} ko^{n+2}X \xrightarrow{\eta} ko^{n+1}X \xrightarrow{c} \cdots$$

(with $n = -(t-s)$). The boundary map in Ext induced by the short exact sequence

$$0 \longleftarrow \mathbb{F}_2 \longleftarrow H^*C\eta \longleftarrow \Sigma^2\mathbb{F}_2 \longleftarrow 0$$

is multiplication by h_1, corresponding to the fact that Sq^2 connects the two classes in $H^*C\eta$.

Figure A.10.19 shows the part of this sequence containing c and R when $X = S$. The maps are forced by the fact that the boundary is multiplication by h_1 and the exactness of the sequence. They are all $\mathrm{Ext}_{\mathcal{A}(1)}(\mathbb{F}_2, \mathbb{F}_2)$-module homomorphisms. We see that $c(a) = h_0v^2$ and $c(b) = v^4$ corresponding to the fact that $c(\alpha) = 2v^2$ and $c(\beta) = v^4$. Similary, $R(v) = h_0$, $R(v^2) = h_1^2$, and $R(v^3) = a$, corresponding to $R(v) = 2$, $R(v^2) = \eta^2$, and $R(v^3) = \alpha$ in homotopy.

Figure A.10.20 shows the same part of the sequence for $X = BC_2 = BO(1)$ which may be illuminating in Section 5.6.A and Chapter 9.

Finally, Figure A.10.18 encapsulates within one diagram the relations between Sq^2, the $\eta - c - R$ sequence, the d^1 differential in the Bockstein spectral sequence and the relation between R and realification r.

FIGURE A.10.18. The $\eta - c - R$ sequence and its relations to various operations

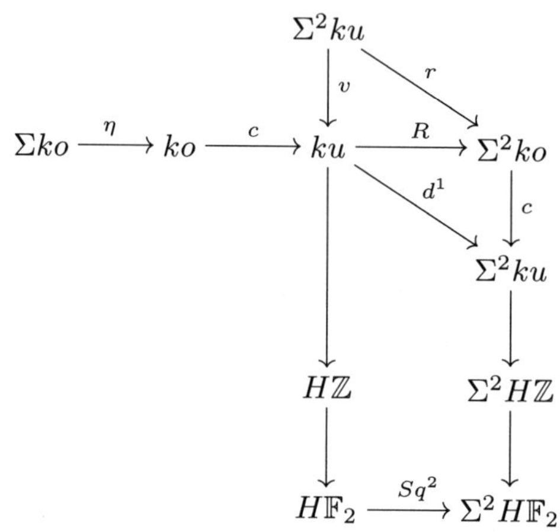

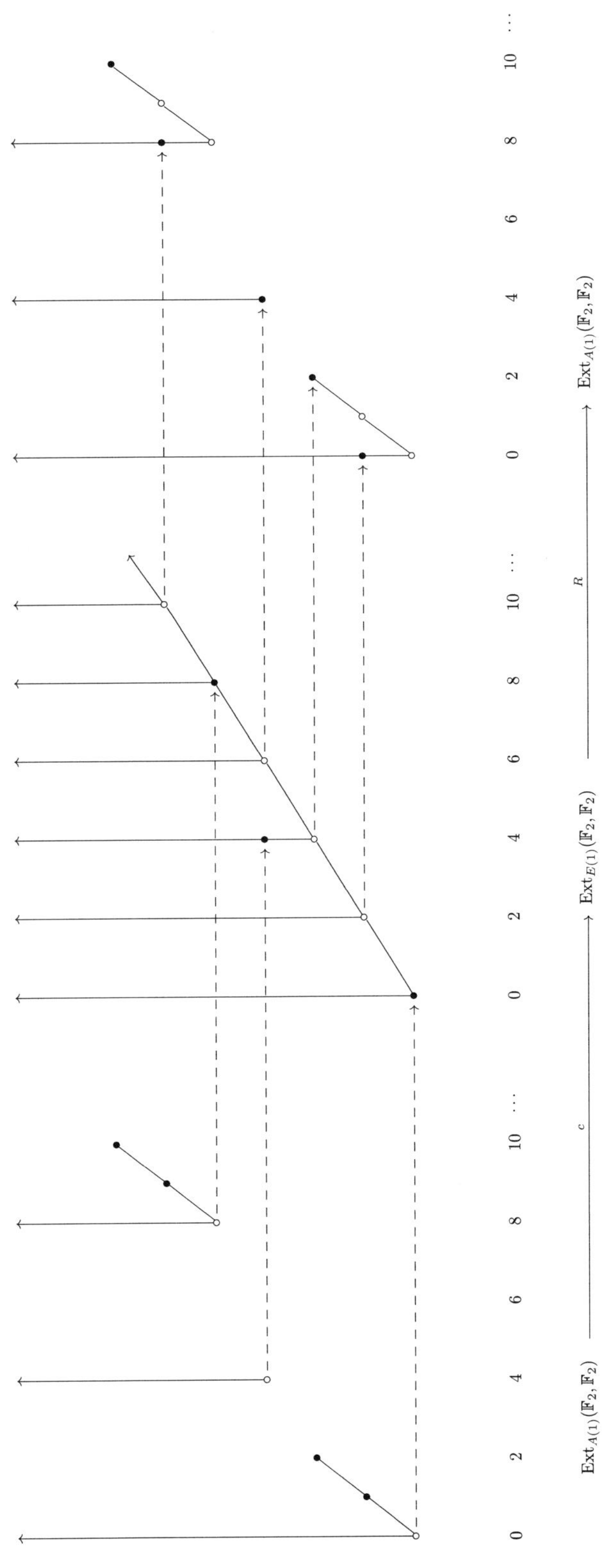

FIGURE A.10.19. The maps c and R in the $\eta - c - R$ sequence in $\mathrm{Ext}(\mathbb{F}_2, \mathbb{F}_2)$

FIGURE A.10.20. The maps c and R in the $\eta - c - R$ sequence in $\mathrm{Ext}(\mathbb{F}_2, P)$

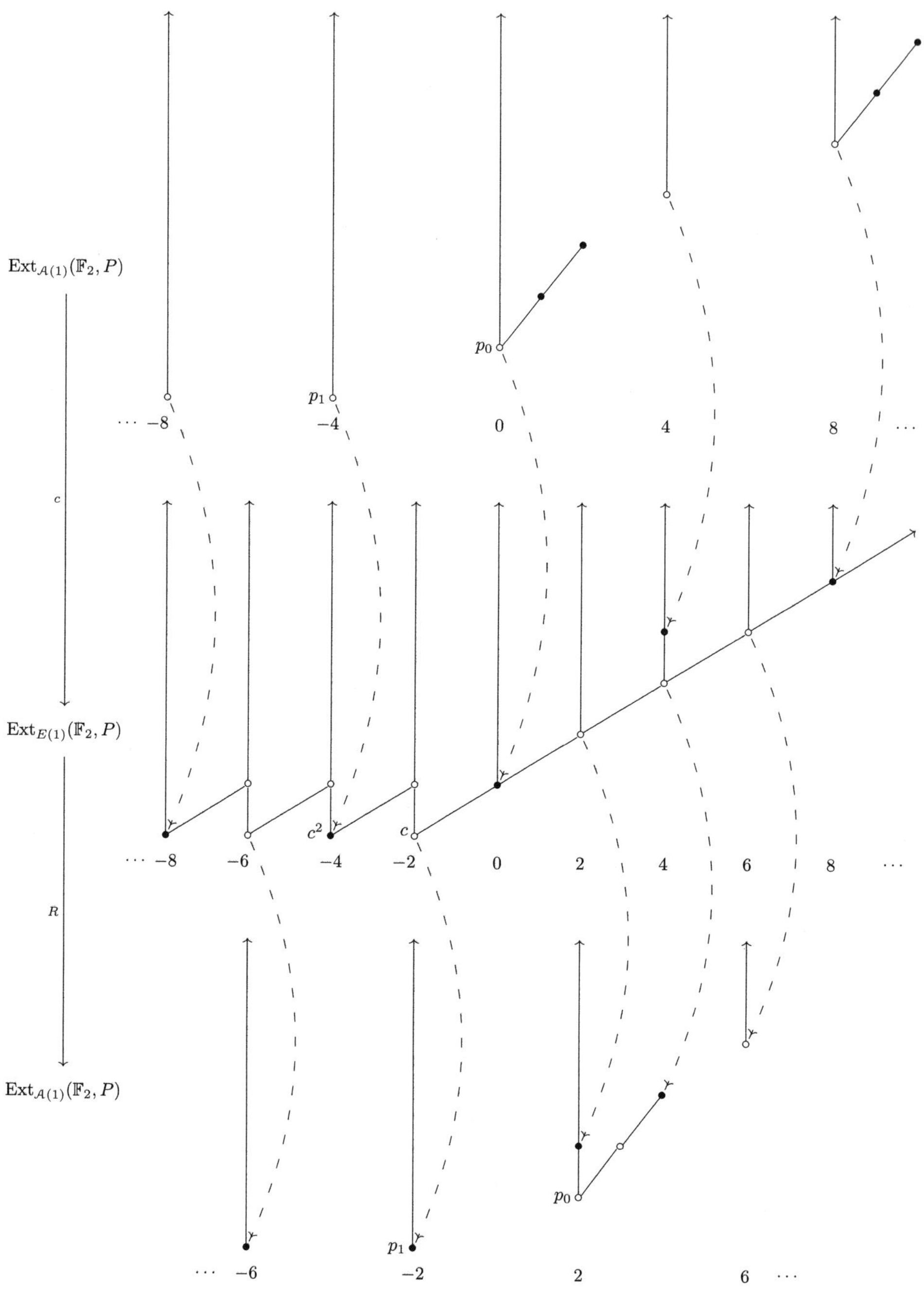

APPENDIX B

Conventions

B.1. General conventions

Unreduced: Homology and cohomology is generally unreduced unless the contrary is indicated by a tilde. The exceptions to this rule are in subsections dealing with spectra, and clearly flagged.

Degree and codegree: All homological (lower) indices are called degrees, and all cohomological (upper) indices are called codegrees. These are related by $M_i = M^{-i}$.

Mod p cohomology: Unless otherwise indicated, coefficients of ordinary cohomology are in $\mathbb{F}_p$ for some prime p.

Suspension: We have indicated suspension in three different ways, according to what is convenient. The algebraic and topological suspensions are related by

$$\Sigma^n M = M(n) = \Sigma_{-n} M.$$

We also write $\mathrm{Start}(n)M$ for the suspension in which the lowest non-zero entry is in degree n. Use of the notation implies that M is non-zero and bounded below. Dually, we write $\mathrm{Finish}(n)M$ for the suspension in which the highest non-zero entry is in degree n. Use of the notation implies that M is non-zero and bounded above.

Groups: We write 2^r or $V(r)$ for an elementary abelian group of rank r, $[n]$ for a cyclic group of order n and $\langle n \rangle$ for an unspecified group of order n. For cyclic groups C_n, dihedral groups D_n or quaternion groups Q_n, the subscript n is the order of the group.

Representations: (i) If H is the normal subgroup generated by s in G, with cyclic quotient G/H, then $\hat{s}$ denotes a one dimensional representation of G with kernel H.

(ii) The trivial one dimensional complex representation is denoted ϵ.

K-theory: The spectra ko and ku were originally denoted bo and bu. We prefer the notation ko and ku to emphasize the relation to KO and KU. We reserve the notation bo and bu for their connective covers, on the grounds that ko and ku have 0th spaces $BO \times \mathbb{Z}$ and $BU \times \mathbb{Z}$, while the connective covers have 0th spaces BO and BU.

Equivariance: (i) The cohomology theory $E_G^*(\cdot)$ and the homology theory $E_*^G(\cdot)$ are generally functors of G-spaces. The rare exception is that to follow classical usage, they may be functors of G-spaces with reality, which is to say of $\widehat{G}$-spaces (see Remark 2.2.6).

(ii) Notation is chosen so that forgetting equivariance requires no change. Thus if X is a G-space, and H is a subgroup, the same notation denotes X as an H-space. This principle applies equally to equivariant spectra. The letter G then appears when we specify that maps or homotopy groups are equivariant:

$$\pi_n^G(X) = [S^n, X]^G = [S^n, X^G] = \pi_n(X^G).$$

APPENDIX C

Indices

C.1. Some recurrent notation

- When discussing structures with reality, we work with an extension

$$1 \longrightarrow G \longrightarrow \hat{G} \longrightarrow Q \longrightarrow 1$$

  where Q is a group of order 2 defining the real involution.
- We have a short exact sequence

$$0 \longrightarrow TU \longrightarrow ku_G^* \longrightarrow QU \longrightarrow 0$$

  where $QU = \mathrm{im}(ku_G^* \longrightarrow KU_G^*)$.
- Similarly, we have short exact sequences

$$0 \longrightarrow T \longrightarrow ko_G^* \longrightarrow \overline{QO} \longrightarrow 0$$

  where $\overline{QO} = \mathrm{im}(ko_G^* \longrightarrow KU_G^*)$ and

$$0 \longrightarrow ST \longrightarrow ko_G^* \longrightarrow QO \longrightarrow 0$$

  where $\overline{QO} = \mathrm{im}(ko_G^* \longrightarrow KO_G^*)$. For the relationship with TO and various collections of η-multiples, $\tau, S\tau$ and $Q\tau$, see Section 6.2.

C.2. Index of calculations

We summarize where to find the calculations for various specific groups.

Group	Name	$ku^*(BG)$	$ku_*(BG)$	$ko^*(BG)$	$ko_*(BG)$
C_n	cyclic	[**28**, 2.2]	[**28**, 3.4.3]	6.3	7.3
$G_{p,q}$	non-abelian pq	[**28**, 2.3.2]	-	-	-
Q_8	quaternion	[**28**, 2.4.6]	[**28**, 3.4.3]	6.4	7.4
Q_{2^n}	quaternion	[**28**, 2.4.5]	[**28**, 3.4.3]	6.4	7.5
$SL_2(3)$	special linear	[**28**, 2.4.12]	-	-	-
$SL_2(\ell)$	special linear	[**28**, 2.4.11]	-	-	-
D_8	dihedral	[**28**, 2.5.5]	[**28**, 3.5.1]	8.2	8.5
D_{2^n}	dihedral	[**28**, 2.5.4]	8.4	8.2	8.5
A_4	alternating	[**28**, 2.6.2]	-	6.6	7.7
Σ_q	symmetric, q prime	[**28**, 2.3.3]	-	-	-
V	elementary abelian	[**28**, 4.2]	[**28**, 4.1.1]	10	12

Group	$ku^*(BG)$	$ku_*(BG)$	$ko^*(BG)$	$ko_*(BG)$
$T(n)$	5.2	5.2	5.2	-
$Sp(1)^n$	5.3	5.3	5.3	5.3
$Sp(n)$	5.3	5.3	5.3	5.3
$SU(n)$	5.4	5.4	-	-
$U(n)$	5.5	5.5	-	-
$O(1)$	5.6.A	[**28**, 4.1.1]	10	12
$O(2)$	5.6.B	-	5.6.B	-
$O(3)$	5.6.C	-	-	-
$SO(3)$	5.7	-	-	-

C.3. Index of symbols

$[n](x)$	$(1-(1-vx)^n)/v$ the multiplicative n-series
$[n]$	Cyclic group of order n
$\langle n \rangle$	Unspecified abelian group of order n
2^a	Elementary abelian 2-group of rank a
$M^\vee$	vector space dual $\mathrm{Hom}_k(M, k)$
$\lvert V \rvert$	V with the trivial action
$W \uparrow^G$	induced representation
$[M]$	Hilbert series
$(1-x)^r_{[i]}$	truncation of $(1-x)^r$
$M(n)$	n-fold suspension of M
$R\{x\}$	Free R-module on one generator x
$R[x]$	Polynomials over R-module on one generator x
$R[[x]]$	Power series over R-module on one generator x
$\hat{x}$	natural representation of the quotient by $\langle x \rangle$
$X^{(n)}$	n-skeleton
$X\langle n \rangle$	$(n-1)$-connected cover
X^{hQ}	Q-homotopy fixed points $Q\text{-map}(EQ_+, X)$ of X.
X_{hQ}	Q-homotopy orbits $EQ_+ \wedge_Q X$ of X.
$\hat{U}$	$\mathbb{R}\hat{G} \otimes_{\mathbb{R}G} U$
$F(n)$	$n \times n$ matrices over F

C.4. Index of notation

A

A_4		alternating group of degree 4
$\mathcal{A}_p$		mod p Steenrod algebra
$\mathcal{A}(0)$		$\langle Sq^1 \rangle \cong E(Sq^1)$ in $\mathcal{A}_2$
$\mathcal{A}(1)$		$\langle Sq^1, Sq^2 \rangle$ in $\mathcal{A}_2$
A		Assembly
$A(m,i)_s$	11.6	Generators in a resolution of TO
a_0		Adams spectral sequence counterpart of p
α		(1) generator of ko_4
		(2) generic one dimensional representation

B

$BSS(a)$	4.1	Bockstein spectral sequence for a
β		(1) Real Bott element of degree 8
		(2) generic one dimensional representation

C

$C_k(F)$	2.3	k-th negative definite Clifford algebra over the field F
$C'_k(F)$	2.3	k-th positive definite Clifford algebra over the field F
$C^0_k(F)$	2.3	Degree 0 part of $C_k(F)$.
$(C'_k)^0(F)$	2.3	Degree 0 part of $C'_k(F)$.
$\mathbb{C}$		Complex numbers
$\mathcal{C}$	8.3	Chern submodule of $RU(D)$
$\mathcal{C}_0$	8.3	
$\mathcal{C}_*$	8.3	
c		Complexification (of real representations)
$\tilde{c}$		Complexification (of quaternionic representations)
c_i^H		Chern class in $H^*(BG)$
c_i^{RU}		Chern class in $RU(G)$
$c_i^{KU_G}$		Chern class in KU_G^*
c_i^{KU}		Chern class in $KU^*(BG)$
c_i^{ku}		Chern class in $ku^*(BG)$
C_n		cyclic group of order n
χ		(1) generic notation for a character
		(2) Euler characteristic

D

D_{2n}		dihedral group of order $2n$
D		(1) Dihedral group
		(2) Double of a graded vector space
		(3)ko-valued index
$D', D'', \overline{D}$		(1) Dihedral subgroups
δ		Determinant representation

E

$E(t_1, \ldots, t_n)$		exterior algebra on generators $t_1, \ldots, t_n$
$E(1)$		exterior algebra $E(Q_0, Q_1)$
EG		contractible free G-space
$\widetilde{E}G$		the join $S^0 * EG$
e_H		Euler class in $H^*(BG)$
e_{RU}		Euler class in $RU(G)$
e_{KU_G}		Euler class in KU_G^*
e_{ku}		Euler class in $ku^*(BG)$
e_{KU}		Euler class in $KU^*(BG)$

$e(\alpha)$	9.5	Euler class for elementary abelian group
$\overline{e}(\alpha)$	9.5	divided Euler class for elementary abelian group
ϵ		the trivial one dimensional representation
η		(1) generator of the stable 1-stem (2) generator of ko_1
$\mathcal{E}(G)$	2.1	ideal generated by Euler classes
$E_{1\frac{1}{2}}$		Some d_1 differentials still to take place

F

F_h	9.7	hth term in resolution of TU
F_h'	11.6	hth term in resolution of TO
$\mathbb{F}_p$		field with p elements

G

G		generic notation for a compact Lie group (usually finite)
$\hat{G}$		An extension of G by a quotient of order 2
$\mathbb{G}_m$		multiplicative group scheme
γ_t		Grothendieck γ operations
$\Gamma_I M$		I-power torsion in M
Γ_B	11.7	Generator in resolution of TO
γ_B	11.7	$d\Gamma_B$

H

$\mathbb{H}$		Quaternions
$H_I^*(R)$		Grothendieck local cohomology of R at the ideal $\mathfrak{m}$
$H\mathbb{F}_p$		mod p Eilenberg-MacLane spectrum
$H\mathbb{Z}$		integral Eilenberg-MacLane spectrum
$H\mathbb{Z}_p^\wedge$		p-adic Eilenberg-MacLane spectrum
H_B	11.5	Generator in resolution of TO
h_B	11.5	dH_B
h_0		the Adams spectral sequence counterpart of 2

I

I		augmentation ideal $I = \ker(ku^*(BG) \longrightarrow ku^*)$
I_n	9.5	Module generated by the n-fold products of divided Euler classes for $V(r)$
I_n	9.6	nth image in the Koszul complex
i		Complex number with $i^2 = -1$
i_V		the inclusion $S^0 \longrightarrow S^V$
$\inf_Q^G$		Inflation functor from Q-equivariant to G-equivariant objects.

J

J		(1) augmentation ideal $J = \ker(RU(G) \longrightarrow \mathbb{Z})$
		(2) conjugate-linear map with $J^2 = \pm 1$
j		non-complex quaternion with $j^2 = -1$

K

k		(1) quaternion ij
		(2) generic notation for a field
		(3) generic notation for a ku_*-algebra
KU		complex periodic K-theory
ku		complex connective K-theory
KO		real periodic K-theory
ko		real connective K-theory
$K\mathbb{R}$		periodic K-theory with reality
$k\mathbb{R}$		connective K-theory with reality
$ko_n^+(BG)$		subgroup of indices of psc spin n-manifolds
$K_R(\underline{x})$		Koszul complex over R on the sequence $\underline{x}$
$K(m,i)_{s,a}$	11.6	Generators of $A(m,i)_s$ in resolution of TO

L

lu		Adams summand of ku
L		string module
L_n	11.5	syzygy in resolution of TU_{r-k}
L_A^m	[**47**]	universal ring for multiplicative A-equivariant formal group laws
Λ^i		i-th exterior power operator
λ_i		i-th exterior power of the defining representation

M

M_{4k-1}	10.3	QU contribution to $ko_{4k-1}(BG)$ at E^1
ModRees(G)		modified Rees ring, generated by Chern classes
$M_G(C_k)$	2.3	G-equivariant graded modules over the Clifford algebra C_k
m_B	11.1.1	$Sq^2(y_B X)$ generator of $ZPC[Sq^2]$
$m_{B,R}$	11.3	$Sq^2(y_B X_R)$ generator of Sq^2 cycles in resolution of TU

N

$N_G(C_k^0)$	2.3	G-equivariant modules over the Clifford algebra C_k^0
$\nu_p(n)$		largest power of p dividing n

O

ω		primitive cube root of unity

P

P		generic Sylow p-subgroup
PW		the polynomial ring $\mathbb{F}_p[x_1, \ldots, x_r]$
PC		the polynomial ring $\mathbb{F}_2[y_1, \ldots, y_r]$ where $y_i = x_i^2$
PP		the polynomial ring $\mathbb{F}_2[z_1, \ldots, z_r]$ where $z_i = x_i^4$
$\mathcal{P}$		family of proper subgroups.
p		(1) a prime (2) inversion of β
p_i		ith Pontryagin class of a quaternionic representation
p^*		generator of modified reduction of JO
p_S	9.6.2	$Sq^2(y_S)$, generator of ZPC
ψ_i	6.4	representation of Q_{2^n}
$\Phi^Q X$		Geometric Q-fixed points of X

Q

Q	2.2	A group of order 2 corresponding to complex conjugation
Q_{2^n}		quaternion group of order 2^n
Q_0		Bockstein, Sq^1 if $p = 2$
Q_1		First higher Milnor Bockstein, $Sq^1Sq^2 + Sq^2Sq^1$ if $p = 2$
QU		the image of $ku_G^* \longrightarrow KU_G^*$, usually the modified Rees ring
QO	6.2	Image of ko_G^* in periodic real K-theory
$\overline{QO}$	6.2	Image of ko_G^* in periodic complex K-theory
Q_T	6.2	η-multiples in QO
Q_R	9.8	equal to Q_R^0; maps to q_R
Q_R^i	9.8	generator of resolution of TU_n
q		quaternionification (of complex represenations)
q_S	9.2	$Q_1Q_0(\prod_{i\in S} x_i)$

R

$RO(G)$		Real representation ring
$RU(G)$		Complex representation ring
$RSp(G)$		Symplectic representation ring
RU_{asc}		Anti-self-conjugate virtual representations.
RU_{sc}		Self-conjugate virtual representations.
R		(1) lift of r along multiplication by v (2) generic notation for a ring, or for $ku^*(BG)$
$\mathbb{R}$		Real numbers
$Rees(R, J)$	1.3	Rees ring of R: subring of $R[v, v^{-1}]$ generated by R, v and $1/v \cdot J$
r_θ	8.1	rotation by an angle θ
$r_{s,n}$	8.1	rotation by an angle $2s\pi/2^{n+1}$
r		(1) realification (of complex representations) (2) p-rank of group

C.5. Index of terminology

K

Koszul complex	9.6	

L

Lewis-May fixed points		right adjoint to inflation; categorical fixed points
lightning flash		certain $E(1)$-module
local cohomology	[**54**]	Grothendieck's method of calculating cohomology with support
LCSS	2.5	Local cohomology spectral sequence
Local cohomology spectral sequence		

M

modified Rees ring		Subring of KU_G^* generated by $1, v$ and Chern classes

P

periodic K-theory		$KU^*(\cdot)$ or $KO^*(\cdot)$
Poincare series		See Hilbert series

Q

quaternion group		Q_{2^n}
quaternionic representation	2.1	representation over $\mathbb{C}$ with conjugate linear J with $J^2 = -1$

R

rank		size of largest elementary abelian subgroup
real representation	2.1	representation over $\mathbb{C}$ with conjugate linear J with $J^2 = 1$
Rees ring		subring of $RU(G)[v, v^{-1}]$ generated by $1, v, 1/v \cdot J$

S

self-conjugate	2.1	
string module		certain $E(1)$-module

T

Tate cohomology		combination of homology and cohomology of BG

Bibliography

[1] J. F. Adams "On Chern characters and the structure of the unitary group" *Proc. Cambridge Philos. Soc.* **57** (1961) 189–199.

[2] J. F. Adams "Lectures on generalised cohomology" *Lecture Notes in Math.* **99** Springer-Verlag (1969) 1–138.

[3] J. F. Adams "Stable homotopy and generalised homology " (Reprint of the 1974 original) Chicago Lectures in Mathematics, University of Chicago Press, Chicago, IL (1995).

[4] J. F. Adams "Lectures on Lie groups" W. A. Benjamin, Inc., New York and Amsterdam (1969).

[5] J. F. Adams and H. R. Margolis "Modules over the Steenrod algebra" *Topology* **10** (1971) 271–282.

[6] J. F. Adams and S. Priddy "Uniqueness of BSO" *Math. Proc. Cambridge Philos. Soc.* **80** (1976) no. 3, 475–509.

[7] D. W. Anderson "The real K-Theory of classifying spaces" *Proc. Natl. Acad. Sci. USA* **51** (1964) 634–636.

[8] S. Araki "Orientations in τ-cohomology theories" *Jpn. J. Math.* (N.S.) **5** (1979) no. 2, 403–430.

[9] S. Araki and M. Murayama "τ-cohomology theories" *Jpn. J. Math.* (N.S.) **4** (1978) no. 2, 363–416.

[10] M. F. Atiyah, "Characters and cohomology of finite groups", *IHES Pub. Math.* **9** (1961) 23–64.

[11] M. F. Atiyah "Bott periodicity and the index of elliptic operators" *Q. J. Math.* **19** (1968) 113–140.

[12] M. F. Atiyah "K-theory and reality" *Q. J. Math.* **17** (1966) 367–386.

[13] M. F. Atiyah, R. Bott and A. Shapiro, "Clifford modules" *Topology* **3** (1964) suppl. 1, 3–38.

[14] M. F. Atiyah and G. B. Segal "Equivariant K-theory and completion" *J. Diff. Geom.* **3** (1969) 1–18.

[15] M. F. Atiyah and D. O. Tall "Group representations, λ-rings and the J-homomorphism" *Topology* **8** (1969) 253–297.

[16] Egidio Barrera-Yanez and Peter B. Gilkey, "The eta invariant and the connective K-theory of the classifying space for cyclic 2 groups", *Ann. Global Anal. Geom.* **17** (1999) no. 3, 289–299.

[17] Dilip Bayen, "Real connective K-theory of finite groups", PhD thesis, Wayne State University (1994).

[18] Dilip Bayen and Robert R. Bruner "Real connective K-theory and the quaternion group" *Trans. Amer. Math. Soc.* **348** (1996) no. 6, 2201–2216.

[19] Raoul Bott "The stable homotopy of the classical groups" *Ann. of Math.* (2) **70** (1959) 313–337.

[20] Raoul Bott "Lectures on $K(X)$" W. A. Benjamin, Inc., New York and Amsterdam (1969).

[21] B. Botvinnik, P. Gilkey and S. Stolz "The Gromov-Lawson-Rosenberg conjecture for groups with periodic cohomology" *J. Differential Geom.* **46** (1997) no. 3, 374–405.

[22] T. Bröcker and T. tom Dieck "Representations of compact Lie groups" Springer-Verlag (1985).

[23] Robert R. Bruner, "Ossa's theorem and Adams covers", *Proc. Amer. Math. Soc.* **127** (1999) no. 8, 2443 – 2447.

[24] Robert R. Bruner "On Ossa's theorem" In preparation.

[25] Robert R. Bruner "An Adams spectral sequence primer" In preparation.

[26] W. Bruns and J. Herzog "Cohen-Macaulay rings" Cambridge Studies in Advanced Mathematics, **39**. Cambridge University Press, Cambridge (1993).

[27] V. M. Buhstaber and A. S. Miscenko "Elements of infinite filtration in K-theory" *Dokl. Akad. Nauk SSSR* **178** (1968) 1234–1237.

[28] R. R. Bruner and J. P. C. Greenlees "Connective K-theory of finite groups", *Mem. Amer. Math. Soc.* **165** (2003).

[29] P. E. Conner and E. E. Floyd "The relation of cobordism to K-theories" *Lecture Notes in Math.* **28** Springer-Verlag, Berlin and New York, (1966).

[30] P. E. Conner and L. Smith "On the complex bordism of finite complexes" *Inst. Hautes Etudes Sci. Publ. Math.* **37** (1970).

[31] Donald Davis "Generalized homology and the generalized vector field problem" PhD thesis, Stanford University (1972).

[32] Donald Davis "Generalized homology and the generalized vector field problem" *Quart. J. Math. Oxford* Ser. (2) **25** (1974) 169–193.

[33] Donald Davis, private conversation (July 2010).

[34] W. G. Dwyer, T. Schick and S. Stolz "Remarks on a conjecture of Gromov and Lawson" High-dimensional manifold topology, 159–176, World Sci. Publ., River Edge, NJ (2003).

[35] Daniel Dugger, Daniel(1-OR) "An Atiyah-Hirzebruch spectral sequence for KR-theory" *K-Theory* **35** (2005), no. 3-4, 213–256.

[36] A. D. Elmendorf, I. Kriz, M. A. Mandell and J. P. May "Rings, modules, and algebras in stable homotopy theory" With an appendix by M. Cole. *Mathematical Surveys and Monographs* **47**. American Mathematical Society, Providence, RI (1997).

[37] L. Fajstrup "Periodic equivariant real K-theories have rational Tate theory" *Manuscripta Math.* **91** (1996) no. 2, 211–221.

[38] H. Fausk "Equivariant homotopy theory for pro-spectra" *Geometry and Topology* **12** (2008) 103–176.

[39] Charles H. Giffen, "Bott periodicity and the Q-construction" Algebraic K-theory (Poznań, 1995) 107–124, *Contemp. Math.* **199**, Amer. Math. Soc., Providence, RI (1996).

[40] S. Gitler, M. Mahowald and R. James Milgram "The nonimmersion problem for RP^n and higher-order cohomology operations" *Proc. Natl. Acad. Sci. USA* **60** no. 2 (1968) 432–437.

[41] J. P. C. Greenlees "K-homology of universal spaces and local cohomology of the representation ring" *Topology* **32** (1993) 295–308.

[42] J. P. C. Greenlees "Augmentation ideals of equivariant cohomology rings" *Topology* **37** (1998) 1313–1323.

[43] J. P. C. Greenlees, "Tate cohomology in commutative algebra" *J. Pure Appl. Algebra* **94** (1994) no. 1, 59–83.

[44] J. P. C. Greenlees "Augmentation ideals of equivariant cohomology rings" *Topology* **37** (1998) 1313–1323.

[45] J. P. C. Greenlees, "Equivariant versions of real and complex connective K-theory" *Homology, Homotopy Appl.* **7** (2005) no. 3, 63–82.

[46] J. P. C. Greenlees "Equivariant connective K-theory for compact Lie groups" *J. Pure Appl. Algebra* **187** (2004) no. 1–3, 129–152.

[47] J. P. C. Greenlees "Multiplicative equivariant formal group laws" *J. Pure and Applied Algebra* **165** (2001) no. 2, 183–200.

[48] J. P. C. Greenlees "Spectra for commutative algebraists" Proceedings of the 2004 Chicago Summer School, *Contemp. Math.* **436** (2007) 149–173.

[49] J. P. C. Greenlees and G. Lyubeznik, "Rings with a local cohomology theorem and applications to cohomology rings of groups", *J. Pure and Applied Algebra* **149** (2000) 267–285.

[50] J. P. C. Greenlees and J. P. May "Derived functors of I-adic completion and local homology" *J. Algebra* **149** (1992) 438–453.

[51] J. P. C. Greenlees and J. P. May, "Equivariant stable homotopy", Handbook of Algebraic Topology, North Holland (1995) 277–323.

[52] J. P. C. Greenlees and J. P. May, "Localization and completion theorems for MU-module spectra", *Annals of Math.* **146** (1997) 509–544.

[53] J. P. C. Greenlees and J. P. May, "Generalized Tate cohomology", *Mem. Amer. Math. Soc.* **113** no. 543 (1995).

[54] A. Grothendieck "Local cohomology" *Lecture Notes in Math.* **41** Springer-Verlag (1967).

[55] Shin Hashimoto, "On the connective K-homology groups of the classifying spaces BZ/p^r", *Publ. Res. Inst. Math. Sci.* **19** (1983) no. 2, 765–771.

[56] M. A. Hill "The 3-local tmf-homology of $B\Sigma_3$" *Proc. Amer. Math. Soc.* **135** (2007) no. 12, 4075–4086.
[57] M. J. Hopkins, N. J. Kuhn, and D. C. Ravenel, "Morava K-theories of classifying spaces and generalised characters for finite groups", Proceedings of 1990 Barcelona Conference on Algebraic Topology, *Lecture Notes in Math.* **1509** Springer-Verlag (1992) 186–209.
[58] M. J. Hopkins, N. J. Kuhn, and D. C. Ravenel, "Generalised group characters and complex oriented cohomology theories", *Amer. J. Math.* **13** (2000) 553–594.
[59] M. Hovey, B. E. Shipley and Jeff Smith "Symmetric spectra" *J. Amer. Math. Soc.* **13** (2000) no. 1, 149–208.
[60] Po Hu "On real-oriented Johnson-Wilson cohomology" *Algebr. Geom. Topol.* **2** (2002) 937–947.
[61] Po Hu and I. Kriz "Real-oriented homotopy theory and an analogue of the Adams-Novikov spectral sequence" *Topology* **40** (2001) no. 2, 317–399.
[62] D. Hussemoller "Fibre bundles" Springer-Verlag (1966).
[63] M. Joachim and A. Malhotra "The Gromov-Lawson-Rosenberg conjecture for manifolds with dihedral fundamental group" (In preparation).
[64] David Copeland Johnson and Larry Smith "On the relation of complex cobordism to connective K-theory" *Proc. Cambridge Philos. Soc.* **70** (1971), 19–22.
[65] David Copeland Johnson and W. Stephen Wilson, "On a theorem of Ossa", *Proc. Amer. Math. Soc.* **125** (1997) 3753–3755.
[66] R. Jung PhD thesis (Mainz).
[67] Richard M. Kane "Operations in connective K-theory" *Mem. Amer. Math. Soc.* **34** (1981) no. 254.
[68] M. Karoubi "Sur la K-théorie équivariante" *Lecture Notes in Math.* **136** Springer-Verlag (1970) 187–253.
[69] M. Karoubi "K-Theory" Grundlehren der mathematischen Wissenschaften **226** Springer-Verlag (1978).
[70] N. Kitchloo and Stephen W. Wilson "On the Hopf ring for $ER(n)$" *Topology Appl.* **154** (2007) no. 8, 1608–1640.
[71] N. Kitchloo and Stephen W. Wilson "The second real Johnson-Wilson theory and nonimmersions of RP^n" *Homology, Homotopy Appl.* **10** (2008) no. 3, 223–268.
[72] N. Kitchloo and Stephen W. Wilson "The second real Johnson-Wilson theory and nonimmersions of RP^n. II" *Homology, Homotopy Appl.* **10** (2008) no. 3, 269–290.
[73] S. P. Lam (unpublished).
[74] P. S. Landweber, "Conjugations on complex manifolds and equivariant homotopy of MU" *Bull. Amer. Math. Soc.* **74** (1968) 271–274.
[75] L. G. Lewis, J. P. May, M. Steinberger (with contributions by J. E. McClure) "Equivariant stable homotopy theory" *Lecture Notes in Math.* **1213** Springer-Verlag (1986).
[76] Mark Mahowald, email communication (February 2010).
[77] M. Mahowald and R. James Milgram "Operations which detect Sq^1 in connective K-theory and their applications" *Quart. J. Math. Oxford* Ser. (2) **27** (1976), no. 108, 415–432.
[78] A. Malhotra "The Gromov-Lawson-Rosenberg conjecture for manifolds with fundamental group elementary abelian of order 8" (In preparation).
[79] A. Malhotra and K. Rodtes "The Gromov-Lawson-Rosenberg conjecture for manifolds with fundamental group semidihedral of order 16" (In preparation).
[80] H. R. Margolis, "Eilenberg-MacLane spectra", *Proc. Amer. Math. Soc.* **43** (1974) 409–415.
[81] H. Matsumura, "Commutative ring theory", *Cambridge Studies in Advanced Mathematics* **8** Cambridge University Press (1986).
[82] J. P. May "A concise course in algebraic topology" Chicago Lectures in Mathematics. University of Chicago Press, Chicago, IL (1999).
[83] J. P. May and R. J. Milgram, "The Bockstein and the Adams spectral sequences" *Proc. Amer. Math. Soc.* **83** (1981) no. 1, 128 –130.
[84] J. McCleary "A user's guide to spectral sequences" Second edition. Cambridge Studies in Advanced Mathematics, **58**. Cambridge University Press, Cambridge (2001).
[85] M. Mimura and H. Toda "Topology of Lie groups. I, II" (Translated from the 1978 Japanese edition by the authors) Translations of Mathematical Monographs **91** American Mathematical Society, Providence, RI (1991).

[86] S. P. Novikov, "Methods of algebraic topology from the point of view of cobordism theory" (Russian) *Izv. Akad. Nauk SSSR* Ser. Mat. **31** (1967) 855–951.

[87] E. Ossa, "Connective K-theory of elementary abelian groups", Proceedings of 1987 Osaka Conference on Transformation groups, *Lecture Notes in Math.* **1375** Springer-Verlag (1989) 269–275.

[88] J. Rosenberg and S. Stolz, "A "stable" version of the Gromov-Lawson conjecture" The Cech centennial (Boston, MA, 1993) 405–418, *Contemp. Math.* **181**, Amer. Math. Soc., Providence, RI (1995).

[89] T. Schick "A counterexample to the (unstable) Gromov-Lawson-Rosenberg conjecture" *Topology* **37** (1998) no. 6, 1165–1168.

[90] Reinhard Schultz, "Positive scalar curvature and odd order abelian fundamental groups" *Proc. Amer. Math. Soc.* **125** (1997) no. 3, 907–915.

[91] G. B. Segal, "The representation ring of a compact Lie group", *IHES Pub. Math.* **34** (1968) 113–128.

[92] G. B. Segal, "K-homology theory and algebraic K theory", *Lecture Notes in Math.* **575**, Springer-Verlag (1977) 113–127.

[93] J.-P. Serre "Linear representations of finite groups" Springer-Verlag (1977).

[94] Larry Smith "On the relation of connective K-theory to homology" *Proc. Cambridge Philos. Soc.* **68** (1970) 637–639.

[95] S. Stolz "Splitting certain $MSpin$-module spectra" *Topology* **33** (1994) no. 1, 159–180.

[96] S. Stolz "Positive scalar curvature metrics—existence and classification questions" Proceedings of the International Congress of Mathematicians, Vol. 1, 2 (Zürich, 1994) 625–636, Birkhäuser, Basle (1995).

[97] Robert E. Stong "Determination of $H^*(BO(k,\ldots,\infty),Z_2)$ and $H^*(BU(k,\ldots,\infty),Z_2)$" *Trans. Amer. Math. Soc.* **107** (1963) 526–544.

[98] N. P. Strickland "The BP$\langle n\rangle$ cohomology of elementary abelian groups" *J. London Math. Soc.* (2) **61** (2000) no. 1, 93–109.

[99] Cherng-Yih Yu "On $ko_*(BV)$ for elemenary abelian 2-groups" PhD thesis, Notre Dame (1995).

Titles in This Series

169 **Robert R. Bruner and J. P. C. Greenlees,** Connective real K-theory of finite groups, 2010

168 **Michiel Hazewinkel, Nadiya Gubareni, and V. V. Kirichenko,** Algebras, rings and modules: Lie algebras and Hopf algebras, 2010

167 **Michael Gekhtman, Michael Shapiro, and Alek Vainshtein,** Cluster algebra and Poisson geometry, 2010

166 **Kyung Bai Lee and Frank Raymond,** Seifert fiberings, 2010

165 **Fuensanta Andreu-Vaillo, José M. Mazón, Julio D. Rossi, and J. Julián Toledo-Melero,** Nonlocal diffusion problems, 2010

164 **Vladimir I. Bogachev,** Differentiable measures and the Malliavin calculus, 2010

163 **Bennett Chow, Sun-Chin Chu, David Glickenstein, Christine Guenther, James Isenberg, Tom Ivey, Dan Knopf, Peng Lu, Feng Luo, and Lei Ni,** The Ricci flow: Techniques and applications, Part III: Geometric-analytic aspects, 2010

162 **Vladimir Maz′ya and Jürgen Rossmann,** Elliptic equations in polyhedral domains, 2010

161 **Kanishka Perera, Ravi P. Agarwal, and Donal O'Regan,** Morse theoretic aspects of p-Laplacian type operators, 2010

160 **Alexander S. Kechris,** Global aspects of ergodic group actions, 2010

159 **Matthew Baker and Robert Rumely,** Potential theory and dynamics on the Berkovich projective line, 2010

158 **D. R. Yafaev,** Mathematical scattering theory: Analytic theory, 2010

157 **Xia Chen,** Random walk intersections: Large deviations and related topics, 2010

156 **Jaime Angulo Pava,** Nonlinear dispersive equations: Existence and stability of solitary and periodic travelling wave solutions, 2009

155 **Yiannis N. Moschovakis,** Descriptive set theory, 2009

154 **Andreas Čap and Jan Slovák,** Parabolic geometries I: Background and general theory, 2009

153 **Habib Ammari, Hyeonbae Kang, and Hyundae Lee,** Layer potential techniques in spectral analysis, 2009

152 **János Pach and Micha Sharir,** Combinatorial geometry and its algorithmic applications: The Alcála lectures, 2009

151 **Ernst Binz and Sonja Pods,** The geometry of Heisenberg groups: With applications in signal theory, optics, quantization, and field quantization, 2008

150 **Bangming Deng, Jie Du, Brian Parshall, and Jianpan Wang,** Finite dimensional algebras and quantum groups, 2008

149 **Gerald B. Folland,** Quantum field theory: A tourist guide for mathematicians, 2008

148 **Patrick Dehornoy with Ivan Dynnikov, Dale Rolfsen, and Bert Wiest,** Ordering braids, 2008

147 **David J. Benson and Stephen D. Smith,** Classifying spaces of sporadic groups, 2008

146 **Murray Marshall,** Positive polynomials and sums of squares, 2008

145 **Tuna Altinel, Alexandre V. Borovik, and Gregory Cherlin,** Simple groups of finite Morley rank, 2008

144 **Bennett Chow, Sun-Chin Chu, David Glickenstein, Christine Guenther, James Isenberg, Tom Ivey, Dan Knopf, Peng Lu, Feng Luo, and Lei Ni,** The Ricci flow: Techniques and applications, Part II: Analytic aspects, 2008

143 **Alexander Molev,** Yangians and classical Lie algebras, 2007

142 **Joseph A. Wolf,** Harmonic analysis on commutative spaces, 2007

141 **Vladimir Maz′ya and Gunther Schmidt,** Approximate approximations, 2007

140 **Elisabetta Barletta, Sorin Dragomir, and Krishan L. Duggal,** Foliations in Cauchy-Riemann geometry, 2007

139 **Michael Tsfasman, Serge Vlăduţ, and Dmitry Nogin,** Algebraic geometric codes: Basic notions, 2007

138 **Kehe Zhu,** Operator theory in function spaces, 2007

137 **Mikhail G. Katz,** Systolic geometry and topology, 2007

136 **Jean-Michel Coron,** Control and nonlinearity, 2007

135 **Bennett Chow, Sun-Chin Chu, David Glickenstein, Christine Guenther, James Isenberg, Tom Ivey, Dan Knopf, Peng Lu, Feng Luo, and Lei Ni,** The Ricci flow: Techniques and applications, Part I: Geometric aspects, 2007

134 **Dana P. Williams,** Crossed products of C^*-algebras, 2007

133 **Andrew Knightly and Charles Li,** Traces of Hecke operators, 2006

132 **J. P. May and J. Sigurdsson,** Parametrized homotopy theory, 2006

131 **Jin Feng and Thomas G. Kurtz,** Large deviations for stochastic processes, 2006

130 **Qing Han and Jia-Xing Hong,** Isometric embedding of Riemannian manifolds in Euclidean spaces, 2006

129 **William M. Singer,** Steenrod squares in spectral sequences, 2006

128 **Athanassios S. Fokas, Alexander R. Its, Andrei A. Kapaev, and Victor Yu. Novokshenov,** Painlevé transcendents, 2006

127 **Nikolai Chernov and Roberto Markarian,** Chaotic billiards, 2006

126 **Sen-Zhong Huang,** Gradient inequalities, 2006

125 **Joseph A. Cima, Alec L. Matheson, and William T. Ross,** The Cauchy Transform, 2006

124 **Ido Efrat, Editor,** Valuations, orderings, and Milnor K-Theory, 2006

123 **Barbara Fantechi, Lothar Göttsche, Luc Illusie, Steven L. Kleiman, Nitin Nitsure, and Angelo Vistoli,** Fundamental algebraic geometry: Grothendieck's FGA explained, 2005

122 **Antonio Giambruno and Mikhail Zaicev, Editors,** Polynomial identities and asymptotic methods, 2005

121 **Anton Zettl,** Sturm-Liouville theory, 2005

120 **Barry Simon,** Trace ideals and their applications, 2005

119 **Tian Ma and Shouhong Wang,** Geometric theory of incompressible flows with applications to fluid dynamics, 2005

118 **Alexandru Buium,** Arithmetic differential equations, 2005

117 **Volodymyr Nekrashevych,** Self-similar groups, 2005

116 **Alexander Koldobsky,** Fourier analysis in convex geometry, 2005

115 **Carlos Julio Moreno,** Advanced analytic number theory: L-functions, 2005

114 **Gregory F. Lawler,** Conformally invariant processes in the plane, 2005

113 **William G. Dwyer, Philip S. Hirschhorn, Daniel M. Kan, and Jeffrey H. Smith,** Homotopy limit functors on model categories and homotopical categories, 2004

112 **Michael Aschbacher and Stephen D. Smith,** The classification of quasithin groups II. Main theorems: The classification of simple QTKE-groups, 2004

111 **Michael Aschbacher and Stephen D. Smith,** The classification of quasithin groups I. Structure of strongly quasithin K-groups, 2004

110 **Bennett Chow and Dan Knopf,** The Ricci flow: An introduction, 2004

109 **Goro Shimura,** Arithmetic and analytic theories of quadratic forms and Clifford groups, 2004

108 **Michael Farber,** Topology of closed one-forms, 2004

107 **Jens Carsten Jantzen,** Representations of algebraic groups, 2003

106 **Hiroyuki Yoshida,** Absolute CM-periods, 2003

105 **Charalambos D. Aliprantis and Owen Burkinshaw,** Locally solid Riesz spaces with applications to economics, second edition, 2003

For a complete list of titles in this series, visit the
AMS Bookstore at **www.ams.org/bookstore/**.